12th International Scientific-Technical Conference on Electromechanics and Robotics "Zavalishin's Readings" (ER (ZR) -2017)

AA004068

MATEC Web of Conferences Volume 113 (2017)

St. Petersburg, Russia
18 - 22 April 2017

Editors:

Andrey Ronzhin
Vladislav Shishlakov

ISBN: 978-1-5108-4328-8

Printed from e-media with permission by:

Curran Associates, Inc.
57 Morehouse Lane
Red Hook, NY 12571

Some format issues inherent in the e-media version may also appear in this print version.

Printed by Curran Associates, Inc. (2017)

For additional information, please contact EDP Sciences – Web of Conferences at the address below.

EDP Sciences – Web of Conferences
17, Avenue du Hoggar
Parc d'Activité de Courtabœuf
BP 112
F-91944 Les Ulis Cedex A
France

Phone: +33 (0) 1 69 18 75 75
Fax: +33 (0) 1 69 28 84 91

contact-edps@webofconferences.org

Additional copies of this publication are available from:

Curran Associates, Inc.
57 Morehouse Lane
Red Hook, NY 12571 USA
Phone: 845-758-0400
Fax: 845-758-2633
Email: curran@proceedings.com
Web: www.proceedings.com

TABLE OF CONTENTS

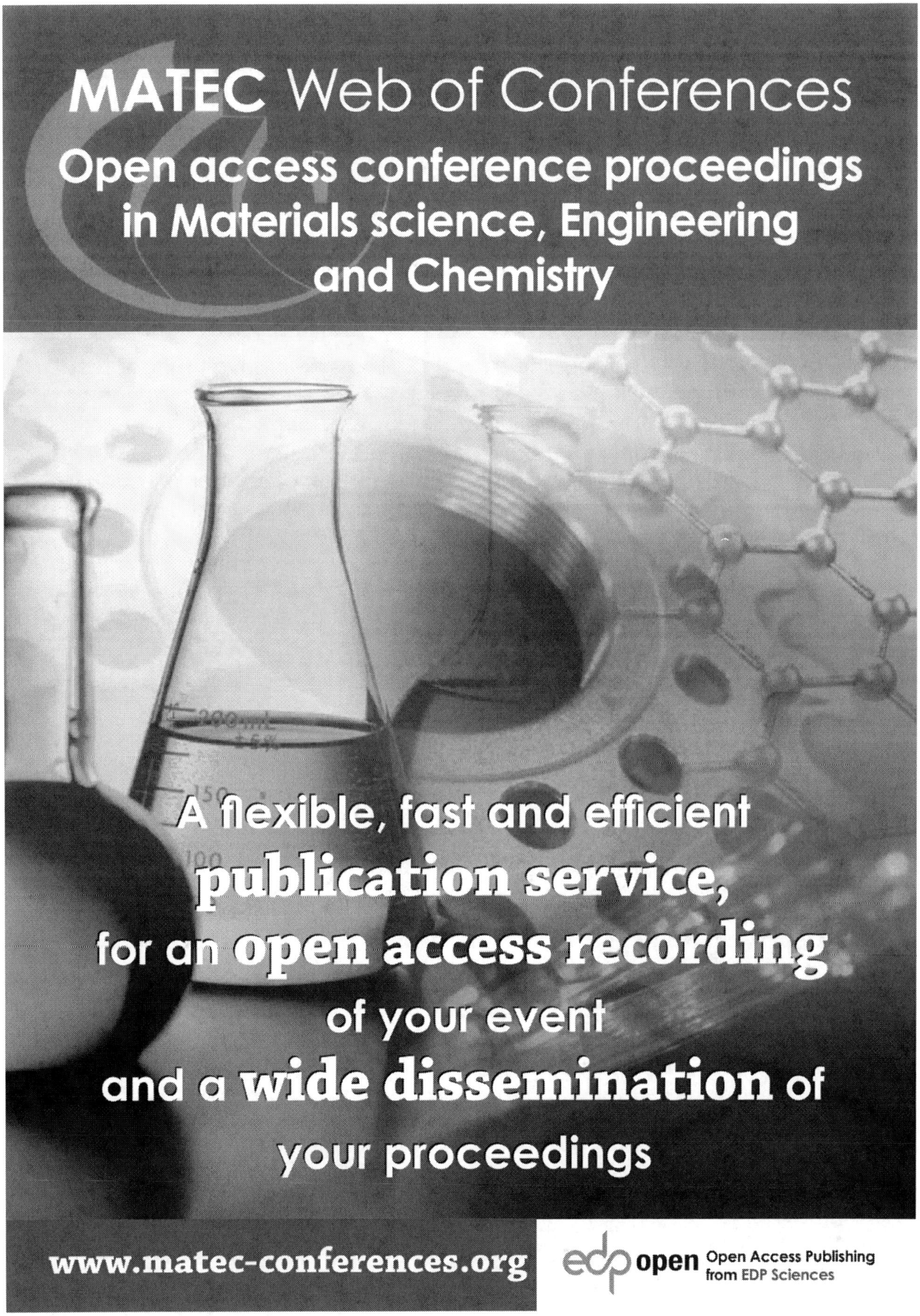
MATEC Web of Conferences
Open access conference proceedings in Materials science, Engineering and Chemistry
A flexible, fast and efficient publication service, for an open access recording of your event and a wide dissemination of your proceedings
www.matec-conferences.org
edp open Open Access Publishing from EDP Sciences

edp sciences

MATEC Web of Conferences

Statement of Peer review

In submitting conference proceedings to MATEC Web of Conferences, I certify to the Publisher that I adhere to the **Policy on Publishing Integrity** of the journal in order to safeguard good scientific practice in publishing.

1. All articles have been subjected to peer review administered by the proceedings editors.
2. Reviews have been conducted by expert referees, who have been requested to provide unbiased and constructive comments aimed, whenever possible, at improving the work.
3. Proceedings editors have taken all reasonable steps to ensure the quality of the materials they publish and their decision to accept or reject a paper for publication has been based only on the merits of the work and the relevance to the journal.

Title, date and place of the conference

12th International Scientific-Technical Conference on Electromechanics and Robotics "Zavalishin's Readings"-2017, April 18-22, 2017, St. Petersburg, Russia

Proceedings editor(s):

Andrey Ronzhin, Vladislav Shishlakov

Date and editor's signature June 09, 2017

Preface

The 12th International Scientific-Technical Conference on Electromechanics and Robotics "Zavalishin's Readings" - 2017, ER(ZR)-2017 was held on 18-22 April 2017 in St. Petersburg, Russia. The conference was established as a tribute to the memory of Dmitry Aleksandrovich Zavalishin (1900-1968) – a Russian scientist, corresponding member of the USSR Academy of Sciences, professor, doctor of technical sciences, honored worker of science and technology of the RSFSR, founder of the school of valve energy converters based on electric machines and valve converters energy. The first conference was organized by the Institute of Innovative Technologies in Electromechanics and Robotics of the St. Petersburg State University of Aerospace Instrumentation in 2006 and is timed to the 60th anniversary of the Institute. Today the conference has become international and be held for the twelfth time in St. Petersburg.

The purpose of the conference is the exchange of information and progressive results of scientific research work of scientific and pedagogical workers, young scientists, graduate students, applicants and students in the field of: automatic control systems, electromechanics, electric power engineering and electrical engineering, mechatronics, robotics, automation, technical physics and management in the electric power industry.

Around 100 papers were submitted to the conference and each paper was reviewed by several scientists. In this year around 30% of best papers were published in English proceedings by EDP Sciences on EI Serial Journal of MATEC web of Conferences indexed in SCOPUS, Thomson Reuters (Web of Science), Inspec, etc.

Due to great efforts of reviewers and scientific editors Prof. Andrey Ronzhin and Prof. Vladislav Shishlakov this issue was carefully prepared and consists of 29 contributions of authors from Russia, Ukraine, Germany, China, Viet Nam, Belarus, and Czech Republic. Special thanks are due to the members of the Local Organizing Committee and secretary Sergey Solyonyj for their tireless effort and enthusiasm during the conference organization.

We express our deepest gratitude to all participants for their valuable contribution to the successful organization of ER(ZR)-2017, hope for and look forward to your attention to the next International Scientific-Technical Conference on Electromechanics and Robotics "Zavalishin's Readings" in 2018. The conference website is located at: http://suai.edu.ru/conference/zav-read/.

General Chair
of 12th International Scientific-Technical Conference
on Electromechanics and Robotics "Zavalishin's Readings" – 2017,
Rector of the St. Petersburg State University of Aerospace Instrumentation
Prof. Yulia A. Antokhina

MATEC Web of Conferences **113**, 01001 (2017) DOI: 10.1051/matecconf/201711301001
12th International Scientific-Technical Conference on Electromechanics and Robotics "Zavalishin's Readings" - 2017

The simulation solution of nonlinear problems of ergodic electric-power processes transformation

Alexander Bulgakov*

Donetsk National Technical University, 83001 Donetsk, Ukraine

Abstract. The algorithm for the simulation solution of the stochastic electric-power processes problems (quadratic inertial smoothing (QIS) and quadratic cumulative averaging (QCA)) has been stated. It includes simulation of initial process realizations, transformation of realizations and ensemble statistical analysis of these processes, as well as simulation quality check by the test problems. Peculiarities of skewness, kurtosis and rated maxima (quantiles) calculations based on the ensemble of the stochastic ergodic process realizations have been considered. Density and cumulative distribution function of the quadratic stochastic processes with normal distribution have been analyzed. Ensemble and single realization errors for rated maxima have been evaluated. It has been proposed to use test check with distribution quantiles when searching for simulation solution of QIS on the stage of quadratic transformation. Approximation of rated maxima using the gamma distribution law is given.

1 Introduction

The current standard for electromagnetic compatibility (EMC) [1] involves the principle of quadratic cumulative averaging [2] to take into account the inertia of objects. The rated maximum of half-hour load [3] also implements a cumulative principle. A simulation solution for a nonlinear problem has been proposed in [4] through the example of quadratic inertial smoothing (QIS) of stochastic electric-power processes. The inertial principle [5] is free from the disadvantages inherent in the cumulative principle of EMC evaluation. The standard [6] is based on the inertial principle of EMC interference evaluation. Some approximate approaches for solution of the problem have been considered in [7].

The detailed definition of numerical characteristics, cumulative function and density distribution for a random variable is given in [8-9] and for stochastic processes in [9-10]. In practice, as a rule, single realizations of stochastic ergodic processes are considered. The simulation solution implies the statistical manipulation of characteristics in the ensemble of stochastic process realizations. The presented material is based on the example of heating an object with the constant of inertia *T*.

2 A heat model

The superheat temperature of the object and the current are related by the linear differential equation

$$T\vartheta_T' + \vartheta_T = c_\vartheta I^2 = c_\vartheta z, \quad (1)$$

where $z = \vartheta_0 = I^2$, and the coefficient c_ϑ is equal to the ratio of the continuous permissible (symbol $\frown$) temperature $\hat{\vartheta}$ in °C to the square admissible continuous current $\hat{I}$ in amperes.

The mean temperature

$$\vartheta_c = c_\vartheta I_e^2 = c_\vartheta z_e \quad (2)$$

is proportional to the squared effective value I_e (in short – effective, rms) or rms z_e of quadratic process $z(t)$. They don't depend on *T*.

In addition to the continuous permissible the maximum permissible temperature is standardized $\hat{\vartheta}_M$.

The rated maximum ϑ_{TM} for stochastic process $\vartheta_{TM}(t)$ is determined by a cumulative distribution function $F_{\vartheta T}(\vartheta_T)$ from the equation

$$F_{\vartheta T}(\vartheta_{TM}) = 1 - E_x. \quad (3)$$

As a rule, a marginal probability E_x is taken in the range from 0.05 to 0.001 (for example, in [1] it is taken equal to 0.05 for the voltage quality indicators). If the rated maximum of the temperature ϑ_{TM} exceeds the $\hat{\vartheta}_M$, then the object must be turned off.

Let us consider a probabilistic model of load variation [11-12]. The normal distribution law of the process $I(t)$ is characterized by the mean I_c and the standard deviation σ_I. An exponential autocorrelation function (CF) for the electric-power processes is proposed in [2]:

$$K_I(\tau) = \sigma_I^2 \exp\{-\alpha|\tau|\}. \quad (4)$$

The parameter α is inverse to the correlation time τ_k. An exponential-cosine CF is used in [12-13]:

* Corresponding author: bulgakov-work@mail.ru

$$K_I(\tau) = \sigma_I^2 \exp\{-\alpha|\tau|\}\cos(\beta\tau). \quad (5)$$

The parameter β characterizes the periodic component of the process. The effective current is $I_e = \left(I_c^2 + \sigma_I^2\right)^{0.5}$.

A per-unit system (pu, symbol $*$) is introduced for generalization. Base values are the correlation time τ_k and standard deviation σ_I. In per-unit system $I_{c*} = I_c/\sigma_I$, $\sigma_{I*} = 1$, $T_* = \alpha T$, $\tau_* = \alpha\tau$, $\alpha_* = 1$. After replaceing in (1) I^2 with I_{c*}^2/σ_{I*}^2 we obtain

$$\vartheta_{T*} = \vartheta_T / c_\vartheta \sigma_I^2.$$

Taking into account that $dt_* = \alpha dt$ and $T_* = \alpha T$ the equation (1) can be expressed as

$$T_* \vartheta'_{T*} + \vartheta_{T*} = I_*^2. \quad (6)$$

The per-unit system allows us to simulate the current for the values $\alpha_* = 1$, $\sigma_{I*} = 1$ and I_{c*} only. The resulting T_*-characteristics are universal. They cover any T, α, I_c and σ_I values in actual values.

The block diagram of a heat model is shown in Figure 1. A stationary stochastic process $I_*(t)$ is an input to the model. QIS transformation is implemented using a squarer (block 1) and the 1-st order inertial link (block 2). A quadratic inertial process $\vartheta_{T*}(t)$ is an output of block 2. In case of need square-rooting is performed to obtain a reduced inertial process $I_{eT*}(t)$.

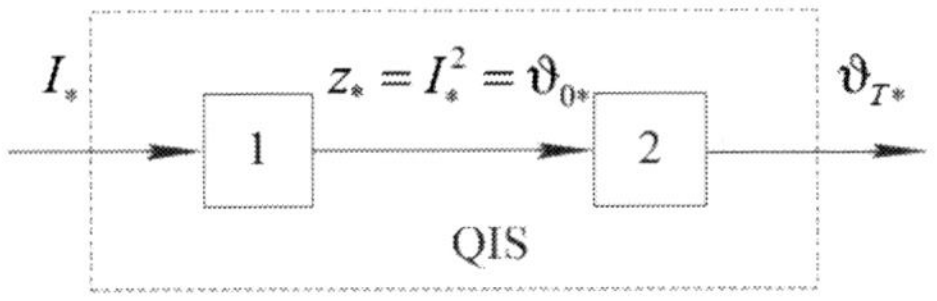

Fig. 1. A heat model.

The exact solution for the distribution density of the process is unknown. Only some numerical characteristics can be determined precisely. For CF of the form (4) mean value is

$$\vartheta_{Tc} = I_e^2 = I_c^2 + \sigma_I^2, \quad (7)$$

the standard deviation is

$$\sigma_{\vartheta T} = \sqrt{2}\sigma_I \sqrt{\frac{\sigma_I^2}{1+2\alpha T} + \frac{2I_c^2}{1+\alpha T}}, \quad (8)$$

in pu

$$\vartheta_{Tc*} = I_{e*}^2 = I_{c*}^2 + \sigma_{I*}^2 = I_{c*}^2 + 1, \quad (9)$$

$$\sigma_{\vartheta T*} = \sqrt{2}\sigma_{I*}\sqrt{\frac{\sigma_{I*}^2}{1+2T_*} + \frac{2I_{c*}^2}{1+T_*}} = \sqrt{2}\sqrt{\frac{1}{1+2T_*} + \frac{2I_{c*}^2}{1+T_*}}. \quad (10)$$

3 Test problems

Subject to the existence of an inertia there can only be a simulation solution of the problem, so it is advisable to set high requirements for the quality of initial stochastic process realizations, as well as to verify the simulation solution on basis of test problems.

A test problem (a check) is the special case, which an exact analytic solution can be found for. In regard to QIS transformation of stochastic electric-power processes the two test problems are given in [4]: a quadratic inertialess transformation and QIS transformation when the mean value of an initial process is equal to 0. In the first case an exact analytic solution is known after transformation (the process distribution law), in the second case – solution is given for moments of distribution.

3.1 Quadratic inertialess transformation

A quadratic process $z(t)$ occurs at the output of block 1 (Fig.1). It is possible to verify the simulation quality at this stage. The calculation is performed for the ensemble simulation density and distribution function to compare them with the exact solution. It is also possible to evaluate errors of distribution moments though it is not necessary in this case.

The distribution density of a quadratic stationary stochastic process $z = I^2$ is given by the exact formula in [10]:

$$f_z(z) = \frac{1}{\sqrt{2\pi}\sigma_I 2\sqrt{z}} \times \left[\exp\left\{-\frac{\left(\sqrt{z} - I_c\right)^2}{2\sigma_I^2}\right\} + \exp\left\{-\frac{\left(-\sqrt{z} - I_c\right)^2}{2\sigma_I^2}\right\}\right]. \quad (11)$$

The distribution density can be expressed through a hyperbolic function

$$f_z(z) = \frac{1}{\sqrt{2\pi}\sigma_I\sqrt{z}} \exp\left\{-\frac{z + I_c^2}{2\sigma_I^2}\right\} \mathrm{ch}\left\{\frac{I_c\sqrt{z}}{\sigma_I^2}\right\}. \quad (12)$$

Considering quadratic transformation $z \geq 0$.

Kolmogorov-Smirnov test gives evaluation in *Y*-direction of the distribution function $F_z(z)$, but evaluation of maximum rated values in accordance with practical confidence principle [14] is performed in *X*-direction. The distribution can have a considerable length (Fig.2) in the area of the rated values (long «tails»). This can introduce a significant error. It is proposed to increase the accuracy in addition to the evaluation with Kolmogorov-Smirnov test further to test simulation solutions for quantiles of inertialess quadratic transformation.

3.2 QIS of zero-mean initial process

The second test check is run for $T > 0$. In this case the exact solution can only be found for the moments of a process distribution $\vartheta_{T*}(t)$. In pu the dispersion

$$D_{\vartheta T*} = D_{\vartheta T}/\sigma_I^4 = \mu_{\vartheta T2*} = \frac{2}{1+2T_*} + \frac{4I_{c*}^2}{1+T_*}, \quad (13)$$

where $\mu_{\vartheta T2*}$ – is the second central moment in pu. The mean value is determined by the formula (9).

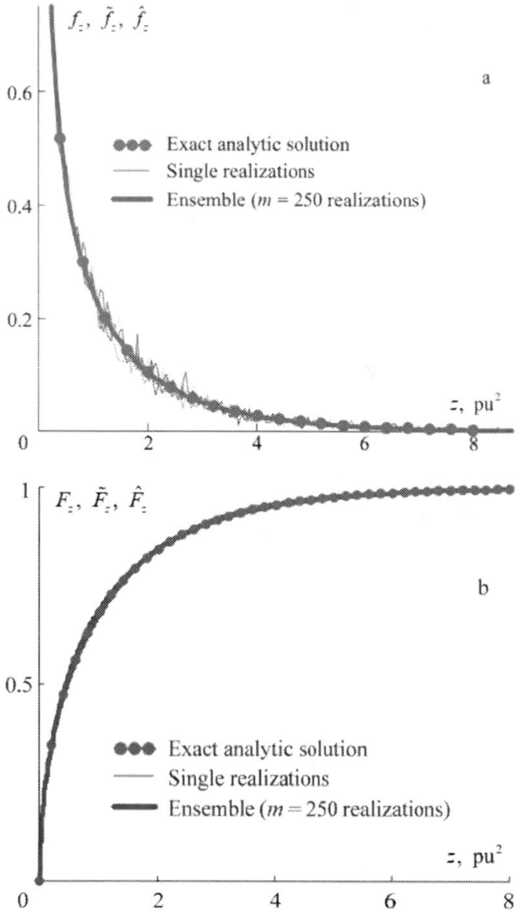

Fig. 2. Distribution law of a quadratic stationary stochastic process ($I_{c*} = 0$ pu): density (a) and cumulative function (b).

In theory the general formulae derived from [10] allow us to define the third and fourth central moments with any mean values of I_c, but the result acceptible for practical application can be received only when $I_c = 0$.

Leaving out simple but cumbersome calculations according to the formulae (34.33) and (34.27), we give the finite expressions:

$$D_{\vartheta T*} = \frac{2}{1+2T_*}, \tag{14}$$

$$\mu_{\vartheta T3*} = \mu_{\vartheta T3} / \sigma_I^6 = \frac{8}{(1+T_*)(1+2T_*)}, \tag{15}$$

$$\mu_{\vartheta T4*} = \mu_{\vartheta T4} / \sigma_I^8 = \frac{12(15+25T_*+2T_*^2)}{(1+2T_*)^2(3+5T_*+2T_*^2)}. \tag{16}$$

The skewness and kurtosis are

$$\mathrm{Sk}_{\vartheta T} = \frac{2\sqrt{2}}{1+T_*}\sqrt{1+2T_*}, \quad \mathrm{Ex}_{\vartheta T} = \frac{36+50T_*}{3+5T_*+2T_*^2}. \tag{17}$$

4 Algorithm for the simulation solution

The main statements of the method to obtain the simulation solution of the QIS problem have been given in [4]. Let us state the algorithm taking into account that it is mostly common to use reduced values which are equidimensional with the initial process, in instances, when power quality factors are evaluated (a generalized model for EMC estimation is given in [15]).

1. The simulation of ensemble initial realizations of a stochastic process.
2. The verification and correction of mean values and standard deviations in the simulated initial realizations.
3. The verification of the correlation function (CF) reproduction precision.
4. The verification of the initial process distribution law.
5. The quadratic transformation of the process realizations.
6. The verification of the ensemble quadratic process distribution law and ensemble quantiles.
7. The inertial smoothing of the quadratic process realizations.
8. Obtaining a reduced inertial process.
9. The calculation of the process realizations numerical characteristics (mean value, standard deviation, variance, 3-d and 4-th order distribution moments).
10. Averaging to obtain the ensemble numerical characteristics.
11. The ensemble skewness and kurtosis calculation.
12. The calculation of the simulation density and cumulative distribution function of process realizations.
13. Averaging to obtain the ensemble simulation density and cumulative distribution function.
14. The quantiles calculation (rated maxima) of quadratic inertial or reduced inertial processes (using the ensemble cumulative distribution function).

5 Ensemble characteristics

The proposed algorithm contains the calculation of process numerical characteristics. The single realizations calculation (symbol ~) and ensemble averaging (symbol ^) are used. Such characteristics as skewness and kurtosis are useful for the test problem solutions, while they are not the ultimate goal of calculations. The calculation of the skewness and kurtosis is performed with ensemble averaged values of standard deviation, 3rd and 4th central moments:

$$\hat{\mathrm{Sk}} = \frac{\hat{\mu}_3}{\hat{\sigma}_z^{1.5}} = \frac{\mathrm{M}\{\tilde{\mu}_3\}}{\mathrm{M}\{\tilde{\sigma}_z\}^{1.5}}, \quad \hat{\mathrm{Ex}} = \frac{\hat{\mu}_4}{\hat{\sigma}_z^2} - 3 = \frac{\mathrm{M}\{\tilde{\mu}_4\}}{\mathrm{M}\{\tilde{\sigma}_z\}^2} - 3. \tag{18}$$

Required rated maxima are determined with an ensemble cumulative distribution function

$$\hat{F}_z(z) = \mathrm{M}\{\tilde{F}_z(z)\} \tag{19}$$

6 Approximation by the gamma distribution law

The density can be approximated by dint of gamma distribution when $T = 0$:

$$f_{\vartheta 0}(\vartheta_0) = \frac{\lambda^{\eta}}{\Gamma(\eta)} \vartheta_0^{\eta-1} \exp\{-\lambda\vartheta_0\}, \tag{20}$$

where $\Gamma(\eta)$ – is complete gamma function determined by a dimensionless shape parameter η. The scale parameter λ has a dimension of $(\mathrm{pu})^{-2}$. Both parameters depend on the effective value I_e of the initial process $I(t)$ and standard deviation σ_z of the quadratic process $z(t)$. In case of the quadratic inertialess transformation:

$$\lambda = \frac{I_e^2}{\sigma_{\vartheta 0}^2} = \frac{I_c^2 + \sigma_I^2}{2\sigma_I^2(\sigma_I^2 + 2I_c^2)},$$
$$\eta = \frac{I_e^4}{\sigma_{\vartheta 0}^2} = \frac{(I_c^2 + \sigma_I^2)^2}{2\sigma_I^2(\sigma_I^2 + 2I_c^2)}. \tag{21}$$

Let us define numerical characteristics of the gamma distribution when approximating. The mean value is

$$\vartheta_{0c} = \frac{\eta}{\lambda} = \frac{I_e^4/\sigma_{\vartheta 0}^2}{I_e^2/\sigma_{\vartheta 0}^2} = I_e^2. \tag{22}$$

The standard deviation is

$$\sigma_{\vartheta 0} = \frac{\sqrt{\eta}}{\lambda} = \sqrt{2}\sigma_I\sqrt{\sigma_I^2 + 2I_c^2}. \tag{23}$$

The skewness and kurtosis are

$$\mathrm{Sk} = \frac{2}{\sqrt{\eta}} = \frac{2}{\sqrt{I_e^4/\sigma_{\vartheta 0}^2}} = \frac{2}{I_e^2/\sigma_{\vartheta 0}} = \frac{2\sqrt{2}\sigma_I\sqrt{\sigma_I^2 + 2I_c^2}}{\sigma_I^2 + I_c^2}, \tag{24}$$

$$\mathrm{Ex} = \frac{6}{\eta} = \frac{6}{I_e^4/\sigma_{\vartheta 0}^2} = \frac{12\sigma_I^2(\sigma_I^2 + 2I_c^2)}{(I_c^2 + \sigma_I^2)^2}. \tag{25}$$

If the mean value of the initial process is zero, the densities calculted according to the formulae (12) and (20) are exactly the same. But when $I_c > 0$, the distribution types are different (Fig.3).

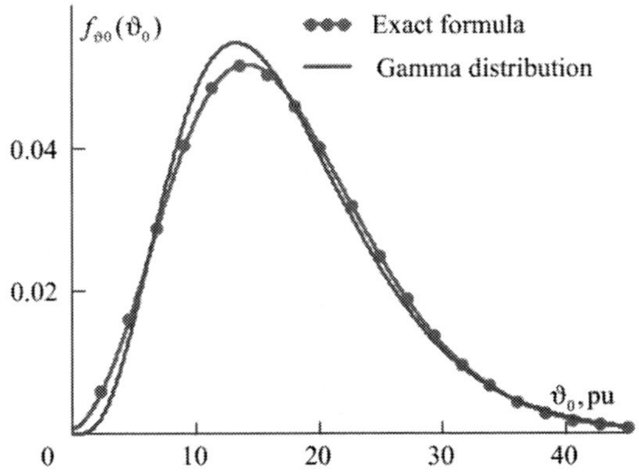

Fig. 3. The distribution density of a quadratic stationary stochastic process and gamma distribution law (I_{c*} = 4 pu).

For the problem under consideration the quality of the approximation is evaluated based on how accurately the inverse distribution of a quadratic process is reproduced when E_x = 0.05. The percentage error is calculated using a formula :

$$\delta = \left(\frac{\vartheta_{0M\Gamma}}{\vartheta_{0M}} - 1\right)100\%,$$

where ϑ_{0M} is an exact solution according to (12), $\vartheta_{0M\Gamma}$ is an approximate solution according to (21). The calculation results for different mean values are given in Table 1.

Table 1. Errors of approximation: $T_* = 0$, $E_x = 0.05$.

I_{c*}, o.e.	ϑ_{0M*}	$\vartheta_{0M\Gamma*}$	δ, %
0	3.841	3.841	0
1	7.002	6.928	−1.057
2	13.285	13.366	0.61
3	21.575	21.804	1.061
4	31.867	32.183	0.992
5	44.154	44.528	0.847

While approximating the error when defining the rated maxima can be both negative and positive.

Let us consider the gamma distribution characteristics in case of a nonzero inertia. The distribution density when T > 0 is

$$f_{\vartheta T}(\vartheta_T) = \frac{\lambda^{\eta}}{\Gamma(\eta)} \vartheta_T^{\eta-1} \exp\{-\lambda\vartheta_T\}. \tag{26}$$

The parameters are defined according to the expressions:

$$\lambda = \frac{I_e^2}{\sigma_{\vartheta T}^2}, \quad \eta = \frac{I_e^4}{\sigma_{\vartheta T}^2}. \tag{27}$$

Taking into account (10), in pu

$$\lambda_* = \frac{I_{c*}^2 + 1}{\dfrac{2}{1+2T_*} + \dfrac{4I_{c*}^2}{1+T_*}},$$
$$\eta_* = \frac{(I_{c*}^2 + 1)^2}{\dfrac{2}{1+2T_*} + \dfrac{4I_{c*}^2}{1+T_*}}. \tag{28}$$

7 Simulation

The initial simulation process $I(t)$ consists of m = 250 realizations with $I_{c*} = 0$ and exponential CF. Single realization minimum and maximum values of z_M (Fig.4, large bright circles) were 3.613 and 4.044 pu^2.

Ensemble value $\hat{z}_M = 3.837$ pu^2 (a large dark circle). The evaluation by formulas (12) gives the exact value equal to 3.841 pu^2.

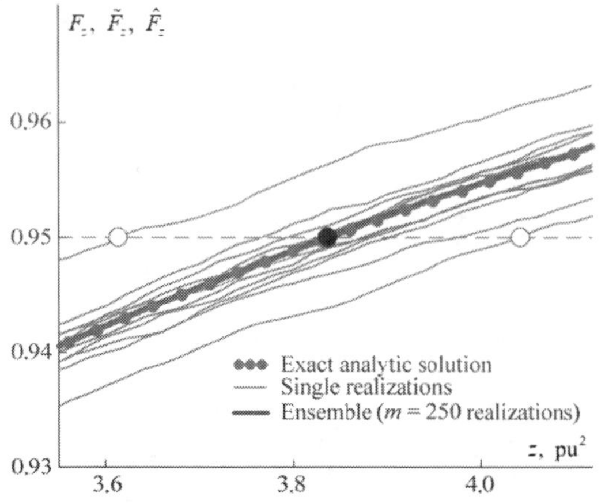

Fig. 4. Cumulative distribution functions of a quadratic stochastic process and rated maxima.

The maximum and minimum ratio errors for the single realization rated maxima z_M were −5.94% and 5.29%. The ratio error for the ensemble was 0.104%. It confirms the necessity to use ensemble of realizations of a stochastic process in the simulation solution of nonlinear transformations of the stochastic power processes, because even inertialess nonlinear transformation of the estimated maxima for individual realizations are reproduced with considerable error.

Simulation skewness and kurtosis of a quadratic inertial process $\vartheta_{T*}(t)$ taken over ensemble and realizations are given in Fig.5. The exact solution is calculated according to (17).

The distribution moments based on the single realizations differ greatly from the exact values. As it can be expected the dispersion for kurtosis is greater than the dispersion for skewness. The ensemble-averaged values of these characteristics are close to the exact values.

The simulation solution of the QIS problem is given in Fig.6 in the form of T_*-characteristic. The simulation is performed for every value of I_{c^*}. The results of the approximation with the help of the gamma distribution based on the formulae (26-28) are marked with circles.

The quantitative meanings of maxima and approximation errors are partially shown in Tables 2 and 3. It is necessary to point out that the approximation errors of the process rated maxima $\vartheta_{T*}(t)$ when E_x = 0.05 do not exceed 1.2% as per their absolute value. At the same time when E_x = 0.001 the approximation quality essentially deteriorates. According to the experiment data there is no steady error reduction.

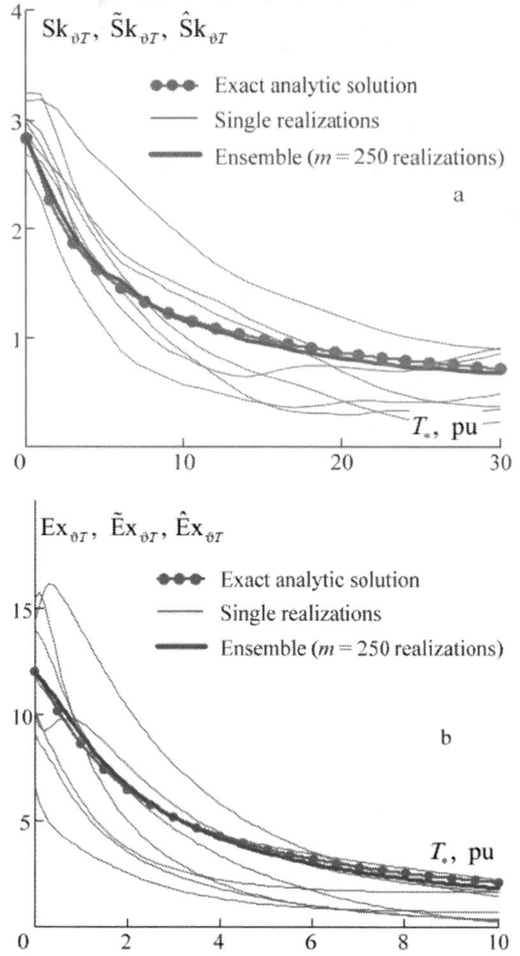

Fig. 5. Numerical characteristics of quadratic inertial process: skewness (a) and kurtosis (b).

The maximum error is 1.176% when E_x = 0.05, and −14.562% when E_x = 0.001.

Table 2. Errors of approximation: I_{c^*} = 1, E_x = 0.05.

T_*	$\hat{\vartheta}_{TM*}$	$\vartheta_{TM\Gamma*}$	δ, %
1	5.272	5.21	−1.176
3	4.195	4.154	−0.977
9	3.309	3.294	−0.453
15	3.001	3.003	0.067
21	2.833	2.845	0.424
30	2.681	2.704	0.698

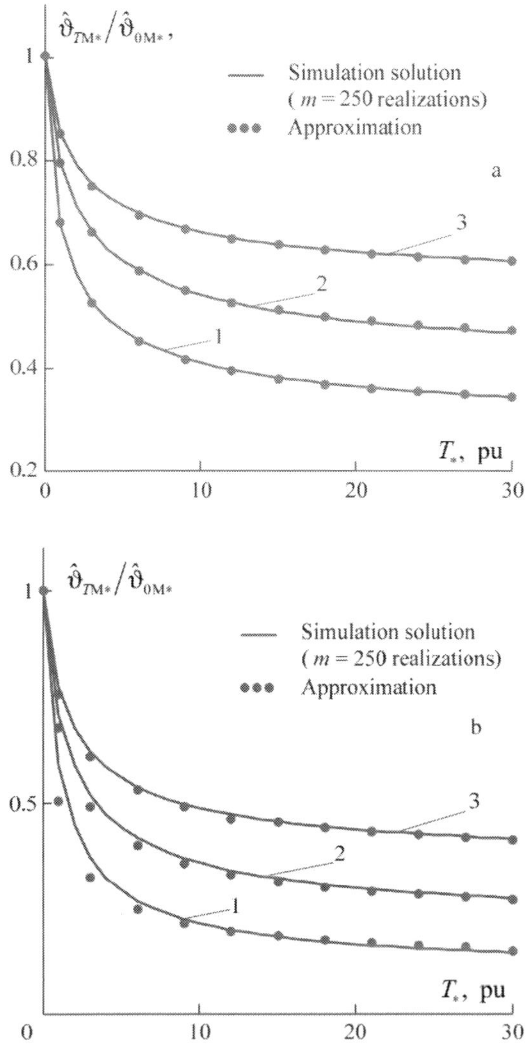

Fig. 6. T_* -characteristics of quadratic inertial process and their approximation when E_x = 0.05 (a) and E_x = 0.001 (b): 1 – I_{c*} = 0 pu; 2 – I_{c*} = 2 pu; 3 – I_{c*} = 4 pu.

Table 3. Errors of approximation: I_{c*} = 0, E_x = 0.001.

T_*	$\hat{\vartheta}_{TM*}$	$\vartheta_{TM\Gamma*}$	δ, %
1	6.353	5.442	−14.652
3	3.983	3.475	−12.749
9	2.403	2.306	−4.039
15	1.962	1.971	0.446
21	1.747	1.8	1.891
30	1.568	1.611	4.768

8 Conclusion

To increase the simulation solution accuracy of the nonlinear problems of ergodic electric-power process transformation it is reasonable to fulfil the statistical analysis of ensemble realizations rather than single realizations.

The rated maxima of the quadratic inertial process can be approximated by gamma distribution with a relatively low error when E_x = 0.05.

References

1. EN 50160:2010 Voltage characteristics of electricity supplied by public distribution networks.
2. A. Shydlovsky, E. Kourennyi, *Introduction to the statistical dynamics of power supply systems* (Kiev, Naukova dumka, 1984)
3. PUE 7th Edition, all sections and exhibits.
4. E. Kourennyi, E. Dmitrieva, A. Bulgakov, Proceedings of the RAS. Power Engineering, **4**, 109-122 (2016)
5. E. Kourennyi, E. Dmitrieva, A. Lyutyi, A. Bulgakov, Electrical Technology Russia (The Electricity), **1**, 12-20 (2016)
6. IEC 61000-4-15:2010, Electromagnetic compatibility (EMC)-Testing and measurement techniques – Flickermeter - Functional and design specifications
7. E. Kourennyi, A. Bulgakov, Russian Electromechanics, **5**, 75-81 (2016)
8. E. Wentzel, *Probability Theory* (Moskow, Knorus, 2010)
9. A. Papoulis, S. Unnikrishna Pillai, *Probability, Random Variables and Stochastic Processes, 4th ed.* (New York, McGraw-Hill, 2002)
10. A. Sveshnikov, *Applied methods of the theory of random functions* (Moskow, Nauka, 1986)
11. Z. Tianshu,S. Wanxing, S. Xiaohui, M. Xiaoli, S. Changkai, *UKSim 15th International Conference on Computer Modelling and Simulation,* 519-524 (2013)
12. N. Malysheva, A. Pavlov, *Dynamics of Systems, Mechanisms and Machines,* 1-5 (2016)
13. A. Saltykov, Russian Electrical Engineering, **8**, 454-456 (2008)
14. E. Dmitrieva, Electrical Technology Russia (The Electricity), **8,** 15-21 (2008)
15. V. Kuznetsov, E. Kourennyi, A. Lyutyi, *Electromagnetic compatibility. The nonsymmetry and nonsinusoidality of voltage* (Donetsk, Nord press, 2005)

MATEC Web of Conferences 113, 01002 (2017) DOI: 10.1051/matecconf/201711301002
12th International Scientific-Technical Conference on Electromechanics and Robotics "Zavalishin's Readings" - 2017

The research of heating efficiency of different induction heating systems

Sergey Konesev[1], *Regina* Khazieva[1*], *Alexander* Konev[1] and *Eduard* Kondratyev[2]

[1]Ufa State Petroleum Technological University, 450062 Ufa, Russia,
[2]Gazprom dobycha Yamburg, 629306, New Urengoy, Russia

Abstract. Computer models of tape and coil inductors are described, and a comparison of the heating efficiency depending on various parameters is made. The developed computer model was made in the ELCUT 6.0. As a result of the simulation, data on the heating characteristics (depending on the various parameters of the heating elements) are obtained. The average statistical data of a series of experiments with a tape inductor are given. It is shown that for the same parameters (values of inductance and number of wires), the tape version inductor heats up a pipe to a higher temperature (by 5.08%) than the inductor in the coil version in 10 minutes.

Introduction

The oil and gas industry requires heating systems: in the extraction and transportation of high-viscosity oil, it is necessary to exert a thermal effect to reduce viscosity, facilitate of the pumping process, and reduce the load on pumps and drives [1, 2].

The authors conducted an analytical review of methods and devices for regulating the rheological properties of high-viscosity oil, as well as the use of induction technologies for heating pipelines in foreign articles with a depth of analysis of 10 years [3, 4]. In response to these problems, different technical solutions such as microwave heating [5] or solar water heating system have been proposed [6].

The most effective and the safest heating systems in this case are electrothermal systems (ETS) [7, 8]. These systems include cable (resistive) systems, inductive-resistive systems to contact temperature maintenance, as well as the induction heating systems operating at industrial and medium-high frequencies [9-12]. Advantages of using such systems are high energy efficiency, the possibility of localizing the area of heating, simplification of installation and exploitation [13-15].

Leading countries in the development and patenting of methods and devices for influencing the rheological properties of viscous, high-viscosity and bituminous oil have been identified [16, 17].

An analysis of the patenting dynamics of methods and devices for regulation of the rheological properties of high-viscosity oil based on international patented bases and patent bases of Russia showed the interest of leading world companies in the implementation of new technologies and devices such as induction heating for regulating the rheological properties of anomalous oils [18, 19]. This interest is due to the development of the deposits with viscous, high-viscosity and bituminous oil against the background of depleted reserves of exploited deposits [20].

The paper presents an analytical comparison of the heating efficiency of various induction systems, such as multi-functional integrated electromagnetic component [21] and the classical inductor in coil design. The results for the analysis were obtained during computer simulation and were confirmed experimentally on laboratory samples.

1 Electrothermal systems based on induction heating

The following requirements are imposed on electrothermal systems (ETS) used at oil and gas facilities:

- providing two modes: maintaining temperature and emergency heating;
- power regulation;
- process automation and staff reduction;
- the operating of the ETS while limiting power supply [22-24].

The application use of induction heating systems operating at medium-high frequencies (IHS) is preferable, since they have better controllability for the transfer of heat power, which is necessary under different conditions of oil transfer compared to systems operating at an industrial frequency [25-28].

Induction heating systems consist of an inductor and a power source with a link of increased frequency [29-31]. By varying the frequency of commutations, it is possible to regulate the power of the IHS, and consequently the temperature and volume of the pumped liquid [32-34].

* Corresponding author: khazievart@mail.ru

Using of IHS allows implementing the following basic methods of pipelines heating: local, associated, locally-associated, locally-stepped.

2 Realization of locally-stepped heating on the basis of multi-functional integrated electromagnetic components

The method of locally-stepped heating, which consists in several intensive heatings of viscous fluid at short sections of the pipeline is the most effective in terms of reliability and simplicity of installation [35, 36].

This method is implemented using the system (device) of locally-stepped heating proposed by the authors, which contains a power source, control system, heating elements placed on the pipeline at intervals determined by the temperature mode and the technological process of pumping.

The heating elements are made in the form of multi-functional integrated electromagnetic components (MIEC) [21] and placed on the pipeline at intervals determined by the temperature mode and the technological process of pumping (the experimental layout is shown in Figure 1). Conductive membranes of MIEC fulfill the role of inductor, so that MIEC can be represented as a tape inductor.

The device works with an accompanying network and can receive power from two wires from the positive and negative poles of the power supply. The positive pole of the power source is connected to beginning of the first winding of the heating element; the negative pole of the power supply is connected to the second winding of the heating element. There is a switch added for each heating element. Switch is connected to the end of the first and to the beginning of the second winding of the heating element. One embodiment is a device in which the negative pole of the power source and the end of the heating element second winding are electrically connected to the pipeline [37].

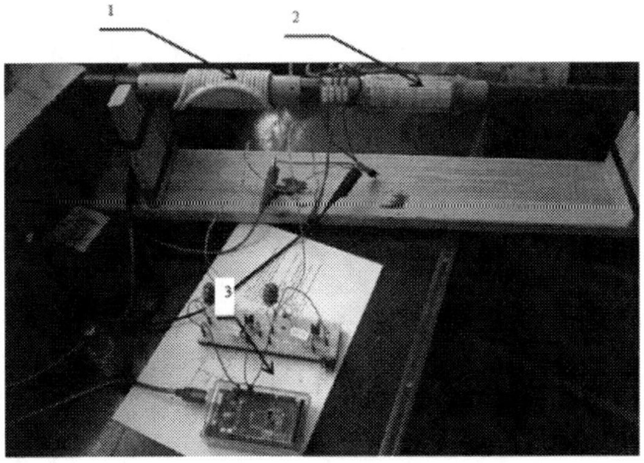

Fig. 1. Coil and tape inductor experimental models (1 – the coil inductor, 2 – the tape inductor, 3 – control system).

To evaluate the effectiveness of the high-viscosity oil rheological properties regulating method, it is necessary to study the thermal, electrodynamic and mechanical processes occurring in IHS heating elements of local-step heating.

The tape inductor is structurally constituted by two layers of copper foil stacked one on top of the other, separated by polyimide film insulation. Inductor is made in the form of a cylinder placed on the pipeline. The sizes of copper foil and polyimide insulation used during the experiment are shown in Table 1. Figure 2 shows images of coil and tape inductors. Figure 3 shows the structure of the tape inductor.

Table 1. Characteristics of copper film and polyimide insulation used during the experiment

	Thickness, μm	Width, cm
Copper film	20	12
Polyimide insulation	10	11

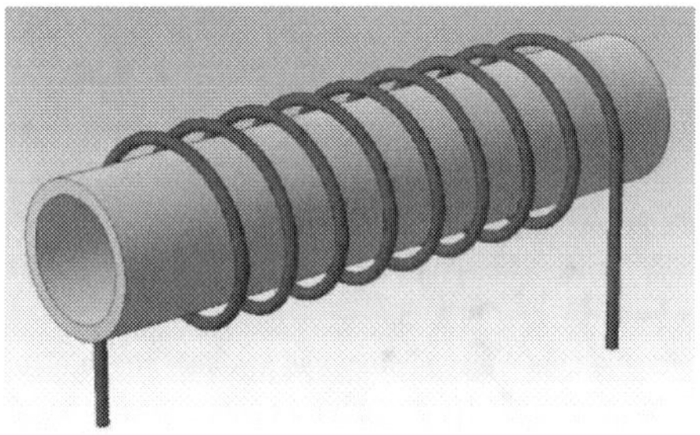

Fig. 2. Coil inductor.

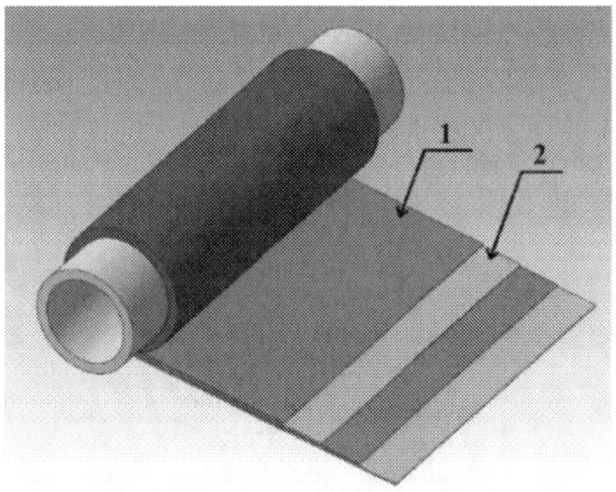

Fig. 3. Tape inductor structure (1 – copper film, 2 – polyimide insulation).

The actual task is to analyze the heating efficiency of single inductors in the coil and tape design, used as the parts for constructing a locally-stepped heating system.

3 Results of experimental and computer modelling

The authors have developed computer models of tape and coil inductors (Fig. 4) and compared their heating efficiency depending on various parameters. Computer model was made in the ELCUT 6.0 program.

For modeling in the program ELCUT 6.0 it is necessary to specify the geometric data of the inductors.

The winding of the inductor in the tape version is made of a copper film with a thickness of 20 mm, a width of 10 cm; a polyimide film with a thickness of 12 mm and a width of 11 cm was used as the insulation. The winding of the coil inductor is made of a copper wire of 2 mm^2 cross section. Figure 4 shows the inductor model in the ELCUT 6.0, and Figure 5 presents the thermogram of the inductor covered by a cover.

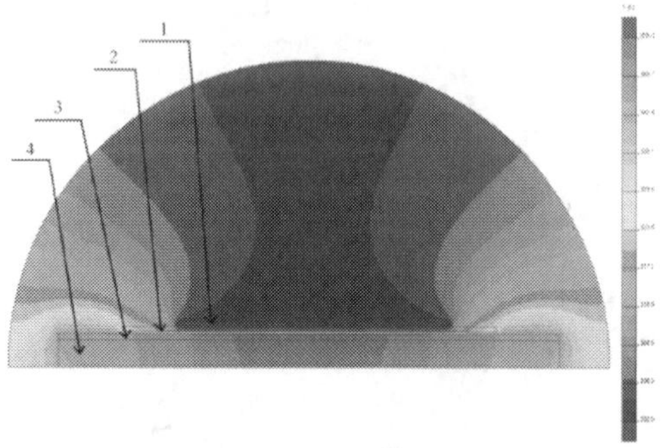

Fig. 4. Tape inductor model in ELCUT 6.0 (1 – tape inductor, 2 – plastic frame, 3 – pipe wall, 4 – empty pipe cavity).

Fig. 5. The thermogram of the tape inductor covered by a cover.

4 The discussion of the results

Table 2 shows the pipe temperature values obtained during the simulation for the given parameters (U = 40 V, f = 100 kHz (for the tape inductor), f = 50 Hz (for the coil inductor), number of turns = 23).

Table 2. Comparison of the data obtained during the simulation at U = 40 V, f = 100 kHz, n = 23.

Time, min	Heating temperature of the pipe under the inductor, ° C		
	Coil inductor model	Tape inductor model	Experiment with tape inductor
0	29	29	29
1	36,78	37,19	35,06
2	43,37	44,16	43,25
3	48,99	50,14	49,13
4	53,79	55,31	53,63
5	57,92	59,82	56,88
6	61,49	63,77	60,13
7	64,59	67,25	63,19
8	67,29	70,35	66,19
9	69,67	73,12	68,56
10	71,76	75,6	71,56

As a result of the simulation, data on the heating characteristics for the coil and tape inductors are obtained (Fig. 6 – 8).

The adequacy of the obtained model was confirmed experimentally. The temperature was derived by the digital dht22 type sensors, located on the tube in the center of each inductor. Data from the sensors were taken within ten minutes. The highest value difference between the results obtained in the experiments compared with the results obtained during the simulation was 6.2%. The experiments were carried out on the prototype IHS pipeline, which is a pipe section with a tape inductor and coil inductor placed on it. Table 2 shows the average statistical data of a series of experiments with a tape inductor.

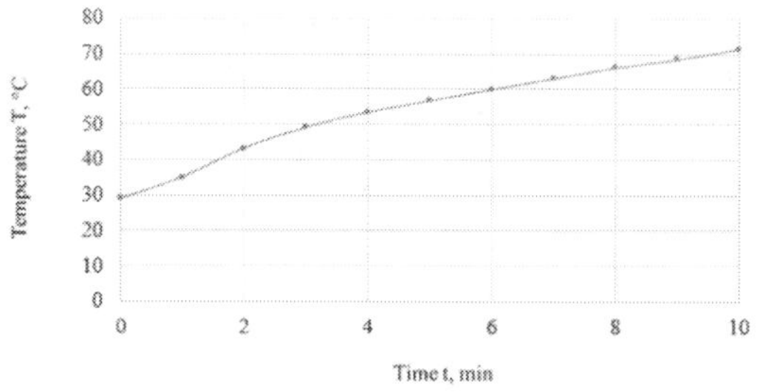

Fig. 6. The results of the efficiency of tube heating, obtained for coil inductor model.

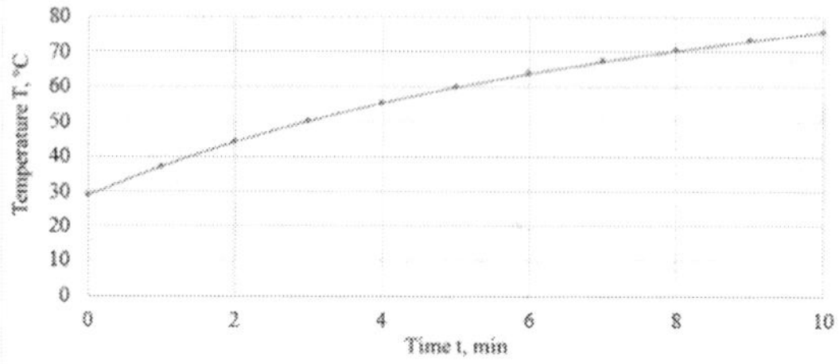

Fig. 7. The results of the efficiency of tube heating, obtained for tape inductor model.

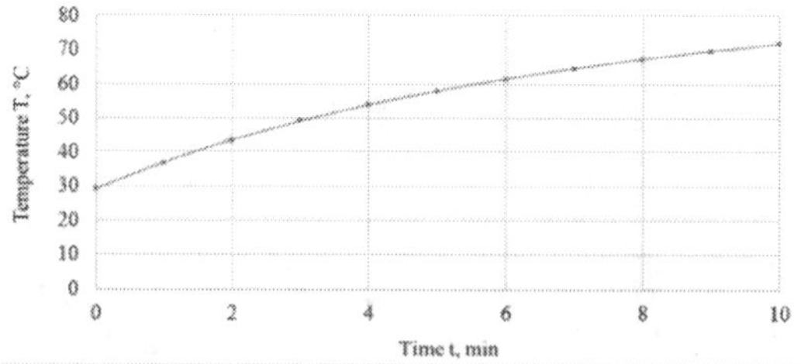

Fig. 8. The results of the efficiency of tube heating, obtained for experiment with tape inductor.

5 Conclusion

With the help of the obtained model, the character of the pipe heating with the tape and coil inductors (depending on various inductors parameters) was evaluated. It is shown that for the same parameters (inductance value, number of turns), the tape inductor made in the form of

MIEK is heated to a higher temperature (by 5.08%) for 10 minutes than the coil version inductor.

References

1. S.G. Konesev, R.T. Khazieva, P.A. Khlupin, E.Y. Kondratyev. Oil and gas business (electronic scientific journal), **5**, 179-189 (2013)
2. S.A. Gorbatkov, A.B. Kuvaldin, V.E. Mineev, V.E. Zhukovsky. Chemistry. **176** (1985)
3. M. Fabbri, A. Morandi, P.L. Ribani. The international journal for computation and mathematics in electrical and electronic engineering, **2**, 480-490 (2008)
4. Y. Pleshivtseva, P. DI Barba, E. Rapoport, B. Nacke, A Nikanorov., S Lupi., E. Sieni, M. Forzan. International journal of microstructure and materials properties, **6**, 532-551 (2014)
5. V. Jakubauskaite, J. Dikun, V. Jankunas, E. Guseinoviene, A. Zukauskaite, V. Kvedaras, *Tenth International Conference on Ecological Vehicles and Renewable Energies (EVER)*, 1-10 (2015)
6. S. Arvas, H. Sajjad, A. Altaf, S. Khan, E. Arvas, *IEEE Asia-Pacific Conference on Applied Electromagnetics (APACE)*, 319-322 (2016)
7. S.G. Konesev, I.A. Makulov, N.M. Mamaev, Oil and Gas Business, **2**, 75-80 (2008)
8. S.G. Konesev, I.A. Makulov, *Electrotechnologies, electric drive and electrical equipment of enterprises*, 117-126 (2007)
9. P. Sturman, R.A. Gray, Composites Part A: Applied Science and Manufacturing, **11**, 1109-1110 (1996)
10. M.M. Semenov, M.M. Batishchev. Patent 2304369 (2007)
11. A.Z. Sahin, *Heat and Mass Transfer*, **1-2**, 133-139 (1996)
12. S.G. Konesev, P.A. Khlupin, I.A. Makulov, *Electric drive, electrotechnology and electrical equipment of enterprises*, 16-20 (2009)
13. S.G. Konesev, V.M. Magzyumov, M.R. Sadikov, *Innovative directions of development of electric drive, electrical technologies and electrical equipment*, 168-177 (2012)
14. S.G. Konesev, P.A. Khlupin, M.R. Sadikov, *Electric drive, electrotechnology and electrical equipment of enterprises*, 212-219 (2011)
15. S.M. Dyviatovsky. Patent 109949 (2011)
16. S. Kazutaka, Y. Hishinuma, A. Nishimura, Journal of the korean physical society, **1** (2006)
17. L.F. Quevedo, C.G. Mancheno, J.A. Morales, F.A. Quizhpi, *IEEE International Autumn Meeting on Power, Electronics and Computing (ROPEC)*, 1-6 (2016)
18. S.G. Konesev, R.T. Khazieva, P.A. Khlupin, E.Y. Kondratyev, Oil and gas business (electronic scientific journal), **5**, 179-189 (2013)
19. S.A. Gorbatkov, A.B. Kuvaldin, V.E. Mineev, V.E. Zhukovsky, Chemistry, 176 (1985)
20. H. Liu, J. Rao, *Second International Workshop on Education Technology and Computer Science*, 174-177 (2010)
21. S.G. Konesev, R.T. Khazieva, R.V. Kirillov, E.Yu. Kondratyev, M.R. Sadikov, P.A. Khlupin, Oil and gas business, **4**, 40-47 (2014)
22. A.I. Elshin, V.M. Kazansky, Patent 21709 (2002)
23. Yu.V. Zverev, A.N. Lapin, Yu.B Tsykarev, Patent 2301378 (2007)
24. Q. Yang, Y. Liu, A. Gu, J. Ding, Z. Shen. Chemical Engineering Science, **6**, 921-931 (2002)
25. V.A. Petrov, P.O. Voronov, Patent 48139 (2005)
26. D. Leutard, G.L. Vignoles, F. Lamouroux, B. Bernard. Journal of Materials Synthesis and Processing, **5**, 259-273 (2001)
27. A.N. Bragin, V.A. Bardokin, Patent 80085 (2009)
28. K. Sadeghipour, J.A. Dopkin, Computers in Industry, **3**, 195-205 (1996)
29. V.A. Cherepanov, Patent 2138137 (1999)
30. C. Mousty, G. Foti, C. Comninellis, V. Reid, *Electrochimica Acta*, **3**, 451-456 (1999)
31. R. Rudolf, P. Mitschang, M. Neitzel, Composites Part A: Applied Science and Manufacturing, **11**, 1191-1202 (2000)
32. P. Mocho, J.C. Bourhis, P. Le Cloirec, Carbon, **7**, 851-856 (1996)
33. D. Pal, V.A. Venkataraman, Sensors and Actuators A: Physical, **1-2**, 151-156 (2002)
34. E.D. Karmanov, S.I. Shaplov, Patent 2263418 (2005)
35. S.G. Konesev, R.T. Khazieva, R.V. Kirillov, E.Yu. Kondratyev, Patent 154768 (2015)
36. S.G. Konesev, Multifunctional integrated electromagnetic component. Patent 2585248 (2016)
37. S.G. Konesev, R.T. Khazieva, R.V. Kirillov, E.Yu. Kondratyev, M.R. Sadikov, Patent 2584137 (2016)

MATEC Web of Conferences **113**, 01003 (2017) DOI: 10.1051/matecconf/201711301003
12th International Scientific-Technical Conference on Electromechanics and Robotics "Zavalishin's Readings" - 2017

Stability markers in control theory

Egor Polilov[1,*] and *Olga* Batiushina[2]

[1]Donbass State Technical University, Department of Automation and Electrical Engineering Systems, Ukraine
[2]Donbass State Technical University, Department of Management, Ukraine

Abstract. The linear dynamic systems stability criterion, allowing one to identify the defective/failed coefficients in the characteristic polynomial at each time quantum, is presented for the first time.

1 Introduction

Is there a great probability to generate Hurwitz's polynomial in a random way? And, fundamentally, is it possible to achieve Hurwitz's features with a selection $(n+1)$ pieces of $\sigma_j, \forall j=\overline{0,n}$ for the high order systems? It is appropriate to concentrate the answer to such questions on the extent/volume assessing of the two clusters, stability and instability, as the interaction of extreme opposites.

It should be admitted that the potential of the so-called stability criteria [1, 2] by Hermite-Biller's theorem about the radicals alternation, Sturm's method for the rational function index calculation, Routh's algorithm, Hurwitz and Lienard-Schipar's determinantal conditions, quadratic forms based on Schur's method, Chebotarev's theorem, a generalization for the entire functions of Hermite-Biler's theorem, Pontryagin's theorem on stable quasipolynomials, *and others* has been exhausted for formulated problems solving and it is quite insignificant despite of the apparent theoretical harmony and algorithmic/computational clutters. Is it necessary to confirm the stability or instability dynamical systems? Out-of-date. With the same success, the step response $h(t)$ step(sys) is the essence of the answer to this question, in addition geometrically. And a map of the radicals pzmap(sys) distribution is more informative (step(sys) and pzmap(sys) the computational procedures of Control System Toolbox in MATLAB). But, in fact, besides the ascertaining of the instability one should clearly understand what to do with non Hurwitz's polynomial? When a patient needs emergency medical care, it is necessary to provide it qualifiedly, but not contemplating the obvious fact to convince yourself that it is really required.

In [3, 4] authors propose a method for computing the distance of a stable polynomial to the set of unstable ones (both in the Hurwitz and in the Schur case). The method is based on the reformulation of the problem as the structured distance to instability of a companion matrix associated to a polynomial.

The ideology of stability markers, which at each time quantum t_k visually identifies the defective/failed $\sigma_k(t), \forall k \in \overline{0,n}$ from a set of the coefficients $\sigma_j(t), \forall j=\overline{0,n}$ of the nonstationary algebraic equation:

$$G(p,t)=\sigma_n(t)p^n+\sigma_{n-1}(t)p^{n-1}+\ldots+\sigma_0(t)=0, \quad (1)$$

and points the recovery ways of each coefficient $\sigma_k(t), \forall k \in \overline{0,n}$, so it is the important issue. There are no similar solutions in the automatic control theory.

2 Material and researches results

Definition. There is the $2\times n$ pieces set spectrum of the geometric averages $\Omega_k(t)$: $\underset{j=\overline{1,n-k}}{\text{gmean}}\left(c_j(t)\right)$, $\forall k \in \overline{0,n}$ locations of the instant dynamic solutions $\langle c_1(t), c_2(t), \ldots, c_{n-k}(t)\rangle_k$, $\forall k \in \overline{0,n}$, fig. 1, formed with a serial dropping of the polynomial (1) highest (2) and youngest (3) degrees. New solutions of the so-called enclosed polynomials of $(n-i,j)$ orders define other combinations of valid and complex conjugated radicals on the complex plane. In their turn, these radicals are geometrically averaged with the own "universe brick" $R_{i(k)}(t)=\sqrt[n-i]{\left|\sigma_0(t)/\sigma_{n-i}(t)\right|}$, $\forall k \in \overline{0,n}$ of the encloses on the right:

$$0=\underbrace{\left|\sigma_n(t)p^n+\underbrace{\left|\sigma_{n-1}(t)p^{n-1}+\underbrace{\left|\ldots+\underbrace{\left|\sigma_1(t)p+\sigma_0(t)\right.}_{R_{i(1)}(t)=\left|\sigma_0(t)/\sigma_1(t)\right|}\right.}_{\ldots}\right.}_{R_{i(n-1)}(t)=\sqrt[n-1]{\left|\sigma_0(t)/\sigma_{n-1}(t)\right|}}\right.}_{R_{i(n)}(t)=\sqrt[n]{\left|\sigma_0(t)/\sigma_n(t)\right|}}, \quad (2)$$

---
[*] Corresponding author: egor.polilov@gmail.com

and $R^*_{i(k)}(t) = \sqrt[n-j]{|\sigma_{n-j}(t)/\sigma_n(t)|}, \forall j \in \overline{0,n}$ of the encloses on the left:

$$0 = \underbrace{\underbrace{\underbrace{\underbrace{\sigma_n(t)p^n + \sigma_{n-1}(t)p^{n-1}}_{R^*_{i(1)}(t)=|\sigma_{n-1}(t)/\sigma_n(t)|} + \ldots}_{\ldots} + \sigma_1(t)p}_{R^*_{i(n-1)}(t)=\sqrt[n-1]{|\sigma_1(t)/\sigma_n(t)|}} + \sigma_0(t)}_{R^*_{i(n)}(t)=\sqrt[n]{|\sigma_0(t)/\sigma_n(t)|}}. \quad (3)$$

Here and further $R_{i(k)}$ and $R^*_{i(k)}$ are the radii of Polilov's- Motchenko's invariants [5, 6].

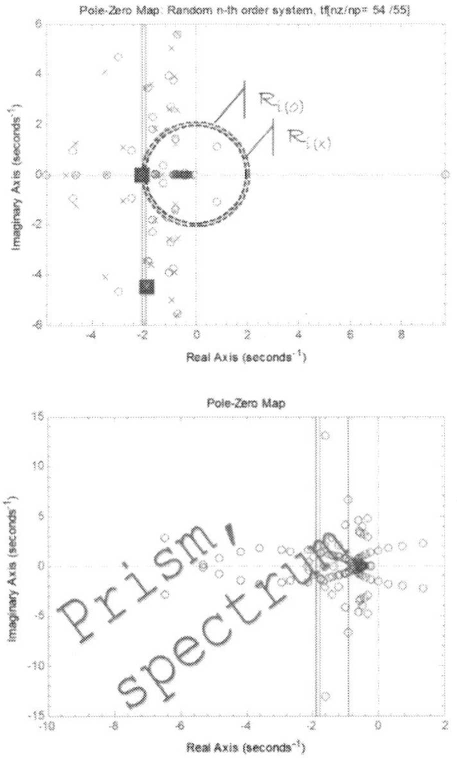

Fig. 1. A distribution map of radicals, geometric averages $R_{i(0)}$ and $R_{i(\times)}$ and spectrum Ω_k $\forall k \in \overline{0, 2\times n}$ of the characteristic polynomial (verticals, 3 of them are shown).

The horizontal and vertical numbering of the $\mathbf{P}(t)$ matrix (4) cells will be associated with the degrees of the polynomial (1), beginning from the highest left top one $\sigma_{n-i}(t)p^{n-i}$ and $\sigma_{n-j}(t)p^{n-j}$ for $i,j = \overline{0,n}$, accordingly, thereby the localization of each cell $\mathbf{P}_{ij}$ will be attached to the spectrum of the geometric averages:

$$\mathbf{P}_{G(p,t)} = \left[\begin{array}{c|c|cccc} & \sigma_n p^n & \sigma_{n-1}p^{n-1} & \ldots & \sigma_1 p & \sigma_0 \\ \hline \sigma_n p^n & 0 & 0 & 0 & 0 & 0 \\ \sigma_{n-1}p^{n-1} & \left|\frac{\sigma_{n-1}}{\sigma_n}\right| & 0 & 0 & 0 & 0 \\ \vdots & \vdots & \vdots & 0 & 0 & 0 \\ \sigma_{n-i}p^{n-i} & \sqrt[i]{\left|\frac{\sigma_{n-i}}{\sigma_n}\right|} & \sqrt[i-1]{\left|\frac{\sigma_{n-i}}{\sigma_{n-1}}\right|} & & 0 & 0 \\ \vdots & \vdots & \vdots & & & \vdots \\ \sigma_1 p & \sqrt[n-1]{\left|\frac{\sigma_1}{\sigma_n}\right|} & \sqrt[n-2]{\left|\frac{\sigma_1}{\sigma_{n-1}}\right|} & \vdots & 0 & 0 \\ \hline \sigma_0 & \sqrt[n]{\left|\frac{\sigma_0}{\sigma_n}\right|} & \sqrt[n-1]{\left|\frac{\sigma_0}{\sigma_{n-1}}\right|} & & \left|\frac{\sigma_0}{\sigma_1}\right| & 0 \end{array}\right], \quad (4)$$

$\sigma_j(t)$, $\forall j = \overline{0,n}$ are still non-stationary in the text here and further, time dependence symbol is omitted because of the matrix overcharging.

The first column and the last row of $\mathbf{P}$-matrices are those investments on the right (2) and on the left (3), which are separated by a dotted line. The other elements $\mathbf{P}_{ij}$ should be associated with the so-called hybrid investments, mentally leaving any adjacent components with k degree in the polynomial $\left(\sigma_{m-j}(t)p^{m-j}\ldots\sigma_i(t)p^i\right)$ during the deliverance from the part of the $j = \overline{1,m}$ oldest and the $i = \overline{0,j}$ youngest degrees p at the same time.

Definition. The linear non-stationary dynamic system is characterized by $m(m-1)/2$ numbers of the geometric averages $\Omega_k(t)$, $\forall k = \overline{0,m}$ of the polynomial numerator $A(p,t)$ zeros and by $n(n-1)/2$ numbers of the geometric averages $\Omega_k(t)$, $\forall k = \overline{0,n}$ of the selected transfer function polynomial denominator $B(p,t)$ poles

$$W(p,t) = \frac{A(p,t)}{B(p,t)} = k_0 \frac{\sum_{i=0}^{m} \alpha_{m-i}(t)p^{m-i}}{\sum_{j=0}^{n} \beta_{n-j}(t)p^{n-j}} = \quad (5)$$

$$= k_0 \frac{\alpha_m(t)p^m + \alpha_{m-1}(t)p^{m-1} + \ldots + \alpha_0(t)}{\beta_n(t)p^n + \beta_{n-1}(t)p^{n-1} + \ldots + \beta_0(t)},$$

forming two lower triangular $\mathbf{P}$ matrices, $(m+1,1)$ and $(n+1,1)$ cells of $\mathbf{P}$ matrices are nothing less than the geometric average $\Omega_k(t)$: $\underset{j=\overline{1,n-k}}{\text{gmean}}\left(c_j(t)\right)$, $\forall k \in \overline{0,n}$ of the instant dynamic solutions $\langle c_1(t),\ c_2(t),\ \ldots,\ c_{n-k}(t)\rangle_k$ location, $\forall k \in \overline{0,n}$ encloses of the numerator $A(p,t)$ polynomial and $\forall k \in \overline{0,n}$ attachments of the denominator $B(p,t)$ polynomial:

$$\mathbf{P}_{A(p,t)} =$$

$$\left[\begin{array}{c|c|cccc}
 & \alpha_m p^m & \alpha_{m-1}p^{m-1} & \ldots & \alpha_1 p & \alpha_0 \\
\hline
\alpha_m p^m & 0 & 0 & 0 & 0 & 0 \\
\alpha_{m-1}p^{m-1} & \left|\dfrac{\alpha_{m-1}}{\alpha_m}\right| & 0 & 0 & 0 & 0 \\
\vdots & \vdots & \vdots & 0 & 0 & 0 \\
\alpha_{m-i}p^{m-i} & \sqrt[i]{\left|\dfrac{\alpha_{m-i}}{\alpha_m}\right|} & \sqrt[i-1]{\left|\dfrac{\alpha_{m-i}}{\alpha_{m-1}}\right|} & & 0 & 0 \\
\vdots & \vdots & \vdots & & & \vdots \\
\alpha_1 p & \sqrt[m-1]{\left|\dfrac{\alpha_1}{\alpha_m}\right|} & \sqrt[m-2]{\left|\dfrac{\alpha_1}{\alpha_{m-1}}\right|} & \vdots & 0 & 0 \\
\hline
\alpha_0 & \underbrace{\sqrt[m]{\left|\dfrac{\alpha_0}{\alpha_m}\right|}}_{\Omega_{0\times}} & \sqrt[m-1]{\left|\dfrac{\alpha_0}{\alpha_{m-1}}\right|} & & \left|\dfrac{\alpha_0}{\alpha_1}\right| & 0
\end{array}\right], \quad (6)$$

$$\mathbf{P}_{B(p,t)} =$$

$$\left[\begin{array}{c|c|cccc}
 & \beta_n p^n & \beta_{n-1}p^{n-1} & \ldots & \beta_1 p & \beta_0 \\
\hline
\beta_n p^n & 0 & 0 & 0 & 0 & 0 \\
\beta_{n-1}p^{n-1} & \left|\dfrac{\beta_{n-1}}{\beta_n}\right| & 0 & 0 & 0 & 0 \\
\vdots & \vdots & \vdots & 0 & 0 & 0 \\
\beta_{n-i}p^{n-i} & \sqrt[i]{\left|\dfrac{\beta_{n-i}}{\beta_n}\right|} & \sqrt[i-1]{\left|\dfrac{\beta_{n-i}}{\beta_{n-1}}\right|} & & 0 & 0 \\
\vdots & \vdots & \vdots & & & \vdots \\
\beta_1 p & \sqrt[n-1]{\left|\dfrac{\beta_1}{\beta_n}\right|} & \sqrt[n-2]{\left|\dfrac{\beta_1}{\beta_{n-1}}\right|} & \vdots & 0 & 0 \\
\hline
\beta_0 & \underbrace{\sqrt[n]{\left|\dfrac{\beta_0}{\beta_n}\right|}}_{\Omega_{0\infty}} & \sqrt[n-1]{\left|\dfrac{\beta_0}{\beta_{n-1}}\right|} & & \left|\dfrac{\beta_0}{\beta_1}\right| & 0
\end{array}\right]. \quad (7)$$

In other words, the scaling constants of cells $\mathbf{P}_{ij} = \sqrt[i-j]{\left|\alpha^*_{m-i} / \alpha^*_{m-j}\right|}$ of $\mathbf{P}$-matrices define the measure of the dominance $\mathbf{P}_{ij} > 1$ or of the subordination $\mathbf{P}_{ij} < 1$ of any enclose from $\left(\alpha_{m-j}(t)p^{m-j} \ldots \alpha_i(t)p^i\right)$ to the invariant $\left(\alpha_m(t)p^m + \alpha_0(t)\right) = 0$.

The dimension of the subdiagonal cells of $\mathbf{P}$-matrices $[rad/\sec]$, all cells of $\mathbf{P}$-matrices (6), (7) are the frequencies, that can be seen obviously, if we present the transfer function (5) with the normalized polynomials $p_{\Omega_{(0)}} = p/\Omega_{(0)}$ and $p_{\Omega_{(\times)}} = p/\Omega_{(\times)}$:

$$W\left(p_{\Omega_{(0,\times)}},t\right) = \frac{A\left(p_{\Omega_{(0)}},t\right)}{B\left(p_{\Omega_{(\times)}},t\right)} = k_0 \frac{\sum_{i=0}^{m} \alpha^*_{m-i}(t)\, p_{\Omega_{(0)}}^{m-i}}{\sum_{j=0}^{n} \beta^*_{n-j}(t)\, p_{\Omega_{(\times)}}^{n-j}} = \quad (8)$$

$$= k_0 \frac{\alpha^*_m(t)\, p^m_{\Omega_{(0)}} + \alpha^*_{m-1}(t)\, p^{m-1}_{\Omega_{(0)}} + \ldots + \alpha_0(t)}{\beta^*_n(t)\, p^n_{\Omega_{(\times)}} + \beta^*_{n-1}(t)\, p^{n-1}_{\Omega_{(\times)}} + \ldots + \beta_0(t)},$$

then, for example, the polynomial $\left(\alpha_{m-1}(t)p^{m-1} \ldots \alpha_1(t)p\right)$ of the hybrid enclose first iteration in $A(p,t)$ determines the corresponding cell of the $\mathbf{P}$-matrix (6): $\sqrt[m-2]{\left|\dfrac{\alpha_1}{\alpha_{m-1}}\right|} = \sqrt[m-2]{\left|\dfrac{\left(\alpha^*_1/\Omega_{(0)}\right)}{\left(\alpha^*_{m-1}/\Omega^{m-1}_{(0)}\right)}\right|} = \underbrace{\sqrt[m-2]{\left|\alpha^*_1/\alpha^*_{m-1}\right|}}_{const,\,[\text{relative units}]} \times \underbrace{\Omega_{(0)}\,[rad/\sec]}_{\text{geometrical mean}}$ etc., and (6) is appropriate to present in the form:

$$\mathbf{P}_{A(p,t)} = \mathbf{P}^*_{A(p,t)}\,[\text{relative units}] \times \Omega_{(0)}\,[rad/\sec] = \quad (9)$$
$$= \mathbf{P}^*_{A(p,t)} \times \sqrt[m]{\left|\alpha_0/\alpha_m\right|},$$

$$\mathbf{P}^*_{A(p,t)} =$$

$$\left[\begin{array}{c|c|cccc}
 & \alpha^*_m p^m_{\Omega_{(0)}} & \alpha^*_{m-1}p^{m-1}_{\Omega_{(0)}} & \ldots & \alpha^*_1 p_{\Omega_{(0)}} & \alpha^*_0 \\
\hline
\alpha^*_m p^m_{\Omega_{(0)}} & 0 & 0 & 0 & 0 & 0 \\
\alpha^*_{m-1}p^{m-1}_{\Omega_{(0)}} & \left|\dfrac{\alpha^*_{m-1}}{\alpha^*_m}\right| & 0 & 0 & 0 & 0 \\
\vdots & \vdots & \vdots & 0 & 0 & 0 \\
\alpha^*_{m-i}p^{m-i}_{\Omega_{(0)}} & \sqrt[i]{\left|\dfrac{\alpha^*_{m-i}}{\alpha^*_m}\right|} & \sqrt[i-1]{\left|\dfrac{\alpha^*_{m-i}}{\alpha^*_{m-1}}\right|} & & 0 & 0 \\
\vdots & \vdots & \vdots & & & \vdots \\
\alpha^*_1 p_{\Omega_{(0)}} & \sqrt[m-1]{\left|\dfrac{\alpha^*_1}{\alpha^*_m}\right|} & \sqrt[m-2]{\left|\dfrac{\alpha^*_1}{\alpha^*_{m-1}}\right|} & \vdots & 0 & 0 \\
\hline
\alpha^*_0 & \underbrace{\sqrt[m]{\left|\dfrac{\alpha^*_0}{\alpha^*_m}\right|}}_{\triangleq 1} & \sqrt[m-1]{\left|\dfrac{\alpha^*_0}{\alpha^*_{m-1}}\right|} & & \left|\dfrac{\alpha^*_0}{\alpha^*_1}\right| & 0
\end{array}\right] -$$

the matrix of the radii scales of enclosed invariants $\left(\alpha_{m-j}(t)p^{m-j} \ldots \alpha_i(t)p^i\right)$ is presented relatively to $\left(\alpha_m(t)p^m + \alpha_0(t)\right) = 0$.

MATLAB $\mathbf{P}$-matrices filling procedure is presented below:

```
function pmatrix=buildpmatrix(den)
[m,n]=size(den);
pmatrix=zeros(n+1);
den_fliplr=fliplr(den);
for i=1:n
 for j=0:n-i
pmatrix(j+1+i,j+1)=nthroot(abs(den_fli
plr(n+1-j-i)/den_fliplr(n+1-j)),i);
 end end
```

Theorem. The nonstationary dynamic system with the transfer function

$$W(p,t)=\frac{A(p,t)}{B(p,t)}=k_0\frac{\sum_{i=0}^{m}\alpha_{m-i}(t)p^{m-i}}{\sum_{j=0}^{n}\beta_{n-j}(t)p^{n-j}}= \tag{10}$$

$$=k_0\frac{\alpha_m(t)p^m+\alpha_{m-1}(t)p^{m-1}+\ldots+\alpha_0(t)}{\beta_n(t)p^n+\beta_{n-1}(t)p^{n-1}+\ldots+\beta_0(t)},$$

is stable at each time quantum t_k if the component $\mathbf{P}_{ij}$ of each subdiagonal $2\times(n-2)$ pieces of the main and secondary diagonals, of each row $n-2$ pieces and each column $n-1$ of the denominator polynomial $B(p,t)$ $\mathbf{P}$ - matrix is under the lack of growth, as well as the same component $\mathbf{P}_{ij}$ of each superdiagonal $n-3$ pieces of her side diagonal, calculated at the same time quantum t_k is under decreasing (11).

$$\mathbf{P}_{B(p,t_k)}=$$

$$\left[\begin{array}{c|cccccc} & \beta_n p^n & \beta_{n-1}p^{n-1} & \cdots & \beta_1 p & \beta_0 & \\ \hline \beta_n p^n & 0 & 0 & 0 & 0 & 0 \\ \beta_{n-1}p^{n-1} & \left|\frac{\beta_{n-1}}{\beta_n}\right| & 0 & 0 & 0 & 0 \\ \vdots & \vdots & \vdots & 0 & 0 & 0 \\ \beta_{n-i}p^{n-i} & \sqrt[i]{\left|\frac{\beta_{n-i}}{\beta_n}\right|} & \sqrt[i-1]{\left|\frac{\beta_{n-i}}{\beta_{n-1}}\right|} & & 0 & 0 \\ \vdots & \vdots & \vdots & & & \vdots \\ \beta_1 p & \sqrt[n-1]{\left|\frac{\beta_1}{\beta_n}\right|} & \sqrt[n-2]{\left|\frac{\beta_1}{\beta_{n-1}}\right|} & \vdots & 0 & 0 \\ \hline \beta_0 & \sqrt[n]{\left|\frac{\beta_0}{\beta_n}\right|} & \sqrt[n-1]{\left|\frac{\beta_0}{\beta_{n-1}}\right|} & & \left|\frac{\beta_0}{\beta_1}\right| & 0 \end{array}\right]. \tag{11}$$

Each element $\mathbf{P}_{ij}$ of the lower triangular matrix $\mathbf{P}_{B(p,t_k)}$ is a linear dynamic system stability marker in the graphical interpretation of the adjacent and neighbouring $\mathbf{P}_{ij\pm1}$ relationship (Fig. 2).

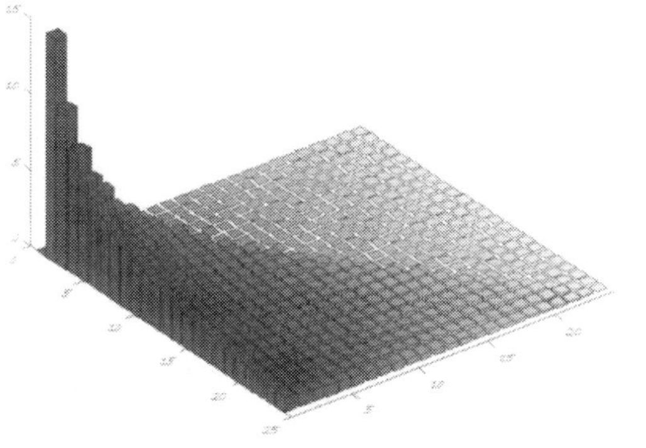

Fig. 2. Probability to generate this harmony in $\mathbf{P}$-matrix of the solutions space with a selection of random samples in the primary space – is equal to 0!

The columns reciprocal geometry, but not their absolute values, is the essential event. The mentioned facts in a formulaic interpretation correspond to:

$$\left|\frac{\beta_{n-1}}{\beta_n}\right|>\left|\frac{\beta_{n-2}}{\beta_{n-1}}\right|>\ldots>\left|\frac{\beta_{n-j-1}}{\beta_{n-j}}\right|>\ldots>\left|\frac{\beta_0}{\beta_1}\right| \text{ for the first,}$$

$$\ldots\sqrt[i]{\left|\frac{\beta_{n-i}}{\beta_n}\right|}>\sqrt[i]{\left|\frac{\beta_{n-1-i}}{\beta_{n-1}}\right|}>\ldots>\sqrt[i]{\left|\frac{\beta_{n-j-i}}{\beta_{n-j}}\right|}>\ldots>\sqrt[i]{\left|\frac{\beta_0}{\beta_i}\right|},$$

for the i-th

$$\ldots\sqrt[n-1]{\left|\frac{\beta_1}{\beta_n}\right|}>\sqrt[n-1]{\left|\frac{\beta_0}{\beta_{n-1}}\right|} \text{ and for the last but one}$$

subdiagonals of the main diagonal.

The same *phenomenon* occurs in component $\mathbf{P}_{ij}$ of each subdiagonal $2\times(n-2)$ pieces of the main and secondary diagonals, of each row $n-2$ pieces and each column $n-1$ of the denominator polynomial $B(p,t)$ $\mathbf{P}$ - matrix is under the lack of growth, as well as the same component $\mathbf{P}_{ij}$ of each superdiagonal $n-3$ pieces of her side diagonal (Fig. 3-7):

$$\sqrt{\left|\frac{\beta_{n-2}}{\beta_n}\right|}>\left|\frac{\beta_{n-2}}{\beta_{n-1}}\right| \text{ for the third,}$$

$$\ldots\sqrt[i]{\left|\frac{\beta_{n-i}}{\beta_n}\right|}>\sqrt[i-1]{\left|\frac{\beta_{n-i}}{\beta_{n-1}}\right|}>\ldots>\sqrt[i-j]{\left|\frac{\beta_{n-i}}{\beta_{n-j}}\right|}>\ldots>\left|\frac{\beta_{n-i}}{\beta_{n-i+1}}\right|,$$

for the i-th

$$\ldots\sqrt[n]{\left|\frac{\beta_0}{\beta_n}\right|}>\sqrt[n-1]{\left|\frac{\beta_0}{\beta_{n-1}}\right|}>\ldots>\sqrt[n-j]{\left|\frac{\beta_0}{\beta_{n-j}}\right|}>\ldots>\left|\frac{\beta_0}{\beta_1}\right|$$

and for the $(n+1)$ rows, and

$$\left|\frac{\beta_{n-1}}{\beta_n}\right| > \sqrt[2]{\left|\frac{\beta_{n-2}}{\beta_n}\right|} > \ldots > \sqrt[i]{\left|\frac{\beta_{n-i}}{\beta_n}\right|} > \ldots > \sqrt[n]{\left|\frac{\beta_0}{\beta_n}\right|}$$ for the first column,

$$\left|\frac{\beta_{n-j-1}}{\beta_{n-j}}\right| > \sqrt[2]{\left|\frac{\beta_{n-j-2}}{\beta_{n-j}}\right|} > \ldots > \sqrt[i-j]{\left|\frac{\beta_{n-i}}{\beta_{n-j}}\right|} > \ldots > \sqrt[n-j]{\left|\frac{\beta_0}{\beta_{n-j}}\right|},$$

j-th

$$\ldots \left|\frac{\beta_1}{\beta_2}\right| > \sqrt{\left|\frac{\beta_0}{\beta_2}\right|}$$ and the $(n-1)$ columns, and also

$$\sqrt[n-1]{\left|\frac{\beta_1}{\beta_n}\right|} > \sqrt[n-3]{\left|\frac{\beta_2}{\beta_{n-1}}\right|} > \ldots > \underbrace{\sqrt{\left|\frac{\beta_{(n-1)/2}}{\beta_{(n+3)/2}}\right|}}_{\text{odd } n} \text{ or } \underbrace{\left|\frac{\beta_{n/2}}{\beta_{(n+2)/2}}\right|}_{\text{even } n}$$ for the first,

$$\sqrt[i]{\left|\frac{\beta_{n-i}}{\beta_n}\right|} > \sqrt[i-2]{\left|\frac{\beta_{n-i+1}}{\beta_{n-1}}\right|} > \ldots > \underbrace{\sqrt{\left|\frac{\beta_{n-i/2-1}}{\beta_{n-i/2+1}}\right|}}_{\text{even } i} \text{ or } \underbrace{\left|\frac{\beta_{(2n-i-1)/2}}{\beta_{(2n-i+1)/2}}\right|}_{\text{odd } i},$$

for the i-th $\sqrt{\left|\frac{\beta_{n-3}}{\beta_n}\right|} > \left|\frac{\beta_{n-2}}{\beta_{n-1}}\right|$ and for the last but one superdiagonals of the secondary diagonal, and

$$\sqrt[n-1]{\left|\frac{\beta_0}{\beta_{n-1}}\right|} < \sqrt[n-3]{\left|\frac{\beta_1}{\beta_{n-2}}\right|} < \ldots < \underbrace{\sqrt{\left|\frac{\beta_{(n-3)/2}}{\beta_{(n+1)/2}}\right|}}_{\text{odd } n} \text{ or } \underbrace{\left|\frac{\beta_{n/2-1}}{\beta_{n/2}}\right|}_{\text{even } n}$$ for the first,

$$\ldots \sqrt[n-j]{\left|\frac{\beta_0}{\beta_{n-j}}\right|} < \sqrt[n-j-2]{\left|\frac{\beta_1}{\beta_{n-j-1}}\right|} < \ldots < \underbrace{\sqrt{\left|\frac{\beta_i}{\beta_{i+2}}\right|}}_{\text{even } i} \text{ or } \underbrace{\left|\frac{\beta_i}{\beta_{i+1}}\right|}_{\text{odd } i},$$ for

j-th $\sqrt[3]{\left|\frac{\beta_0}{\beta_3}\right|} < \left|\frac{\beta_1}{\beta_2}\right|$ and for the last but one subdiagonals of the secondary diagonal.

The apparent awkwardness of the interdependent conditions, there are $2n(n-1)$ inequations, demonstrates how much the balance of phase transition and getting to the cluster stability is delicate. Is there any probability to randomly generate a polynomial $\left(R_{i(x)} : \beta_n^* p^n, \beta_{n-1}^* p^{n-1} \ldots \beta_0^*\right)$? That is a stable according to Hurwitz' criteria? This probability is equal to zero with the growth of the polynomial $B(p,t)$ order! It is similar to the probability of life appearance in the Universe. As, for example, can fifty-one randomly assigned component $\beta_j(t), \forall j = \overline{0,n}$ in $B(p,t)$ satisfy 2*50*49 interdependent locations of the markers?

The violation of any one surely results the dynamic system in the dominant instability cluster. The limits/prohibitions plenty of the each marker $\mathbf{P}_{ij}$ location in the $\mathbf{P}$ matrix guarantees the adjustment mechanism existence of shallow and more profound system. The adjustment mechanism of shallow system is identified with the conditions of rows, columns and subdiagonals of the main diagonal, the adjustment mechanism of more profound system is defined with the subdiagonals conditions of its secondary diagonal.

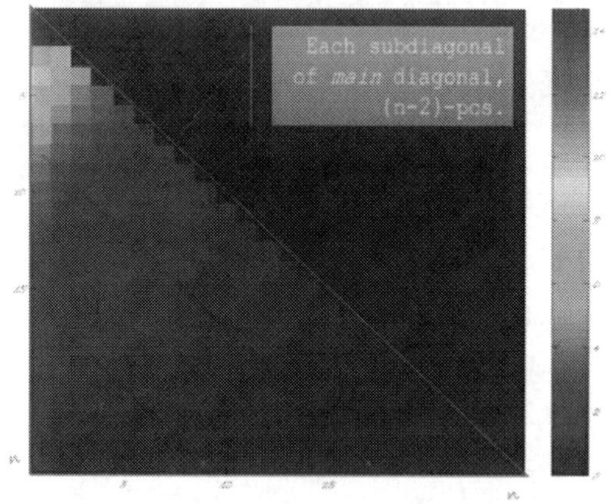

Fig. 3. About the lack of growth of the components $\mathbf{P}_{ij}$ of each subdiagonal of the lower triangular $\mathbf{P}_{B(p,t_k)}$ main diagonal.

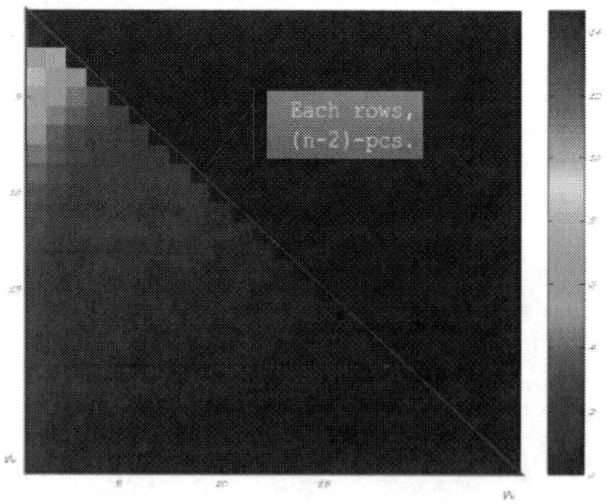

Fig. 4. About the lack of growth of the components $\mathbf{P}_{ij}$ of the lower triangular $\mathbf{P}_{B(p,t_k)}$ each row.

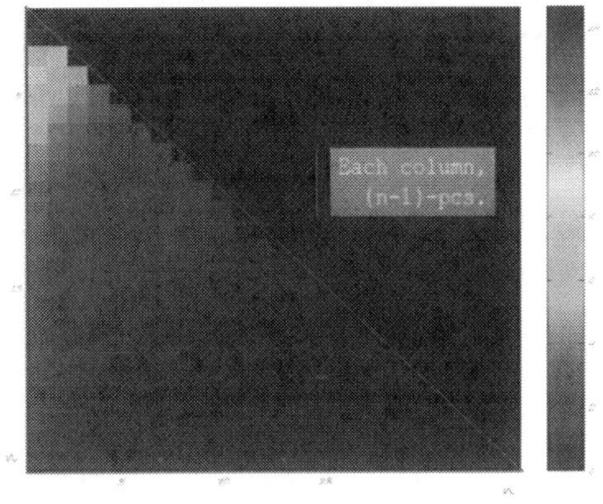

Fig. 5. About the lack of growth of the components $\mathbf{P}_{ij}$ of the lower triangular $\mathbf{P}_{B(p,t_k)}$ each column.

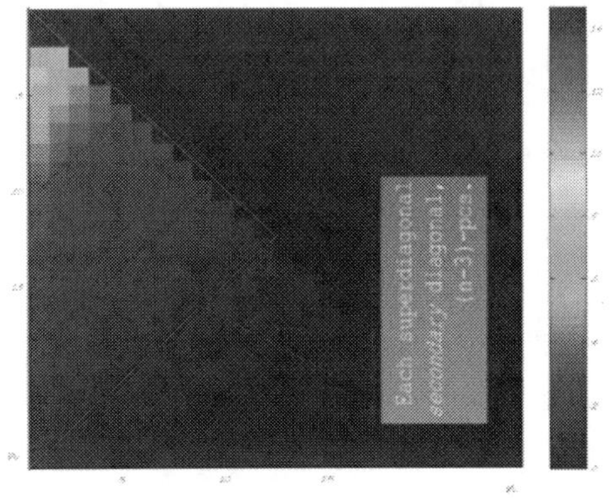

Fig. 6. About the non-decreasing of the components $\mathbf{P}_{ij}$ of each superdiagonal of the lower triangular $\mathbf{P}_{B(p,t_k)}$ secondary diagonal.

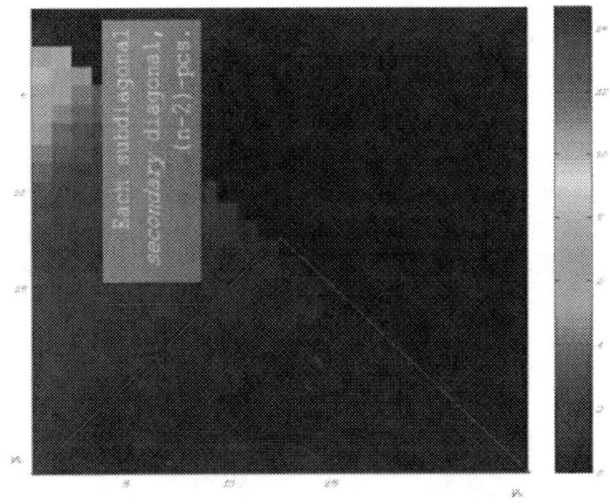

Fig. 7. About the lack of growth the components $\mathbf{P}_{ij}$ of each subdiagonal of the lower triangular $\mathbf{P}_{B(p,t_k)}$ secondary diagonal.

MATLAB procedure of check the columns, rows and subdiagonals lack of growth:

```
%Polilov's theorem about a polynomial
stability according to Hurwitz
for j=1:n
 validateattributes(diag(pmatrix,-j),
{'numeric'},{'decreasing'}) %main
subdiagonals
 validateattributes(pmatrix(j+1,1:j),
{'numeric'},{'decreasing'}) %rows

validateattributes(pmatrix(j+1:n+1,j),
{'numeric'},{'decreasing'}) %columns
validateattributes(nonzeros(diag(flipl
r(pmatrix),j)),
{'numeric'},{'increasing'}) %secondary
subdiagonals
validateattributes(nonzeros(diag(flipl
r(pmatrix),-j)),
{'numeric'},{'decreasing'}) %
secondary subdiagonals
end
```

MATLAB procedure of plot 2D and 3D viewpoint of $\mathbf{P}$ -matrix is presented here:

```
%theorem geometric interpretation,
stability markers
figure, imagesc(pmatrix),
colorbar, set(gcf,'Color','w');
figure, bar3(pmatrix),
colormap(summer)
set(gcf,'Color','w'); grid off
```

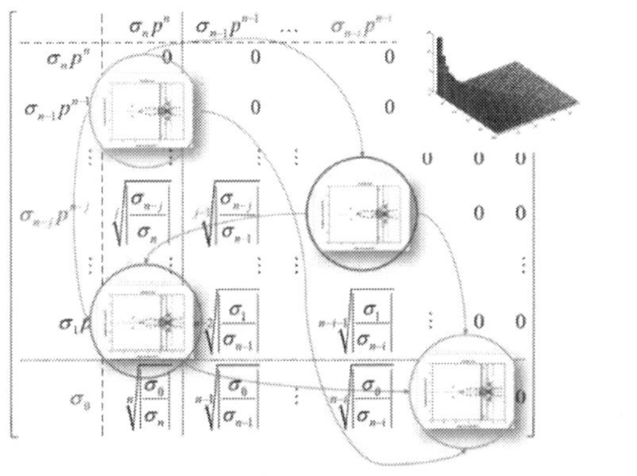

Fig. 8. There is a special radicals geometry in each cell of $\mathbf{P}$ -matrix. Dominant of the neighboring components $\mathbf{P}_{ij\pm1}$ allows to assign the *stability markers*.

3 Conclusion

The linear dynamic systems stability criterion, allowing one to identify the defective/failed $\sigma_k(t), \forall k \in \overline{0,n}$ from a set of non-stationary $\sigma_j(t), \forall j = \overline{0,n}$ in the characteristic polynomial $G(p,t) = 0$ at each time quantum t_k is presented for the first time. $\mathbf{P}$ - matrix, similar to the dispersive prism and the luminous flux, factorizes the dynamic system in the spectrum of geometric averages. The special order of the adjacent neighboring $\mathbf{P}_{ij\pm1}$ alternation as in a strictly regulated sequence of the light spectrum color array shades is substantial and thus it allows assigning the stability markers.

Hurwitz' indicator of more than a hundred polynomials of the high orders with the radicals distribution $c_j(t), \forall j=\overline{1,n}$ according to Butterworth, Bessel, Graham- Lathrop etc., of various geometric averages $\Omega_k(t)=\underset{j=\overline{1,n}}{\text{gmean}}(c_j(t))$ has been geometrically confirmed. The nowlingly unstable dynamic systems, formed from Hurwitz' systems $G(p,t)=0$ with the intentional change of any component from $\sigma_j, \forall j\in\overline{0,n}$ that is refleëcted on Fig. 9, were being studied. In the $\mathbf{P}$ matrix the condition of the markers lack of growth of the corresponding subdiagonals, rows and columns, obviously these are some errors in β_{15} and β_{20} , is disturbed.

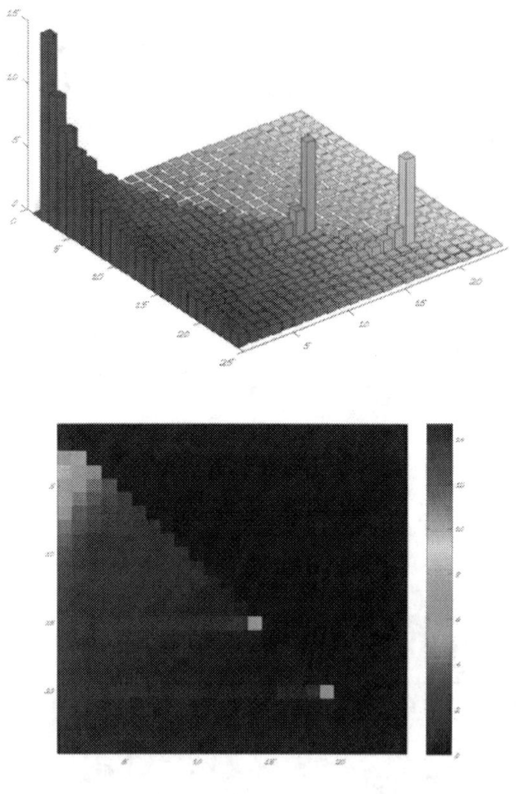

Fig. 9. A fault is in the solutions $\beta_{15}\times 15,0$ and $\beta_{20}\times 20,0$.

And, perhaps, the key moment, for what all this has been started is that it can be seen in $\mathbf{P}$ matrix how much the defective β_{15} and β_{20} have to be changed in order to fit them harmonically to the line with the adjacent "right" markers. The retention of non-stationary system in a very small stability cluster should be assigned on the adaptive computing potential, for example, on the mechanism of the artificial neural networks, which forcibly orders the markers $\mathbf{P}_{ij}$ in the $\mathbf{P}$ matrix at each time quantum t_k in the current cycle and searchless criterion. A generation of the own Hurwitz' polynomials $G_v(p,t)=0$ in the modal control classical problems should be recognized as one of the niches of the $\mathbf{P}$ - matrices effective application.

It is appropriate to speak about finding of the acceptable intervals $\left[\underline{\sigma}_j;\ \overline{\sigma}_j\right]$ of each $\sigma_j(t), \forall j=\overline{0,n}$ variation in the characteristic polynomial $G(p,t)=0$ with the $\mathbf{P}$ - matrices terms. It is a constructive alternative to the Kharitonov' theorem [7]. We can four corner polynomials, four $\mathbf{P}$ - matrices, markers, with the terms of interval polynomials:

$$G_1(p)=\overline{\sigma}_0+\underline{\sigma}_1 p+\underline{\sigma}_2 p^2+\overline{\sigma}_3 p^3+\overline{\sigma}_4 p^4+\ldots,$$

$$G_2(p)=\underline{\sigma}_0+\overline{\sigma}_1 p+\overline{\sigma}_2 p^2+\underline{\sigma}_3 p^3+\underline{\sigma}_4 p^4+\ldots,$$

$$G_3(p)=\overline{\sigma}_0+\overline{\sigma}_1 p+\underline{\sigma}_2 p^2+\underline{\sigma}_3 p^3+\overline{\sigma}_4 p^4+\ldots,$$

$$G_4(p)=\underline{\sigma}_0+\underline{\sigma}_1 p+\overline{\sigma}_2 p^2+\overline{\sigma}_3 p^3+\underline{\sigma}_4 p^4+\ldots$$

References

1. M.M. Postnikov, *Stable polynomials* (Editorial URSS, Moscow, 2004)
2. Q.I. Rahman, G. Schmeisser, *Analytic Theory of Polynomials: Critical Points, Zeros and Extremal Properties* (Oxford: Clarendon Press, 2002)
3. N. Guglielmi, M. Manettac, Journal of Computational and Applied Mathematics, **292**, 638-653 (2016)
4. N. Guglielmi, M. Gürbüzbalaban, M.L. Overton, SIAM J. Matrix Anal. Appl., **34**(2), 709-737 (2013)
5. E.V. Polilov, Collection of Scientific Papers of Donbass State Technical University, **46**, 102–121 (2016)
6. E.V. Polilov, Collection of scientific works "Bulletin of NTU "KhPI": The Problems of automated electric. Theory and Practice, **12**(1121), 72-82 (2015)
7. V.L. Kharitonov, Diff. Eq., **1**(11), 2086-2088 (1978)

Evaluation of functional reliability indices for DC-link capacitors in pulse-width modulation converters

Alexander Gorpinich*

Pryazovskyi State Technical University, Department of Electrical Engineering, 87500 Mariupol, Ukraine

Abstract. The impact of kHz range harmonics on the power losses, thermal stress, and lifetime reduction of the dc-link capacitors in pulse-width modulation (PWM) converters was investigated. Expressions to evaluate the mean time to failure, survival probability, and unavailability of aging failure for dc-link capacitors are presented. The dc-link capacitors failures due to accelerated insulation aging were modeled using Weibull and normal distribution. A case study with the Siemens SINAMICS S120 frequency converter for driving of rolling mill leveler shows that the failures of motor modules due to breakdown of electrolytic dc-link capacitors registered for the time frame from May 2012 to October 2012 can be caused by the increased ambient temperature and additional heating due to high-frequency components of the ripple current. As a possible solution to improve reliability of motor modules, the four AVX FFVE4I0227K film capacitors instead of nine EPCOS B43564 electrolytic capacitors in dc-link were recommended.

1 Introduction

A widespread proliferation of the PWM converters with dc-link in the electric vehicles, uninterruptible power supply, wind turbine and photovoltaic systems, adjustable speed motor drives, etc., causes the increase in the level of current and voltage components with frequencies in the kHz range. The high-frequency harmonic distortion of the current and voltage waveforms in PWM converters is produced by the voltage-source inverter using fast electronic switches which are commutated with very high rate of voltage rise (up to 50 kV/μs) and pulse repetition frequency (about 10-50 kHz) [1, 2]. As an example, Figures 1 and 2 present the distorted current and voltage waveforms generated by LS800-42K2-TD frequency converter with dc-link and registered by digital oscilloscope during laboratory measurements [3]. According to LS800-42K2-TD specification, the space vector PWM (SVPWM) switching frequency for insulated-gate bipolar transistor (IGBT) module can vary from 2 to 16 kHz. The amplitude-time-frequency representation of the distorted current and voltage waveforms registered on the output of the LS800-42K2-TD frequency converter was realized in Wolfram Mathematica using the simplified complex Morlet wavelet [4]. The high-frequency components in the 0.76-97.31 kHz range with the voltage peak values up to 199% and current peak values up to 78% (relative to peak value of the fundamental sinusoidal component) were identified. It was determined that the duration of these high-frequency components reaches up to tens microseconds-only a few milliseconds whereas their pulse edge is similar to square-wave signal one [3].

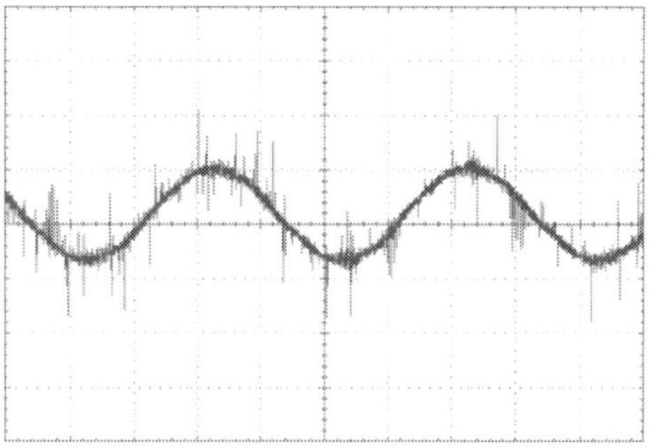

Fig. 1. Measured output phase current waveform.

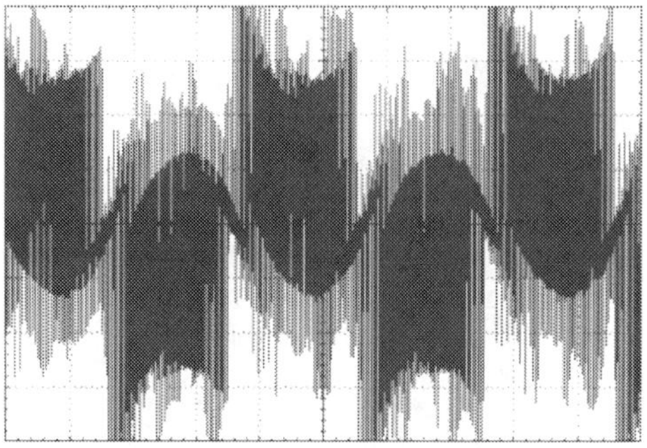

Fig. 2. Measured phase-to-phase voltage waveform at the driven motor terminals.

Obtained peak values and durations of the high-frequency components are inherent in the current and voltage switching impulses limited by the International Electrotechnical Commission (IEC) standards. For

* Corresponding author: gorpinich@ieee.org

instance, IEC 61800 requires the consideration of overvoltages due to commutation oscillations of up to 150% of the peak value of the fundamental voltage (periodically) or even 250% (non-periodically) for the design of low-voltage converters [5]. The high-frequency components with significant peak values cause accelerated thermal and electrical insulation aging of the driven motor, connecting cable, and dc-link capacitors. Furthermore, the resonant oscillations in the kHz range can appear originating failures of the PWM converter and driven motor elements. In the last few years, the practical examples with problems linked to kHz range harmonics were experienced mainly in the industry. Mostly variable speed drives in industrial environments were found stopping due to dc-link overvoltage. At least in one case investigated, a match between an internal rectifier resonance frequency with the disturbing network frequency contributed to the failure [1, 5].

From the author's experience, two cases with failures of motor modules due to breakdown of electrolytic dc-link capacitors in the Siemens SINAMICS S120 frequency converter for driving of the rolling mill leveler of the PJSC "Azovstal Iron & Steel Works" (Ukraine) were registered for the time frame from May 2012 to October 2012. The power rating of the driven induction motors is 1.5 MW whereas dc-link, consisting of the three separate modules with nine EPCOS B43564 electrolytic capacitors in each one, is an integral part of the motor module. It should be noted that the electrolytic capacitors are sharing 60% of the failure distribution for PWM converter elements as represented in Figure 3 [6].

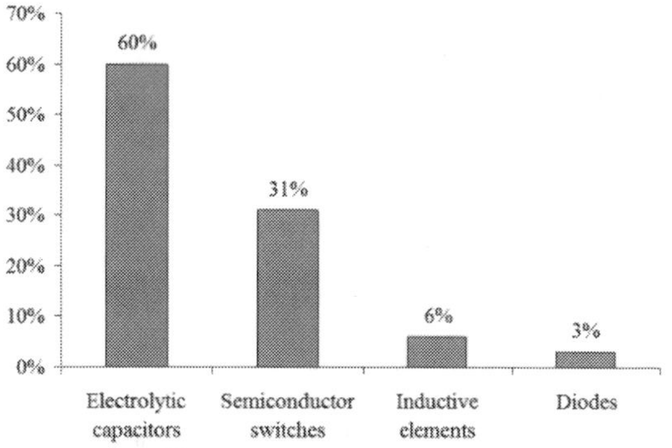

Fig. 3. Failure distribution for PWM converter elements.

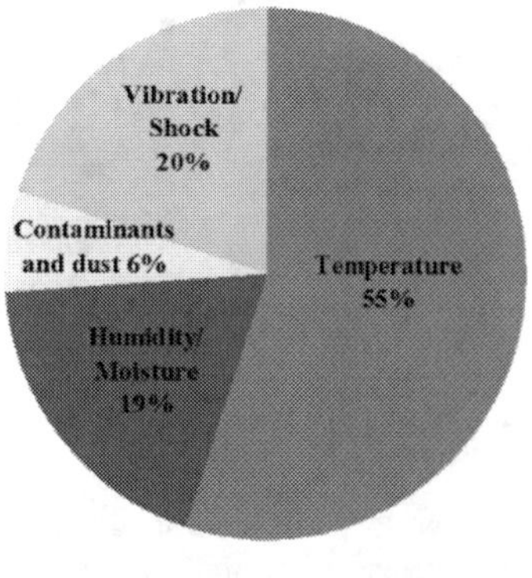

Fig. 4. Failure sources for electronic equipment.

The lifetime of electrolytic capacitors depends on both the rated lifetime at nominal conditions and the actual experienced stresses in the field operation. Long life could be achieved with a large design margin in terms of voltage, ripple current, and temperature. Temperature, vibration, and humidity are the three major stressors that directly or indirectly induce failure in power electronic components. The U.S. Air Force Avionics Integrity Program conducted an investigation into the failure sources of electronic equipment in 1980s and reached the conclusion that temperature is the most dominant stressor as shown in Figure 4 [7].

Methods for predicting the effect of ripple current stress on the power losses, temperature rise, and expected lifetime of the dc-link capacitors used in the variable-frequency drives, power converters, rectifiers, electric vehicle drive systems, photovoltaic inverters, wind turbine and fuel cell power converters have been presented in [8-17]. But so far none of the recognized approaches has been applied for evaluating the functional reliability indices of the dc-link capacitors, such as the probability to aging failure (a conditional probability that capacitor will fail in a subsequent period t after having survived for T operating hours) and the unavailability of aging failure (average probability that capacitor is found not available due to aging failure).

This paper presents an approach to evaluate the effect of high-frequency components of the ripple current on the additional heating, loss of lifetime, and functional reliability indices reduction of the dc-link capacitors in PWM converters. The effectiveness of the proposed approach was demonstrated using case study with the dc-link capacitors of the Siemens SINAMICS S120 frequency converter for driving of the rolling mill leveler of the PJSC "Azovstal Iron & Steel Works" (Ukraine). Based on the comparative analysis, the four AVX FFVE4I0227K film capacitors instead of nine EPCOS B43564 electrolytic capacitors in dc-link were recommended to mitigate the negative impact introduced by the increased ambient temperature and additional heating due to kHz range harmonics of the ripple current.

2 Capacitor models and characteristics

The electrical model parameters, a ripple current spectrum, and the thermal model parameters are needed to evaluate the effect of high-frequency components of the ripple current on the reduction of dc-link capacitor functional reliability indices due to accelerated thermal insulation aging. Equivalent series resistance (ESR) and ripple current are the main parameters for electrical model. Ripple current is the root-mean-square (RMS) value of the alternating current (AC) caused by the ripple voltage. The ripple current value depends on dissipation factor (also known as the tangent of the loss angle $\tan\delta$) which in turn is defined by ESR. Dissipation factor depends on the supply voltage frequency, therefore a ripple current depends on frequency also. The permissible maximum ripple current values are commonly given by the manufacturer datasheets for frequencies 100 or 120 Hz (sometimes for frequencies

10 or 20 kHz). In addition, manufacturer often provides generic curves to determine the ripple current for other operating frequencies.

The ESR model of an electrolytic capacitor is shown in Figure 5 [8, 11], where the resistance R_0 accounts for the sum of resistances of foil, tabs, and terminals, R_1 represents the electrolyte resistance, C_1 represents the terminal capacitance, and the parallel combination of R_2 and C_2 represents the dielectric losses. According to Figure 5, the complex impedance of the capacitor is described as

$$Z_{\text{cap}} = \frac{1}{\dfrac{1}{R_2} - j2\pi f C_2} + R_1 + R_0 - \frac{j}{2\pi f C_1}, \quad (1)$$

where f is the frequency.

Only the real part of this impedance is needed to estimate the additional power losses and heating caused by the ripple current:

$$\text{ESR} = \text{Real}(Z_{\text{cap}}), \quad (2)$$

$$\text{Real}(Z_{\text{cap}}) = R_1 + R_0 + \frac{R_2}{1+(2\pi f R_2 C_2)^2}. \quad (3)$$

Increasing temperature causes a decrease in ESR because the resistance R_1 is reduced due to the increased conductivity of electrolyte. The effect of temperature on R_1 is described by the following expression [8, 11]:

$$R_1 = R_{1\text{base}} \exp\left(\frac{T_{\text{base}} - T_{\text{core}}}{F}\right), \quad (4)$$

where $R_{1\text{base}}$ represents the value of R_1 at the base temperature T_{base} (27 °C), T_{core} is the capacitor core temperature, °C, and F is the temperature sensitivity factor, K^{-1}.

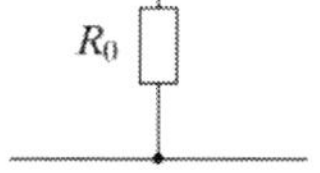

Fig. 5. ESR model of an electrolytic capacitor.

Parameters of the EPCOS B43564 electrolytic capacitor used in dc-link of the Siemens SINAMICS S120 frequency converter driving the rolling mill leveler are given in Table 1 [11, 18].

Table 1. Parameters of the EPCOS B43564 electrolytic capacitor.

Parameters	Value
ESR at 100 Hz (mΩ)	23
Maximum impedance at 10 kHz (mΩ)	28
Resistance of foil, tabs, and terminals R_0 (mΩ)	5.03
Resistance of electrolyte at the base temperature $R_{1\text{base}}$ (mΩ)	6
Temperature sensitivity factor F (K^{-1})	21
Dielectric losses resistance R_2 (mΩ)	20
Terminal capacitance C_1 (μF)	4700
Dielectric losses capacitance C_2 (μF)	11.6
Capacitor core temperature T_{core} (°C)	40
Maximum permissible current (A)	37

Let's consider a possibility to use the film capacitors instead electrolytic ones in dc-link of the Siemens SINAMICS S120 frequency converter. For this purpose, the four AVX FFVE4I0227K film capacitors instead of nine EPCOS B43564 electrolytic capacitors were selected in each dc-link module.

When the operating frequency is higher than 1 kHz, the ESR value of the film capacitor can be modeled as a function of frequency by means of such equation [11]:

$$\text{ESR}(f) = (R_s - A_s) + K(f) \cdot A_s, \quad (5)$$

where R_s is the base resistance, A_s is a given by the manufacturer value depending on the capacitor size, $K(f)$ is the predefined function showing the effect of frequency which is usually given by plots in manufacturer datasheets. $K(f)$ is mathematically expressed as a cubic polynomial function as follows

$$K(f) = k_3 \cdot f^3 + k_2 \cdot f^2 + k_1 \cdot f + k_0, \quad (6)$$

where the coefficients k_0-k_3 can be established by using the least square curve fitting.

In contrast with the electrolytic capacitor, the ESR value of film capacitor increases with the frequency. Its temperature-sensitivity is less than that of electrolytic capacitor because the ESR of film capacitor results mainly from the tabs and contact resistances, but not from the electrolyte [11]. Parameters of the AVX FFVE4I0227K film capacitor are presented in Table 2 [19].

Since the ESR values of electrolytic and film capacitors show frequency-dependent characteristics, a ripple current multiplier M_f is defined in (7) to characterize the ripple current at a given frequency [11]:

$$M_f = \frac{I_f}{I_{100}} = \sqrt{\frac{\text{ESR}_{100}}{\text{ESR}_f}}, \quad (7)$$

where I_{100} and ESR_{100} are the rated ripple current and ESR value at 100 Hz, respectively, I_f and ESR_f represent ripple current and ESR at frequency f.

Table 2. Parameters of the AVX FFVE4I0227K film capacitor.

Parameters	Value
ESR at 100 Hz (mΩ)	0.9
Base resistance R_s (mΩ)	0.9
Terminal capacitance C_1 (μF)	220
Parameter defined by height A_s	0.23
Coefficient k_3	$3.173 \cdot 10^{-7}$
Coefficient k_2	-0.000124
Coefficient k_1	0.02369
Coefficient k_0	1.014
Maximum permissible current (A)	100

The dc-link capacitor ripple current frequency spectra analysis realized in [11] for different operating conditions and modulation strategies shows that SVPWM gives the largest kHz range harmonics in mode when modulation index $M = 0.625$ and load power factor $\cos\varphi = 0.954$. In this case, amplitude of 20 kHz component is equal to 52% from the output phase current amplitude, amplitude of 40 kHz component – 14%, amplitude of 60 kHz component – 9.9%, and amplitude of 80 kHz component – 9%. Let's define the power losses in the considered electrolytic and film capacitors for this ripple current frequency spectrum because SVPWM is used for the Siemens SINAMICS S120 frequency converter. These power losses can be represented as the sum of power losses from the individual frequency current components:

$$P_{\mathrm{cap}} = \sum \mathrm{ESR}_{f_i} \cdot (I_{\mathrm{cap}\, f_i})^2 = \mathrm{ESR}_{100} \sum \left(\frac{I_{\mathrm{cap}\, f_i}}{M_f} \right)^2, \quad (8)$$

where ESR_{f_i} is the ESR value at frequency f_i, $I_{\mathrm{cap}\, f_i}$ is the RMS current of the frequency f_i component.

Table 3 shows the power losses for considered electrolytic and film capacitors. Capacitor hot-spot temperature

$$T_{h\,\mathrm{core}} = P_{\mathrm{cap}} R_{th} + T_a, \quad (9)$$

where T_a is the ambient temperature, °C, R_{th} is the capacitor thermal resistance (for film capacitor $R_{th} = 7.2$ °C/W [19], for electrolytic one pressed against a large metal plate or chassis $R_{th} = 1.02$ °C/W [20]).

For Siemens SINAMICS S120 frequency converter $T_a = 40$ °C so $T_{h\,\mathrm{core}} = 52.3$ °C for electrolytic capacitor and $T_{h\,\mathrm{core}} = 62.9$ °C for film one were obtained by (9).

Table 3. Power losses for electrolytic and film capacitors.

f (kHz)	$I_{\mathrm{cap}\, f_i}$ (%)	**EPCOS B43564**	**AVX FFVE4I0227K**
		$I_{\mathrm{cap}\, f_i}$ (A)	$I_{\mathrm{cap}\, f_i}$ (A)
10	2.1	0.777	2.1
20	52	19.24	52
30	5	1.85	5
40	14	5.18	14
50	3	1.11	3
60	9.9	3.663	9.9
70	1.5	0.555	1.5
80	9	3.33	9
90	1.5	0.555	1.5
100	3	1.11	3
P_{cap} (W)		12.099	3.185

3 Capacitor functional reliability indices

The lifetime for both electrolytic and film capacitors can be derived from the "lifetime doubles in every 10 °C" rule as follows [9, 11, 14]:

$$L = L_0 \cdot 2^{-\frac{\Delta\tau}{10}} = L_0 \cdot \exp(-\beta\Delta\tau), \quad (10)$$

where L and L_0 are the lifetimes under the real operating conditions and maximum permissible temperature of the hottest spot, respectively, $\Delta\tau$ is the additional insulation heating, °C, $\beta = 0.0693$ °C^{-1} is the insulation aging factor.

Based on the assumption of normally distributed random variable $\Delta\tau$, the mean value and standard deviation of the capacitor lifetime, respectively

$$m_L = L_0 \exp(-\beta m_{\Delta\tau} + 0.5\beta^2\sigma_{\Delta\tau}^2), \quad (11)$$

$$\sigma_L = m_L \sqrt{\exp(\beta^2\sigma_{\Delta\tau}^2) - 1}, \quad (12)$$

where $m_{\Delta\tau}$ and $\sigma_{\Delta\tau}$ are the mean value and standard deviation of the additional insulation heating.

Weibull or normal distribution can be used to model the aging failures of dc-link capacitors caused by accelerated thermal insulation aging. Practical experience shows that the same historical reliability data can be successfully approximated by the both distributions.

In the case of Weibull distribution, the reliability function is

$$R(t) = \exp\left[-\left(\frac{t}{a} \right)^b \right], \quad (13)$$

where a and b are the scale and shape parameters.

The probability to aging failure (a conditional probability that capacitor will fail in a subsequent period t after having survived for T operating hours) is

$$P_f = F(t \mid T) = \frac{\exp\left[-\left(\frac{T}{a}\right)^b\right] - \exp\left[-\left(\frac{t+T}{a}\right)^b\right]}{\exp\left[-\left(\frac{T}{a}\right)^b\right]} . \quad (14)$$

The corresponding survival probability is

$$P_r = R(t \mid T) = 1 - F(t \mid T) . \quad (15)$$

The unavailability of aging failure (average probability that capacitor is found not available due to aging failure) can be obtained using the concept proposed in [21, 22]. For this purpose, the period t is divided into N equal intervals, each having a length $\Delta t = t/N$. The following equations can be expressed:

$$U_a = \sum_{i=1}^{N}\left(1 - \frac{2i-1}{2N}\right)P_i , \quad (16)$$

$$P_i = \frac{\exp\left[-\left(\frac{(i-1)\frac{t}{N}+T}{a}\right)^b\right] - \exp\left[-\left(\frac{i\frac{t}{N}+T}{a}\right)^b\right]}{\exp\left[-\left(\frac{T}{a}\right)^b\right]} , \quad (17)$$

where U_a is the unavailability of capacitor aging failure. The availability is

$$A_a = 1 - U_a . \quad (18)$$

From the mathematical expression of the Weibull density distribution function, the following equations can be obtained for evaluating the mean value and standard deviation of the capacitor lifetime:

$$m_L = a\Gamma(1 + 1/b) , \quad (19)$$

$$\sigma_L^2 = a^2[\Gamma(1 + 2/b) - \Gamma^2(1 + 1/b)] , \quad (20)$$

where $\Gamma(1+1/b)$ and $\Gamma(1+2/b)$ are the gamma functions depending on the shape parameter b .
As only m_L and σ_L values were defined by using (11) and (12), the shape parameter b can be calculated by

$$\frac{\Gamma(1+2/b)}{\Gamma^2(1+1/b)} = 1 + \left(\frac{\sigma_L}{m_L}\right)^2 . \quad (21)$$

Then scale parameter a is found from (19) using a known shape parameter b .

The differences between the normal distribution and Weibull distribution are not significant if the shape parameter $b \geq 3.9$. In case of normal distribution the reliability function is given by

$$R(t) = \frac{0.5 + P\left(\frac{m_L - t}{\sigma_L}\right)}{0.5 + P\left(\frac{m_L}{\sigma_L}\right)} , \quad (22)$$

where $P(t) = \frac{1}{\sqrt{2\pi}}\int_0^t e^{-t^2/2}dt$ is the probability integral.

The probability to aging failure

$$P_f = F(t \mid T) = \frac{P\left(\frac{t+T-m_L}{\sigma_L}\right) + P\left(\frac{m_L - T}{\sigma_L}\right)}{0.5 + \Phi\left(\frac{m_L - T}{\sigma_L}\right)} . \quad (23)$$

The corresponding survival probability is found from (15).

The unavailability of aging failure can be derived from (16) where P_i is defined as

$$P_i = \frac{P\left(\frac{i\frac{t}{N}+T-m_L}{\sigma_L}\right) - P\left(\frac{(i-1)\frac{t}{N}+T-m_L}{\sigma_L}\right)}{0.5 + P\left(\frac{m_L - T}{\sigma_L}\right)} . \quad (24)$$

The availability can be obtained from (18). To demonstrate the effectiveness of the proposed approach, it has been assumed that $m_{\Delta\tau} = 12.3$ °C, $\sigma_{\Delta\tau} = 3$ °C for electrolytic capacitor whereas $m_{\Delta\tau} = 22.9$ °C, $\sigma_{\Delta\tau} = 5$ °C for film capacitor (these values are based on the results of $T_{h\,core}$ estimation). Table 4 presents the performance comparison and functional reliability indices for both electrolytic and film capacitors at the $T = 4000$ hrs, $t = 1000$ hrs.

Based on the obtained results it can be concluded that the functional reliability indices for both electrolytic and film capacitors are approximately similar if the aging failures are modeled by either Weibull or normal distribution. Both manufacturers have designed quite competitive capacitors with good characteristics and high reliability. However, it is better to use the four AVX FFVE4I0227K film capacitors instead of nine EPCOS B43564 electrolytic capacitors in dc-link because the film capacitors have the maximum permissible temperature which is 20 °C higher than that for electrolytic ones. At the increased ambient temperature, such solution mitigates the negative impact introduced by the additional heating due to kHz range harmonics of the ripple current. Furthermore, the quantity, dimensions, and cost of the film capacitors are much lower.

Table 4. Performance comparison and functional reliability indices.

Parameters	EPCOS B43564	AVX FFVE4I0227K
Capacitor type	Electrolytic	Film
Quantity	9	4
Capacitance (μF)	4700	220
Total capacitance (μF)	42300	880
Dimensions (d×h, mm)	76.9×105.7	84.5×51
Maximum permissible current (A)	37	100
Rated voltage (V)	400	400
Maximum permissible temperature (°C)	85	105
Lifetime at 85 °C (hrs)	15000	35000
Cost ($)	112	86
Total cost ($)	1008	344
Functional reliability indices (Weibull distribution)		
P_f	0.101	0.091
P_r	0.899	0.909
U_a	0.043	0.043
A_a	0.957	0.957
Functional reliability indices (normal distribution)		
P_f	0.104	0.082
P_r	0.896	0.918
U_a	0.043	0.038
A_a	0.957	0.962
Mean time to failure, hrs	6500	7600

4 Conclusions

The main goal of the paper was to present an approach to evaluate the functional reliability indices for dc-link capacitors in PWM converters. A case study with the dc-link capacitors of the Siemens SINAMICS S120 frequency converter for driving of rolling mill leveler was used to demonstrate the effectiveness of the proposed approach. The dc-link capacitors failures due to accelerated thermal insulation aging were modeled using both Weibull and normal distribution. The results indicate that the failures of motor modules due to breakdown of electrolytic dc-link capacitors, registered for the time frame from May 2012 to October 2012, can be caused by the increased ambient temperature and additional heating due to high-frequency components of the ripple current. Based on the comparative analysis, the four AVX FFVE4I0227K film capacitors instead of nine EPCOS B43564 electrolytic capacitors in dc-link were recommended to improve reliability of motor modules.

The presented approach can be used to evaluate the functional reliability indices reduction of the driven motor and connecting cable, which is caused by the accelerated thermal and electrical insulation aging due to high-frequency components. Other practical applications of the proposed methodology include the estimation of the economical damage due to the kHz range harmonics of the ripple current, and comparative analysis of cost-effectiveness of the passive and active filters to suppress ones.

References

1. C. Unger, *Proc. IEEE/IPEC Int. Conf.* (2010)
2. A. Cavallini, D. Fabiani, G.C. Montanari, IEEE Electr. Ins. Mag., **26**, 3 (2010)
3. A.V. Gorpinich, I.A. Taranenko, Electromagn. Compatib. Saf. Rail. Transp., 7 (2014)
4. N.C.F. Tse, *Proc. IEEE Power Eng. Soc. Gen. Meet.* (PES GM 2006)
5. C. Unger, K. Krueger, M. Sonnenschein, R. Zurowski, *Proc. Int. Conf. Exhib. Electr. Distr.* (CIRED 2005)
6. H. Soliman, H. Wang, B. Gadalla, F. Blaabjerg, J. Renew. Energy Sust. Dev. (RESD), **1**, 2 (2015)
7. H. Wang, M. Liserre, F. Blaabjerg, IEEE Ind. Electr. Mag., **7**, 2 (2013)
8. M.L. Gasperi, IEEE Trans. Ind. Appl., **41**, 6 (2005)
9. H.M. Pang, P.M.H. Bryan, *Proc. IEEE Ann. Appl. Power Electron. Conf. Expos.* (APEC 2010)
10. K. Zhao, P. Ciufo, S. Perera, *Proc. IEEE Int. Conf. Harm. Qual. Power* (ICHQP 2010)
11. H. Wen, W. Xiao, X. Wen, P. Armstrong, IEEE Trans. Veh. Tech., **61**, 7 (2012)
12. H. Wang, Y. Yang, F. Blaabjerg, *Proc. IEEE Ann. Appl. Power Electron. Conf. Expos.* (APEC 2013)
13. H. Wang, M. Liserre, F. Blaabjerg, P.P. Rimmen, J.B. Jacobsen, T. Kvisgaard, J. Landkildehus, IEEE J. Emerg. Select. Top. Power Electron., **2**, 1 (2014)
14. H. Wang, F. Blaabjerg, IEEE Trans. Ind. Appl., **50**, 5 (2014)
15. K. Ma, H. Wang, F. Blaabjerg, IEEE Power Electron. Mag., **3**, 4 (2016)
16. S. Xue, Q. Zhou, J. Li, C. Xiang, S. Chen, *Proc. IEEE Int. Conf. High Volt. Eng. Appl.* (ICHVE 2016)
17. D. Zhou, H. Wang, F. Blaabjerg, *Proc. IEEE Ann. South. Power Electron. Conf.* (SPEC 2016)
18. Aluminum electrolytic capacitors, Capacitors with screw terminals, EPCOS AG, 2012. [Online]. Available: https://ferrite.ru/uploads/pdf/products/passive/capacitors/electrolytic/B43564_B43584.pdf
19. AVX medium power film capacitors for power applications, AVX Corporation, 2015. [Online]. Available: http://catalogs.avx.com/MediumPowerFilm.pdf
20. Aluminum electrolytic capacitors application guide, CDM Cornell Dubilier, 2013. [Online]. Available: http://www.cde.com/resources/catalogs/AEappGUIDE.pdf
21. W. Li, S. Pai, IEEE Power Eng. Rev., **22**, 2 (2002)
22. W. Li, IEEE Trans. Power Syst., **17**, 3 (2002)

MATEC Web of Conferences 113, 01005 (2017) DOI: 10.1051/matecconf/201711301005
12th International Scientific-Technical Conference on Electromechanics and Robotics "Zavalishin's Readings" - 2017

Power factor corrector with a high-order harmonics filtering function

Vladimir Burlaka* and *Sergey* Gulakov

Pryazovskyi State Technical University, Welding Department, 87500 Mariupol, Ukraine

Abstract. The article presents an original method and a control system of the active power factor corrector (PFC), which allows integrating the function of a parallel active filter of high-order harmonics into it. This makes it possible to reduce the total harmonic distortion of the grid voltage at the point of connection of the PFC with the proposed control system, thereby improving the working conditions of other electrical equipment in this grid. The presented control method is applicable to all PFC topologies employing direct control of the input currents.

1 Introduction

The adverse effects of harmonics on electrical power systems are well known [1, 2]. Harmonics current generated by any non-linear load flows from the load into the power system. These harmonic currents degrade the power system performance and reliability and can also cause safety problems. Harmonics need to be clearly located, sources identified and corrective measures taken to prevent them.

Introduction of modern electromagnetic compatibility standards that limit harmonic currents emission (IEC 61000-3-2, IEC 61000-3-4, IEC 61000-3-12) determines the intensive development of technical means that prevent deterioration of the power quality due to the presence of harmonics. These means are passive harmonic filters, harmonic mitigating transformers, active power filters [3-5], power quality conditioners [6, 7], unified [8] and interline [9] power flow controllers. Another way to improve electromagnetic compatibility of power supplies is to equip them with power factor correctors (PFC).

Most widespread among single-phase systems is the PFC based on a boost converter (the so-called Boost-type PFC, Fig. 1).

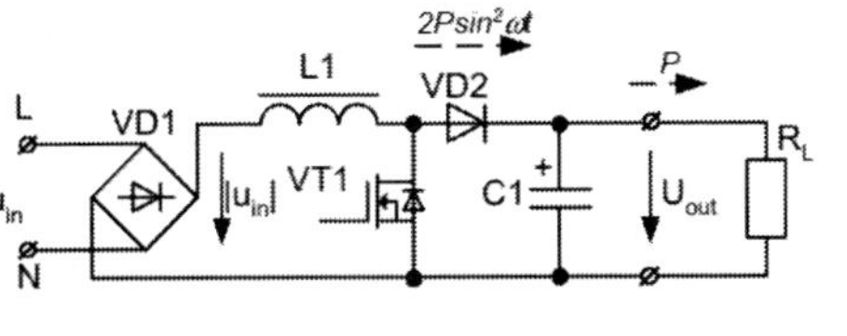

Fig. 1. Typical schematic of the single-phase boost-type PFC power stage.

The vast majority of such PFC behave as a linear active resistance in relation to the supply network (Fig. 2a). There is also a small number of PFC (for example, based on the M81012FP control chip from Mitsubishi) [10], behaving as sources of the sinusoidal current with fundamental frequency (Fig. 2b).

In both cases, the PFC does not have an active influence on the mains voltage spectrum (and in the case shown in Figure 2b, the voltage total harmonic distortion THD_U may even increase due to a voltage fundamental amplitude decrease).

As modern networks are flooded with non-linear loads, the problem of compensation of their negative impact on power quality arises.

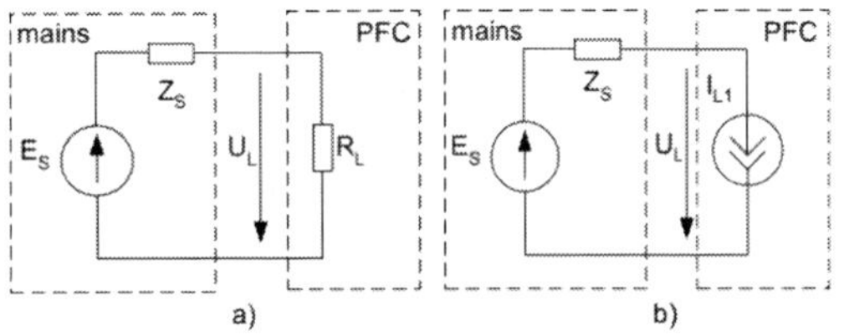

Fig. 2. PFC emulating: a) active resistance; b) sinusoidal current source.

As for today, there is a large number of electric loads that use intermediate conversion of alternating supply voltage to DC (switching power supplies, welding inverters, variable frequency drives, etc.). Typically, this conversion is carried out by means of a diode bridge and a capacitive filter. The input current of such a load has high content of odd harmonics, mainly 3rd and 5th, and typical power factor (PF) lies in the range of 0.5 to 0.7. The presence of a large number of such loads leads to a distortion of the mains voltage waveform, an overload of the neutral wire, a significant deterioration in the operating conditions of power supplies, especially with thyristor regulators with phase angle control, etc.

The negative impact of the non-linear loads on the mains voltage can be sufficiently reduced by means of installing the active power filters (APF) and/or

* Corresponding author: vburlaka@rambler.ru

equipping the loads with power factor correctors. These measures allow bringing the power quality parameters up to modern requirements.

At present, APFs are built using four-quadrant inverters. This is dictated by the need to provide a bi-directional instantaneous active power flow and leads to a complication of the power stage and its control system.

The power stage of the PFC, in its turn, is optimized for operation only with positive instantaneous active power (in the 1st and 3rd quadrants), which makes it impossible to use classical APF control algorithms for it.

In addition to the power stage, parallel APF usually requires installation of current sensors for compensated non-linear loads. However, the need for current sensors can be eliminated by operating the parallel APF inverter in a voltage source mode, as shown in [11].

2 Proposed PFC control strategy

The authors proposed the PFC [12] with a control system which allows for integration of parallel APF functions into it while ensuring regulation of the PFC output voltage, working with a positive input instantaneous active power, and limitation of the PFC input current. The control system operates in the conditions of absence of information about the currents of non-linear loads connected to the mains.

The essence of the idea is to change the algorithm of the PFC control system in order to reduce the emulated input resistance.

For this purpose, the input current of the PFC $i_L(t)$ is formed in accordance with the equivalent circuit shown in Figure 3.

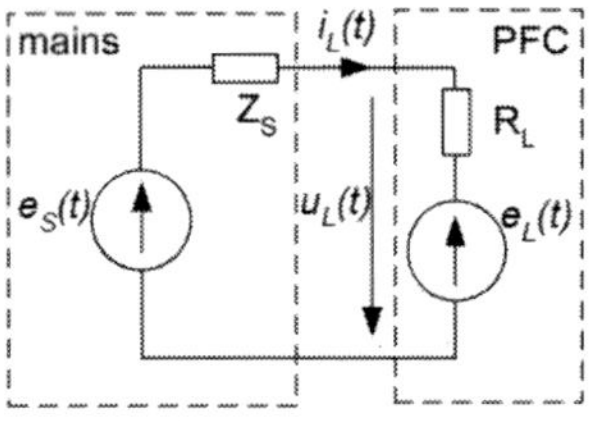

Fig. 3. Proposed PFC equivalent schematic.

As it was said before, the PFC with respect to mains behaves like an active resistance R_L, but now it has its own EMF $e_L(t)$, which contains only the fundamental mains frequency in its spectrum.

The source of the EMF $e_L(t)$ is synchronized with the mains voltage by means of phase locked loop (PLL):

$$e_L(t) = E_R sin(\omega t), \qquad (1)$$

where ω– mains voltage angular frequency; E_R – amplitude reference.

In this case, the input current of the PFC is

$$i_L(t) = [u_L(t) - e_L(t)]/R_L , \qquad (2)$$

where R_L is the emulated input resistance of the PFC.

The magnitude of the E_R amplitude reference is chosen so that the instantaneous active power of the PFC is always positive: $i_L(t)\cdot u_L(t)>0$ (the condition of the correct operation of the PFC power stage, see above), which allows using the developed control method for almost all existing power stage topologies of active PFC with direct input current control.

The value of R_L is determined from the power balance condition (the necessary condition for regulating the PFC output voltage).

$$(\omega/2\pi)\cdot\int u_L(t)\cdot[u_L(t)-e_L(t)/R_L]dt = P, \qquad (3)$$

where P is the power drawn from the network.

E_R and R_L signal conditioning circuits should not allow rapid changes in these signals, otherwise the performance of the system will deteriorate.

In comparison with the conventional PFC (Fig. 2a), the R_L value obtained for the same output power will be smaller. Thus, because of the reduced input resistance, the PFC will have an increased shunting effect for the high-order harmonics currents, similar to parallel APF.

The block diagram of the control system that implements the developed method is shown in Figure 4.

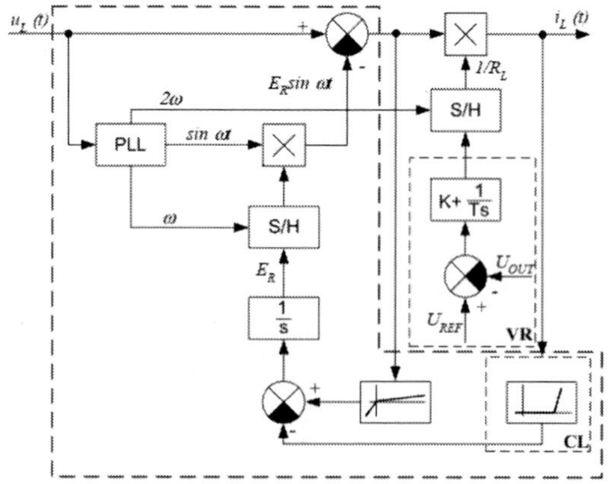

Fig. 4. Block diagram of the proposed PFC control system.

The reference signal of the fundamental harmonic of the mains voltage is obtained using the PLL (Fig. 4) and together with the amplitude reference (E_R) is fed to a multiplier whose output is connected to an adder that generates a difference signal. This signal is fed to the second multiplier, which works in conjunction with the PI controller of the PFC output voltage (voltage regulator VR, Fig. 4), and generates a PFC input current reference signal $i_L(t)$.

The E_R signal is generated by a local negative feedback circuit comprising a stage with a transmission coefficient that is switched depending on the polarity of the input signal, and an integrator. The time constant of the E_R circuit must significantly exceed the mains voltage period (i.e., E_R should not vary significantly during the period).

The PFC input current limiter (CL, Fig. 4) with the threshold stage reduces the E_R amplitude reference in case of an unacceptable increase in the PFC input current. Since this current limiter has a delay due to the E_R integrator, the high-speed hardware current limitation must also be performed directly in the power stage control circuits.

To prevent changing of the emulated resistance R_L and amplitude reference E_R during the mains period, Sample-and-Hold (S/H) stages synchronized with the mains voltage are used in control system. Thus, in the E_R amplitude reference signal, all harmonics of the mains frequency are suppressed, as a result of which it contains only a slowly varying DC component, which is necessary for the correct operation of the PFC. Similarly, in the U_{OUT} signal, all even harmonics of the mains frequency are suppressed, which eliminates the influence of output voltage ripple on the spectrum of the PFC input current [13].

3 Simulation

Figure 5 shows the results of simulating the work of the developed PFC control algorithm under conditions of a distorted mains voltage (the secondary side voltage of one of the Mariupol city substations: U_{RMS} = 187 V, THD_U = 9 %, with the 3rd harmonic coefficient of 8.2 %). The PFC output power is P = 2 kW. Simulation gives the developed PFC power factor PF_{PFC} = 0.818.

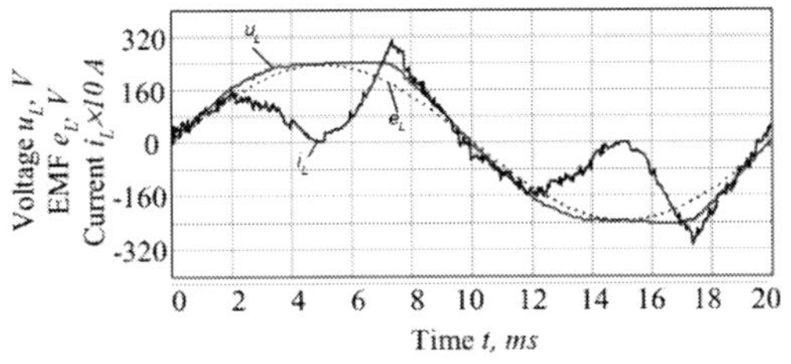

Fig. 5. Simulation results of the input current (i_L) of the proposed PFC under distorted mains voltage waveform (u_L).

It should be noted that the voltage quality parameters of this substation do not comply with modern power quality standards at least due to high THD_U.

In the mains voltage (u_L), high-order harmonics are expressed due to the presence of a large amount of nonlinear loads with uncontrolled single-phase rectifiers in the power supply network. The PFC reacts on this distorted voltage by the appearance of high-order harmonics in its input current (i_L) with the phases of these currents being opposite to the phases of the currents that caused distortion of the voltage waveform.

The analysis of the PFC input current spectrum shows a significant level of the high-order harmonics, significantly above the level of the harmonics in the mains voltage (for the given case, THD_U = 9.1 %, THD_I = 82.5 %). This is caused by the developed PFC control system attempt to increase the active power transferred by high-order harmonics. This fraction of the power can be defined as:

$$P_H^* = 1 - U_{1m} \cdot (U_{1m} - E_R)/(2R_L \cdot P), \quad (4)$$

where U_{1m} is the amplitude of the fundamental harmonic of the mains voltage.

Calculation by formula (4) for the given case shows that 7.9 % of the PFC input power is provided by higher voltage harmonics. While with a PFC emulating purely active load (as shown in Fig. 2a), this fraction will be determined as

$$P_H^* = 1 - 1/(1 + THD_U^2), \quad (5)$$

which, for the case under consideration, is only 0.8 %, and for the case of PFC shown in Fig. 2b, the active power transferred on the high-order harmonics is zero.

The effect of increasing the power fraction of high-order harmonics is associated with a decrease in the PFC emulated input resistance (R_L). In the case under consideration, the developed PFC operates with parameters E_R = 235 V, R_L = 2.03 Ohm. At the same time, the conventional PFC (Fig. 2a) would have R_L = 17.48 Ohm under similar conditions.

Figure 6 shows an example of the effect of replacing the conventional PFC (Fig. 2a) with the PFC employing the described control system (Fig. 3). Replacement of the conventional PFC (Fig. 6a) with a PFC control system, developed by the authors (Fig. 6b), leads to an increase in the total power factor of the system: $PF_2 > PF_1$.

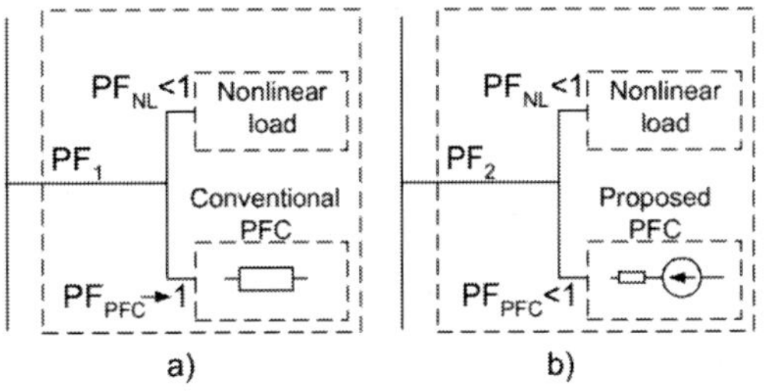

Fig. 6. Effect of using the proposed PFC with nonlinear load: $PF_2 > PF_1$.

Thus, the power factor of the system "non-linear load + developed PFC" will be higher than that of the "non-linear load + conventional PFC" system, despite the fact that the power factor of the developed PFC will be noticeably below unity. This effect is explained by the fact that in the case shown in Fig. 6b, a part of the high-order harmonics of the non-linear load current is supplied by the PFC, and not by the mains.

In case of a "clean" mains voltage (with an insignificant level of high-order harmonics), the operation of the developed PFC is completely analogous to the conventional one, i.e. its input current will contain only the fundamental harmonic component.

4 Experiments

For experimental verification of theoretical results, a modified PFC based on the MC33262 control IC was created (Fig. 7). The instantaneous mains voltage processing circuit of the PFC has been modified to implement the active filtering function with the possibility of switching it off.

The processing of the mains voltage signal was carried out using a single-chip microcontroller

ATTiny13; the generation of a current reference signal for the MC33262 was performed by filtering the microcontroller-generated pulse width modulated (PWM) signal using the low settling time filter developed by the authors [14]. The appearance of the signal processing unit is shown in Figure 8.

Measurement of power quality parameters and oscillography of voltages and currents was performed using the power quality analyzer developed by the authors [15].

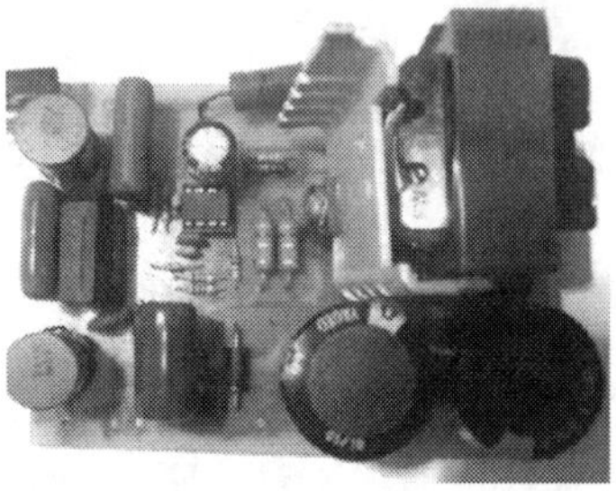

Fig. 7. Experimental PFC prototype.

Fig. 8. Signal processing unit for an experimental PFC.

The PFC operated together with a nonlinear load, which was presented by a bridge rectifier with a 330 µF smoothing capacitor, loaded with an active resistance and consuming 100 watts at a mains voltage of 193 V_{RMS}.

The output power of the experimental PFC was 140 – 160 W.

Figures 9 and 10 show the oscillograms of the mains voltage and current when the active filtering function is turned off and on, respectively. For the convenience of representation, the signals are normalized to their amplitudes.

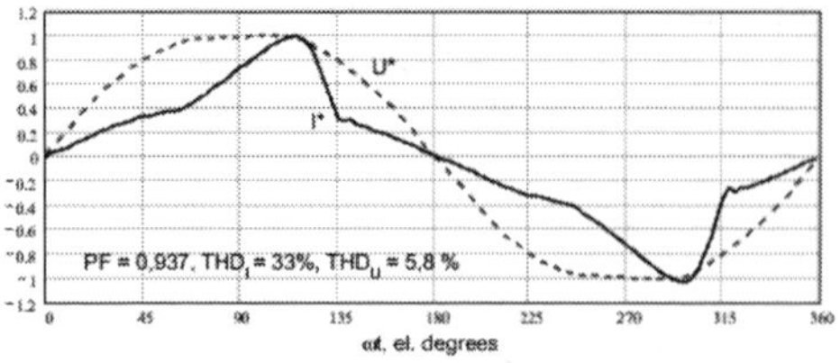

Fig. 9. Mains voltage and current with PFC emulating active resistance.

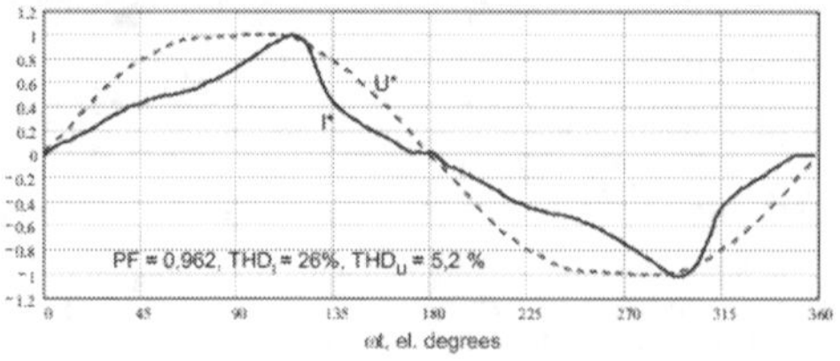

Fig. 10. Mains voltage and current with PFC in active filtering mode.

As can be seen from the presented oscillograms, activation of the active filtering function in the PFC resulted in an increase in system PF from 0.937 to 0.962; the reduction of THD_I from 33 % to 26 % and the reduction of THD_U from 5.8 % to 5.2 %. These results confirm the operability of the described control method of the active PFC.

Figures 11 and 12 show the oscillograms of the mains voltage and the input current of the experimental PFC when operating under conditions of a "clean" mains voltage (Fig. 11) which is simulated using an apparatus described in [16], and under conditions of a "real" mains voltage (Fig. 12).

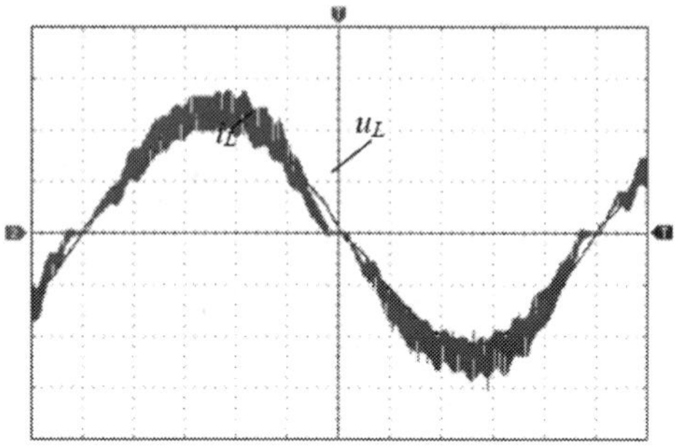

Fig. 11. Mains voltage (u_L) and input current (i_L) of the proposed PFC under "clean" mains voltage (THD_U is less than 2 %).

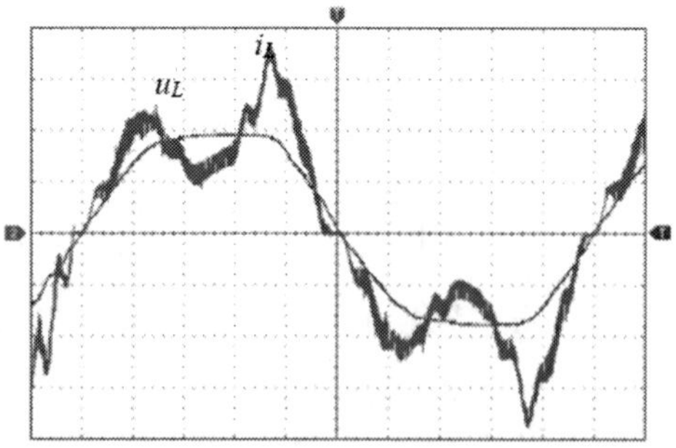

Fig. 12. Mains voltage (u_L) and input current (i_L) of the proposed PFC under "real" mains voltage (THD_U is about 6 %).

In case of low THD of the mains voltage, the proposed PFC works in the same way as conventional PFC emulating active resistance (Fig. 2a). But in case of distorted mains voltage, the proposed PFC starts absorbing harmonic currents in such a way as to lower voltage distortion. In Figure 12, flat-topped voltage waveform appears due to the presence of loads drawing current near the peaks of voltage. Unlike these loads, proposed PFC reacts to such voltage distortion by reducing its current draw near the peaks, thus lowering overall system current THD. A good agreement with the simulation results can be seen (waveforms shown in Figure 12 and Figure 5).

5 Conclusion

Thus, the PFC with the described control system allows the functions of a current sensorless parallel active filter to be partially integrated into it by increasing the ability to absorb high-order harmonic currents generated by the non-linear loads present in the power network. This leads to improving the spectral composition (reducing THD) of the mains voltage. The proposed control method can be used for practically all known PFC topologies with direct control of the input current.

A similar method and control system can be used for 3-phase active PFC, as well as for direct conversion type switchmode power supplies, in which there is a possibility of direct control of the input currents. In this case, one should only take into account that in the case of a three-wire connection to a three-phase network, it is necessary to filter out the zero sequence component of the phase voltages before processing them in accordance with the algorithm described above.

The use of electric apparatus equipped with PFC having the functions of high-order harmonics filtering will ease the achievement of compliance with the standards of electromagnetic compatibility of technical equipment, improve the mains electric power quality and improve the working conditions of other equipment connected to the same power supply network.

References

1. M.K. Soni, N. Soni, International Journal of Science, Engineering and Technology Research (IJSETR), **3**, 214-220 (2014)
2. M. J. Ghorbani, H. Mokhtari, International Journal of Electrical and Computer Engineering (IJECE), **5**, 214-220 (2015).
3. A. Jacob, B.T. Abraham, N. Prakash, R. Philip, International Journal of Advanced Research in Electrical, Electronics and Instrumentation Engineering, **3**, 10253-10261 (2014)
4. S. Gupta, P. Gupta, International Journal of Advanced Computer Research, **3**, 116-120 (2013)
5. R.S. Rani, C.S. Rao, M.V. Kumar, *International Conference on Electrical, Electronics, and Optimization Techniques (ICEEOT)*, 1438-1446 (2016)
6. N. R. Varma, K. Neelima, International Journal of Modern Engineering Research (IJMER), **2**, 268-272 (2012)
7. P.C. Babu, S.S. Dash, International Journal of Computer and Electrical Engineering, **4**, 60-64 (2012)
8. S. Khanchi, V.K. Garg, International Journal of Engineering Research and Applications (IJERA), **3**, 1430-1435 (2013)
9. A. Ghorbani, S.Y. Ebrahimi, M. Ghorbani, J. Electr. Syst. Inform. Technol. (2017)
10. DIP-PFC Application Note: Mitsubishi Electric Corporation, Power Semiconductor Device Division. Doc. No. DPH 2635e (2004)
11. M. Routimo, M. Salo, H. Tuusa, *APEC'2005*, **3**, 1696-1702 (2005)
12. V. Burlaka, S. Gulakov, Patent 100449 UA (2012)
13. V. Burlaka, S. Gulakov, *Modern Issues In Radio Engineering and Telecommunications "RT-2010"*, 54 (2010)
14. V. Burlaka, S. Gulakov, Patent 106021 UA (2013)
15. V. Burlaka, S. Kravchuk, S. Gulakov, V. Leonov, *Donbass 2020: perspectives of developing by eyes of young scientists*, 76-80 (2014)
16. V. Burlaka, S. Gulakov, *15th International Youth Forum "Radioelectronics and Youth in the 21st Century"*, **1**, 109-110 (2011)

Fuzzy-modal control for the ball and beam system

Mikhail Burakov*

State University of Aerospace Instrumentation, B. Morskaya St. 67, 190000, Saint-Petersburg, Russia

Abstract. The article is devoted to the development of a fuzzy-modal controller for the ball-beam position control system. This nonlinear dynamic plant is often considered when developing various control strategies. One option here is to use modal controls based on the linearized system model. However, the peculiarity of the linear controller is that the duration of the transient process does not depend on the initial state of the system. The proposed nonlinear control algorithm is based on the use of a set of modal control laws synthesized for different eigenvalues of a closed loop system. The signals of the modal controllers are matched by a fuzzy inference circuit using an a priori linguistic description of the states of the system. The simulation results show that fuzzy-modal control provides a transient time proportional to the initial deviation of the system.

1 Introduction

For the modern stage of the development of the theory of automatic control (TAU), the interaction of methods of intellectual control and classical sections of the TAU is characteristic. One of the variants of such interaction is the fuzzy-modal control considered in this paper.

The simulation of the "ball and beam" system is of great practical importance, because the transient processes here are similar to the dynamics of the aircraft during takeoff and landing, as well as when moving in the turbulent zone [1, 2]. To solve this problem, various approaches can be used, including modal control and fuzzy logic.

The space-state method is an effective design tool that ensures high accuracy of control systems [3, 4]. However, the modal control law is linear, and the reaction time of the system does not depend on its initial state. For essentially nonlinear objects such an idealization is unsatisfactory, which stimulates the use of fuzzy [5, 6] and neural network [7] methods for solving this problem.

Fuzzy logic allows you to manage complex objects using nonlinear algorithms [8, 9]. But the task of constructing fuzzy rules is not strictly formalized, and the accuracy of control may not be high enough.

In this paper, we investigate the option of organizing a control system for a nonlinear object, based on the use of a set of modal regulators whose signals are matched by a fuzzy logic inference scheme. This variant resembles a fuzzy logic controller (FLC) of the Takagi-Sugeno type [10, 11], in which the set of rules corresponds to a set of linear models, each of which describes the local area of the object's phase space. The construction of Takagi-Sugeno type FLCs for nonlinear objects was considered in [12, 13], where PID regulators were used as local ones. The advantage of modal control lies in the possibility of an analytic calculation of parameters that allows the poles of a closed system to be placed in a preselected position, thus ensuring the necessary characteristics of transient processes.

2 Mathematical model of control plant

Consider a mathematical model of the control object. We introduce the following notation: θ is the tilt angle of the trough, m is the ball mass, r is the radius of the ball, p is the coordinate along the beam axis, l is the length of the beam, J_1 and J_2 are the moments of inertia of the beam and ball, and τ is the control moment. The scheme of the forces acting in the system is shown in Figure 1.

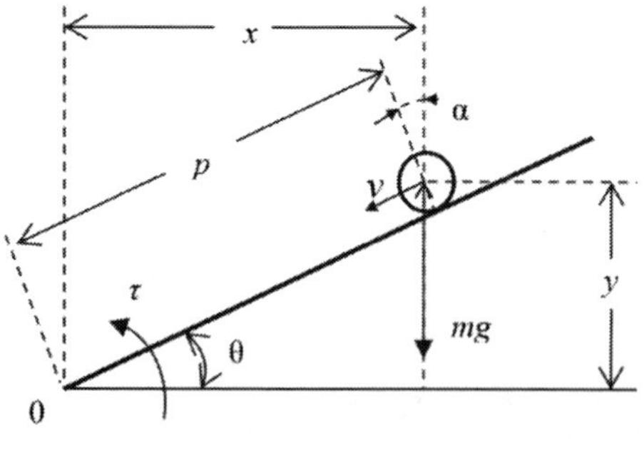

Fig. 1. Schematic representation of the ball-beam system.

A nonlinear model of the system dynamics was obtained, for example, in [1] on the basis of the Lagrange equations, it has the form:

$$\begin{cases} \left(mp^2 + J_1\right)\ddot{\theta} + 2mp\dot{p}\dot{\theta} + mgp\cos\theta = \tau, \\ \left(\frac{J_2}{r^2} + m\right)\ddot{p} + mg\sin\theta - mp\dot{\theta}^2. \end{cases} \quad (1)$$

* Corresponding author: bmv@sknt.ru

We introduce the state variables:

$$X = \begin{bmatrix} x_1 \\ x_2 \\ x_3 \\ x_4 \end{bmatrix} = \begin{bmatrix} p \\ \dot{p} \\ \theta \\ \dot{\theta} \end{bmatrix}.$$

Then it follows from (1) that

$$\dot{X} = \begin{bmatrix} \dot{x}_1 \\ \dot{x}_2 \\ \dot{x}_3 \\ \dot{x}_4 \end{bmatrix} = \begin{bmatrix} x_2 \\ \dfrac{m\left(x_1 x_4^2 - g \sin x_3\right)}{\dfrac{J_2}{r^2} + m} \\ x_4 \\ \dfrac{-2m x_1 x_2 x_4 - m g x_1 \cos x_3 + \tau}{m x_1^2 + J_1} \end{bmatrix}. \quad (2)$$

Let us consider linearization (2):

$$\dot{X} = F(X,\tau), \quad A = \frac{\partial F(X,\tau)}{\partial X}; \quad B = \frac{\partial F(X,\tau)}{\partial \tau}.$$

Let the working point have the coordinates of the state $X_0 = [p_0\ 0\ 0\ 0]^T$, then from (2) it follows that

$$A = \begin{bmatrix} 0 & 1 & 0 & 0 \\ 0 & 0 & \dfrac{-mg}{\dfrac{J_2}{r^2} + m} & 0 \\ 0 & 0 & 0 & 1 \\ \dfrac{-mg}{m p_0^2 + J_1} & 0 & 0 & 0 \end{bmatrix},$$

$$B = \begin{bmatrix} 0 \\ 0 \\ 0 \\ \dfrac{1}{m p_0^2 + J_1} \end{bmatrix}, \quad C = \begin{bmatrix} 1 \\ 0 \\ 0 \\ 0 \end{bmatrix}^T. \quad (3)$$

Thus, a linearized description of the dynamics of the system is obtained, which depends on the operating point along the length of the beam.

3 Modal control of the ball-beam system

Consider the system with parameters: m = 0.1 kg, r = 0.015 m, J_1 = 0.00001 $kg{\cdot}m^2$, J_2 = 0.025 $kg{\cdot}m^2$ and operating point $X_0 = [0\ 0\ 0\ 0]^T$, then substituting in (3) we have

$$A = \begin{bmatrix} 0 & 1 & 0 & 0 \\ 0 & 0 & -6.8 & 0 \\ 0 & 0 & 0 & 1 \\ -39.2 & 0 & 0 & 0 \end{bmatrix}, \quad B = \begin{bmatrix} 0 \\ 0 \\ 0 \\ 40 \end{bmatrix},$$

We select the position of the poles of the closed loop system: P = [–3 –3 –3 –3], then by the Ackerman formula we can obtain the vector of feedback coefficients K=[–1.28, –0.4, 1.35,0.3].

In Figure 2 is a block diagram of the control system simulation, in Figure 3 and 4 - transient processes with different initial deviations of the ball.

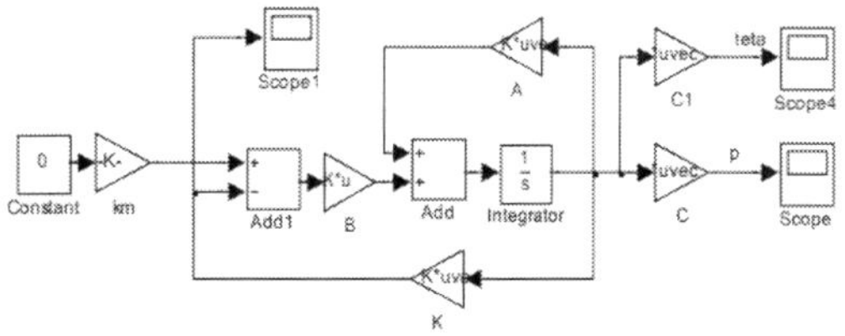

Fig. 2. The scheme of modal control in Simulink MatLab.

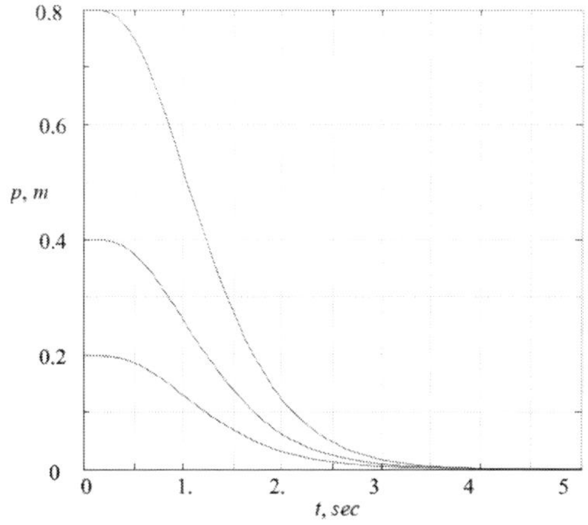

Fig. 3. Transient processes with respect to the position of the ball for various initial deviations.

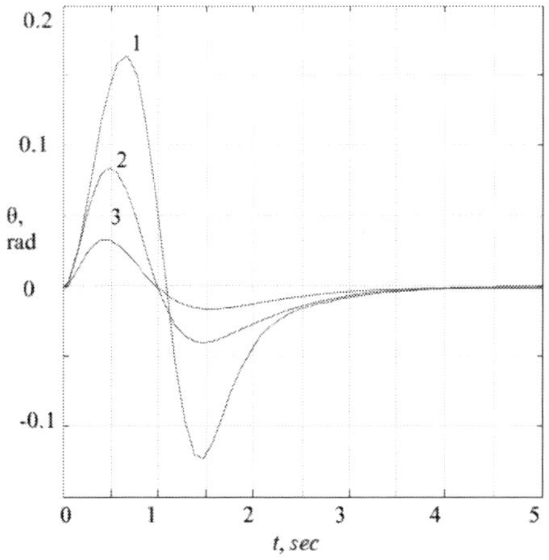

Fig. 4. Transient processes on the slope of the bar for various initial deviations of the ball: 1 - 0.8 m, 2 - 0.4 m, 3 - 0.2 m.

As Figure 3 shows, the time of the transient process is the same regardless of the initial position of the ball.

Consider the use of standard Newtonian polynomials for the choice of the position of the poles of a closed system [3]. Each pole of λ must be negative, and the magnitude of its modulus λ_0 is determined by the speed requirements - the larger λ_0, the shorter the transient time.

Table 1 presents 5 variants of the arrangement of the poles of the closed loop system and the corresponding feedback coefficients of the modal controller. Controller no. 5 is the most slow, and controller no. 1, respectively, the fastest.

Table 1. Parameters of local modal regulators.

No	Vector of poles	Vector of feedbacks gain
1	[−18 −18 −18 −18]	[−387, −85.87, 48.6, 1.8]
2	[−12 −12 −12 −12]	[−77.3, −25.44, 21.6, 1.2]
3	[−9 −9 −9 −9]	[−25.13, −10.73, 12.15, 0.9]
4	[−6 −6 −6 −6]	[−5.75, −3.18, 5.4, 0.6]
5	[−3 −3 −3 −3]	[−1.28, −0.4, 1.35, 0.3]

Examples of transient processes under the control of various regulators are shown in Figure 5. When modeling, the nonlinear model of the plant specified by the system (1) is used.

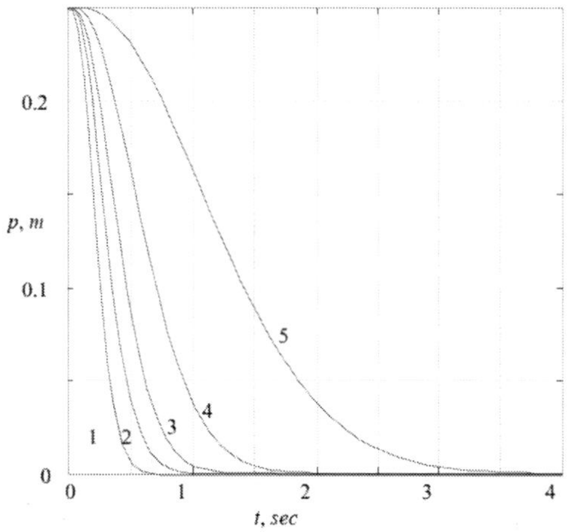

Fig. 5. Illustration of the work of controllers 1 – 5.

Let us formulate the control problem as follows: the duration of the transient process must be directly proportional to the initial deviation of the ball while maintaining the shape of the transient process. In other words, the average velocity of the ball should be approximately constant, regardless of the magnitude of the initial deviation. To solve this problem, it is proposed to simultaneously use a variety of modal controllers whose signals are matched by a fuzzy inference system.

4 Structure of fuzzy modal controller

The use of Takagi-Sugeno's FLC suggests that the fuzzy control law is described by a set of rules in which the conclusion is some analytic dependence on the input variables:

$$if\ \ Y(t) = C_i(X),\ \ then\ u_i = f_i(X),$$

where Y is the output vector of the plant, C is a fuzzy set, f is a function, usually linear, and X is a state vector.

In the problem under consideration

$$u_i = K_i X(t) = k_1^i p(t) + k_2^i \dot{p}(t) + k_3^i \theta(t) + k_4^i \dot{\theta}(t),\ i = \overline{1,5}.$$

A fuzzy set C describes a certain region of the state space of a plant, so it must be defined in a 4-dimensional space:

$$C_i = C_i(p, \dot{p}, \theta, \dot{\theta}).$$

Consider a simplified version when $C_i = C_i(p(t))$. In this case, the fuzzy partitioning option shown in Figure 6 (where T_i are fuzzy sets corresponding to the range of the individual regulators, μ is the degree of membership, Δp is the distance between the current point and the set point).

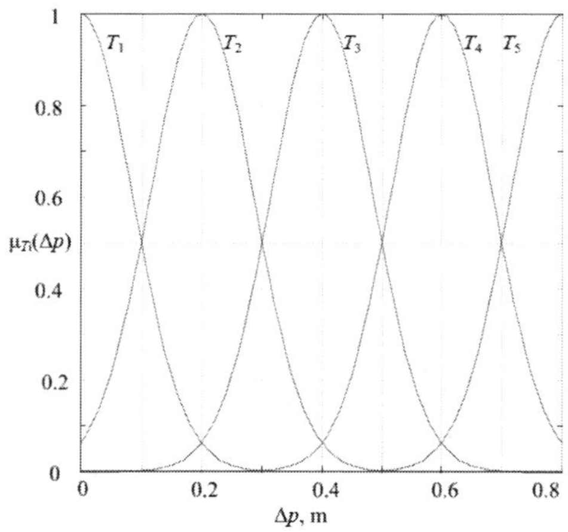

Fig. 6. Variant of fuzzy description of controllers activity areas.

The output of the fuzzy system at each time is calculated by the formula:

$$u(t) = \frac{\sum_{i=1}^{5} \mu_{T_i}(p(t)) u_i}{\sum_{i=1}^{5} \mu_{T_i}(p(t))} = w_1 u_1 + w_2 u_2 + \ldots + w_5 u_5, \quad (3)$$

where $\mu_{Ti}(p(t))$ is the degree of membership of the input value to the i-th fuzzy region; w_i is the weighting factor of the i-th fuzzy controller.

As shown in (3), the output of the FLC here is the weighted sum of the outputs of the linear controllers. The structure of the FLC is shown in Figure 7, where MC is the modal controller, p^* is the setpoint.

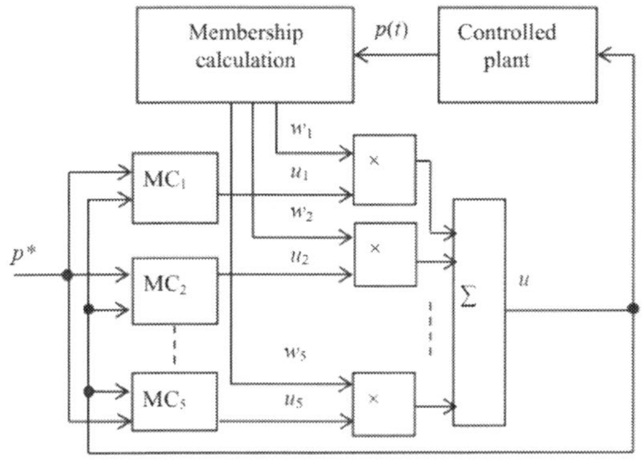

Fig. 7. Structure of fuzzy modal controller.

Thus, each local controller is responsible for its range of deviation of the ball from the target position. Regions overlap, since fuzzy sets are used to describe them.

Figures 8 and 9 show an example of modeling the operation of a fuzzy modal controller under various initial conditions.

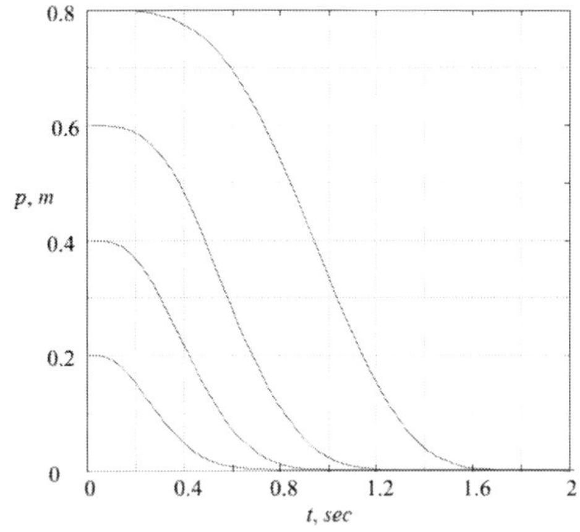

Fig. 8. Dynamics of a ball under a fuzzy-modal control.

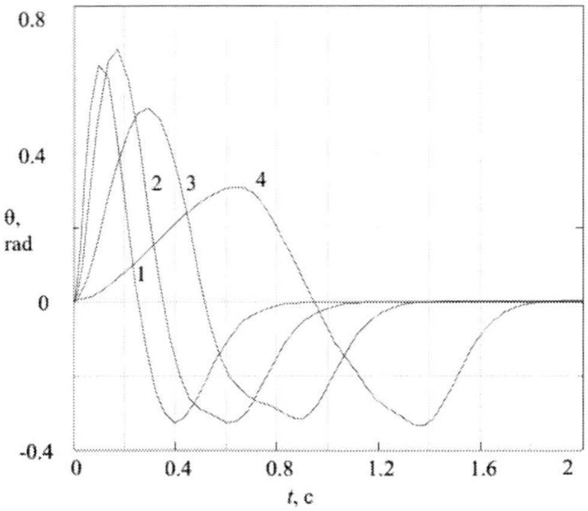

Fig. 9. Transient processes on the slope of the bar for various initial deviations of the ball: 1 - 0.2 *m*, 2 - 0.4 *m*, 3 - 0.6 *m*, 4 - 0.8 *m*.

As Figure 9, the angle of inclination of the strip during the transient process does not exceed the permissible value.

In contrast to the linear modal regulator (Figure 3), here the time of the transient process depends on the initial deviation of the ball.

Figure 10 shows an example of changing the value of the membership of the current position of the ball to the controller activity areas when moving on the beam from 0.8 m to 0 m, and Figure 11 - the control signal generated at the same time.

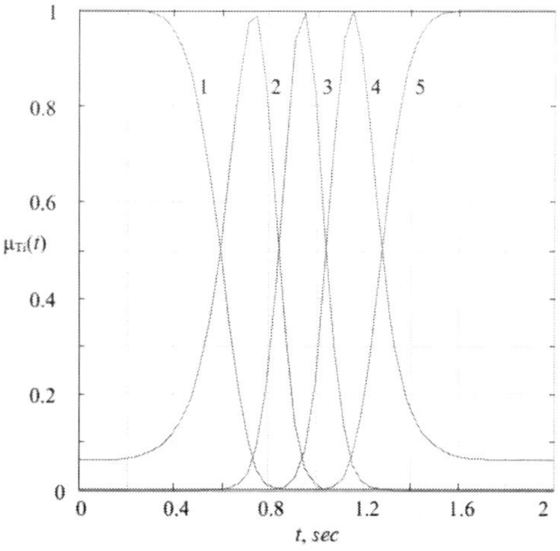

Fig. 10. Dynamics of the degree of belonging of the coordinate of the ball when moving from the point 0.8 m to zero.

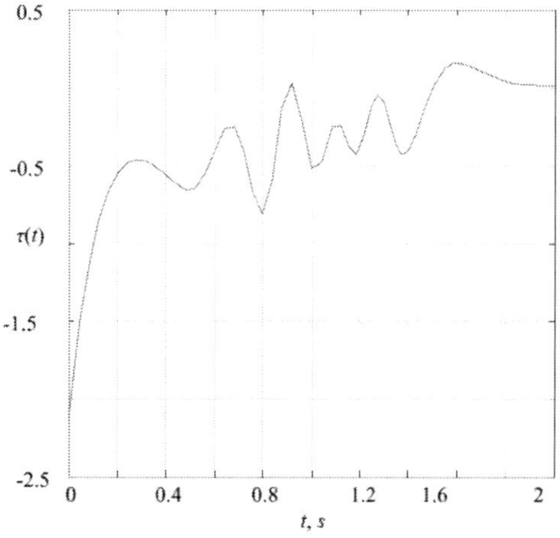

Fig. 11. Control moment of fuzzy modal controller.

Figure 12 compares the velocities of ball motion under various control laws. The fuzzy modal regulator provides an insignificant increase in the time of the transient process in comparison with the fast modal regulator (Fig. 12). However, this results in a smooth change in speed, which does not cause the risk of loss of stability.

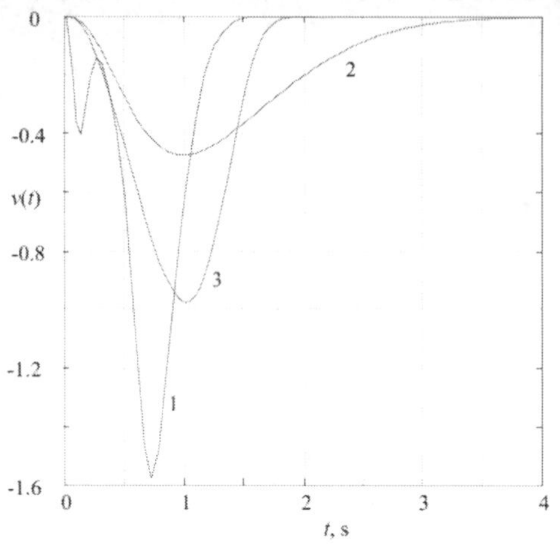

Fig. 12. Velocity of the ball: 1 - fast modal regulator (no. 2), 2 - slow modal regulator (no. 5), 3 - fuzzy modal regulator.

5 Conclusion

The structure of the fuzzy modal controller for controlling a nonlinear dynamic object, a ball on the bar, is proposed in the article.

The proposed scheme allows you to get the following options:

– Perform analytical analysis of local linear regulators for specified quality indicators.

– To provide different speed of local controllers at different points in the state space of the object.

– Use a limited number of local controllers by matching their signals using a fuzzy output circuit.

– Eliminate the drawback of linear control law, in which small deviations of the object are compensated for at the same time as large deviations.

The last property of the fuzzy modal controller is not rigidly established, since it is possible to describe the desired speed for an arbitrary point in the state space. This is the main advantage of the proposed approach, which makes it possible to recommend its use for controlling nonlinear plants of various physical nature.

References

1. J. Hauser, S. Sastry, P. Kokotovic, IEEE Trans. Autom. Control. **37**, 3 (1992)
2. F.A. Salem, Control Theory and Informatics. **3**(4) (2013)
3. R.C. Dorf, R.H. Bishop, *Modern Control Systems* (Upper Saddle River, NJ: Prentice Hall, 2001)
4. K. Ogata, *MATLAB for Control Engineers* (Upper Saddle River, NJ: Pearson Prentice Hall, 2008).
5. A.A. Emhemed, IJFLS, **3**(1) (2013)
6. S.K. Oh, H.J. Jang, W. Pedrycz, Eng. Appl. Artif. Intell., **22** (2009)
7. X. Li, W. Yu, Int. J. Control Autom. Syst., **8**(3) (2010)
8. K.M. Passino, S. Yurkovich, *Fuzzy control* (Addison Wesley Longman, Inc. 1998)
9. Z. Kovacic, S. Bogdan, *Fuzzy Controller Design: Theory and Applications* (CRC Press, 2006)
10. T. Takagi, M. Sugeno, IEEE Trans. Systems Man Cybernet. **15**(116) (1985)
11. M.V. Burakov, M.S. Brunov, SPIIRAS Proceedings, **3**(34), 232-246 (2014)
12. S. Ramesh, S.A. Lincon, IJERA, **3**(6) (2013)
13. M.V. Burakov, A.P. Kirpichnikov, Bulletin of the Kazan Technological University, **19**(4) (2016)

Calculation features of magnetic coupling in non-magnetic structures

Vladimir Kosulin and *Irina* Salova*

SUAI, 190000, St. Petersburg, Russia

Abstract. This paper presents analytical approach for computing the torque transmitted by magnetic coupling with radially polarized permanent magnets. The method for calculating magnetic couplings in nonmagnetic constructions is proposed. 2D models of magnetic couplings for finite element calculation are presented.

1 Introduction

Magnetic couplings are widely used in circulating pumping systems for pumping environmentally hazardous and corrosive liquids in chemical, oil and gas processing, food, pharmaceutical and other industries. In them, the magnetic field is created by permanent magnets (NdFeB and SmCo) installed in the leading and driven couplings. A lot of work has been devoted to the calculation of magnetic couplings, for example, [1-7], where constructions with steel magnetic cores are considered. In [1, 2], the influence of the permanent magnets geometry on the coupling characteristics is investigated. In [3], the influence of the pole pairs number and the air gap size on the magnitude of the torque, developed by the coupling is considered. In [6], the results of optimization of the radial magnetic couplings geometry are presented. As the optimization criterion, the volume of permanent magnets is assumed. To study and calculate coupling parameters, the authors use methods of magnetic fields mathematical modeling, which are based on solving the Laplace or Poisson equations with respect to the scalar or vector potential of the magnetic field in the investigated region.

The last time for the analysis of magnetic systems, the finite element method is widely used [3, 4, 5]. This method gives good agreement with experiment. In a number of cases, the application of these methods is difficult because of their complexity and cumbersomeness of the field theory equations. In addition, the calculation of each version of the magnetic system requires considerable time. In engineering calculations it is required to study the design of the coupling in combination with the main device. At this stage it is required to estimate the possibility of realizing a given moment for various overall dimensions of the structure as a whole. Therefore, a technique is required that makes it possible to quickly calculate the magnetic coupling maximum torque for several design variants. Such method is based on the construction and calculation of circuits for the replacement of magnetic circuits [7,10].

There are designs in which the steel magnetic circuits cannot be used for operating conditions. In such designs, there is no saturation of the magnetic cores. Therefore, such constructions are most easily calculated. In such designs magnets are mounted in a holder, made of non-magnetic material. Since the magnetic permeability of the armature approaches to the magnetic conductivity of air, the amount of useful magnetic flux is significantly influenced by scattering fluxes, which should be taken into account when calculating the magnetic flux in the air gap. The calculation of such systems is connected with the determination of the magnetic resistances of the air gaps, both between the working surfaces of the magnets and along the paths of the scattering fluxes. In this case, the problem becomes linear, since there are no sections with steel.

2 Method content

2.1 Physical basics of the method

Let us consider this method using the example of a magnetic coupling with a radial arrangement of permanent magnets. The element of the magnetic coupling system with the radial arrangement of permanent magnets is shown in Figure 1,a. Permanent magnets, magnetized radially, are shown by hatching, and the structural elements of the coupling are omitted and replaced by air gaps. In Figure 1 also indicates: h_m – the height of a permanent magnet; b_1,b_2 – width of permanent magnets of outer and inner holders, respectively; l – axial dimension of the coupling; $\Delta_1, \Delta_1', \Delta_2, \Delta_2'$ - the distance between the sides of the magnets of the upper (1) and lower (2) holders. The expression for the torque of magnetic coupling as a function of the relative angle between the inner and outer holders has the form [8]:

* Corresponding author: salova_irine@mail.ru

$$M = \frac{p\pi D l \delta}{2\mu_0} B_{\delta m}^2 \sin\Theta, \quad (1)$$

where D – the diameter of the average circle of the working non-magnetic gap δ; Q-the rotating electric angle of the inner holder relative to the outer ($Q=pQ'$, here Q' - rotating space angle); $B_{\delta m}$ – the maximum value of induction in the middle of the air gap δ when the magnet coupling is in the idle mode (Q=0); p – the number of poles pairs; l– the length of the magnet in the axial direction; $\mu_0 = 4\pi \cdot 10^{-7} H/m$. This expression contains the maximum value of induction, to determine which it is required to know the induction distribution with respect to the angle $B(\Theta')$, which requires the solution of the field problem.

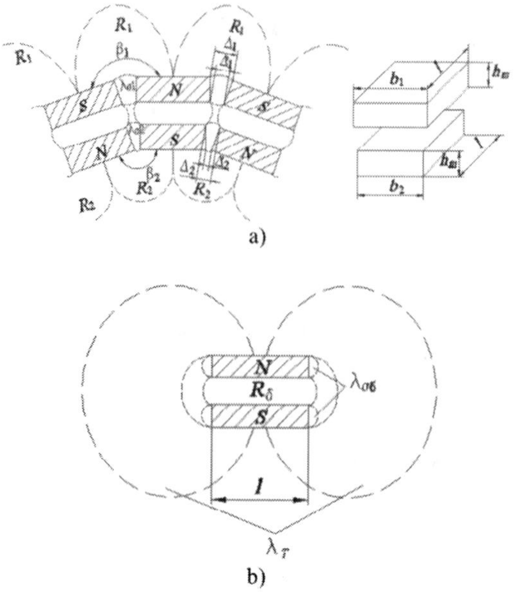

Fig. 1. The distribution of magnetic fluxes of mutually perpendicular planes.

For engineering practice, this calculation meets certain difficulties because it requires the application of software packages, which is not always possible and appropriate. Let us consider the method based on the using of equivalent magnetic circuits, which will allow pre-calculation of magnetic coupling to estimate with a known degree of accuracy the magnitude of the coupling failure torque use the integral characteristics of the magnetic field – magnetic flux. To do this, in the expression for the magnetic coupling torque, we pass to the induction average value and, accordingly, to the magnetic flux F_δ, which closes between the poles of the half couplings in the idle mode:

$$M = \frac{2p^3 \Phi_\delta^2 \delta}{\mu_0 D\pi l \alpha_t^2} \sin\Theta, \quad (2)$$

where $\alpha \approx b_{mid}/t$– coefficient of pole overlap; here b_{mid} - the permanent magnets arc average value of the inner and outer holders; t- pole division ($t=\pi D/2p$).

To calculate the magnetic flux, we construct an equivalent circuit for replacing the magnetic circuit and carry out its calculation.

On the constructive scheme of the device magnetic circuit we determine the probable path of the magnets fluxes, distinguish scattering flows and main work flow. For each flux, we calculate the magnetic conductivity (or magnetic resistance) and construct an equivalent circuit. In this case, the equivalent circuit is compiled for the main working flux, and the influence of the scattering fluxes is taken into account by the reduction factors for the magnets parameters.

In Figure 1, for the model under consideration, the dashed lines conventionally denote the boundaries of the possible paths of the corresponding magnetic fluxes in the system.

The magnetic flux of each permanent magnet consists of two components: a useful magnetic flux passing through an air gap with magnetic resistance R_δ, and scattering fluxes along paths with magnetic conductivities $\lambda_{\sigma 1}$, $\lambda_{\sigma 2}$ and $\lambda_{\sigma b}$. Scattering fluxes do not participate in creating a useful flow and can be excluded from consideration by introducing the reduction coefficients k_σ. The calculation of the air gap magnetic conductivities for most common configurations is performed by known methods [9]. The main working magnetic flux on the outer faces of the outer and inner holders is closed along the paths of air gaps with a magnetic resistance R_1, R_2, and $R_T=1/\lambda_T$ (Fig.1).

Then we can assume that the useful magnetic flux of each pair of magnets, passes through the air gap, consists of two components:

- the flow in the plane perpendicular to the axis of the coupling (Fig.1,a);
- the flow in the plane passing through the coupling axis (Fig. 1,b).

The geometry of the coupling is such that the resistances R_1, R_2 and $R_T=1/\lambda_T$ are comparable, this means that both magnetic flux (transverse and longitudinal) are involved in the creation of a working magnetic flux. This situation is not typical for systems with steel magnetic cores, since practically all the useful magnetic flux is closed in the transverse plane.

Consider the principle of calculating the reduction coefficient k_σ, which takes into account the change in the calculated magnets parameters, with the elimination of scattering fluxes.

If the equivalent scheme is presented for any magnet in the simplest magnetic system (Fig.2,a), then it will include four elements:

- actual MMF source (F_m) - the permanent magnet;
- magnetic resistance R_m of a permanent magnet;
- magnetic resistance R_σ of scattering;
- magnetic resistance R_δ of the air gap for the main (useful) flow.

There will be two obvious transformations:

- MMF source is converted into a current source and we combine the resistances R_m and R_σ into one as connected in parallel: $R' = \frac{R_m R_\sigma}{R_m + R_\sigma}$;
- current source is converted back into the MMF source F'_m with internal resistance $R' = \frac{R_m R_\sigma}{R_m + R_\sigma}$.

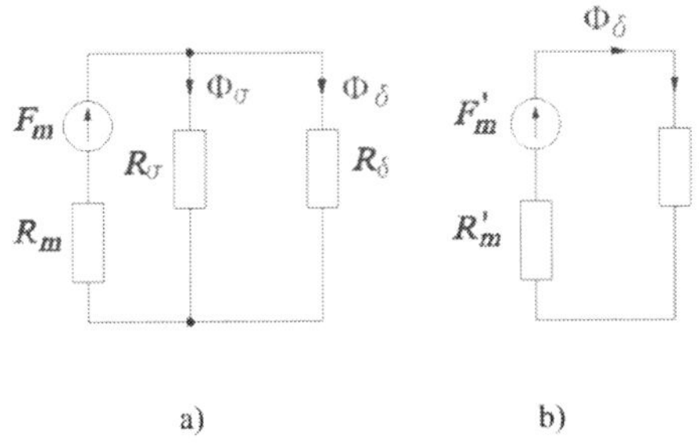

Fig. 2. An equivalent magnetic circuit of the permanent magnet.

We introduce the coefficient of reduction $k_\sigma = 1 + \frac{R_m}{R_\sigma}$. Then the following relations are valid: $F'_m = \frac{F_m}{k_\sigma}$; $R'_m = \frac{R_m}{k_\sigma}$, and the transformed scheme takes the form, as in Figure 2,b.

This transformation made it possible to eliminate the scattering conductivity when the preparation of the equivalent magnetic circuit, replacing the parameters of the permanent magnet by the reduced ones. Let us note, that

$$F_m = h_m H_C; R_m = \frac{h_m H_C}{S_m B_r} = \frac{h_m}{S_m \mu_r \mu_0}, \tag{3}$$

where h_m – the length of the magnet in the direction of magnetization; S_m- the area of magnet neutral section; B_r - the residual induction of the magnet, H_c- coercive force; μ_r – relative permeability of a permanent magnet material.

Taking into account the relationship between the magnetic resistance and magnetic conductivity coefficient of reduction can be calculated, and through the corresponding magnetic conductivities:

$$k_\sigma = 1 + \frac{\lambda_\sigma}{\lambda_m}. \tag{4}$$

For the magnetic circuit in Figure 2b, as a single-circuit one, the magnetic flux F_δ is found from the obvious relation:

$$\Phi_\delta = \frac{F'_m}{R_\delta + R'_m}. \tag{5}$$

Taking into account the distribution of the magnetic fluxes in the magnetic system (Fig.1) equivalent circuit has the form as in Figure 3.

Due to the symmetry of the coupling design and based on the first Kirchhoff's law for magnetic circuits, the relation is valid:

$$\Phi_1 = \frac{\Phi_\delta - \Phi_3}{2}. \tag{6}$$

Let us compose the equations according to the second Kirchhoff's law for this scheme:

$$\begin{cases} \Phi_\delta \left(R'_{m1} + R'_{m2} + R_\delta\right) + \Phi_3 R_T = F'_{m1} + F'_{m2} \\ \Phi_1 \left(R_1 + R_2\right) - 2\Phi_3 R_T = 0 \end{cases}. \tag{7}$$

Solving the obtained equation relative to the flow F_δ, we obtain:

$$\Phi_\delta = \frac{4\left(F'_{m1} + F'_{m2}\right)}{4\left(R'_{m1} + R'_{m2} + R_\delta\right) + \left(1 - k_T\right)\left(R_1 + R_2\right)}, \tag{8}$$

where $k_T = \frac{R_1 + R_2}{R_1 + R_2 + 4R_T}$.

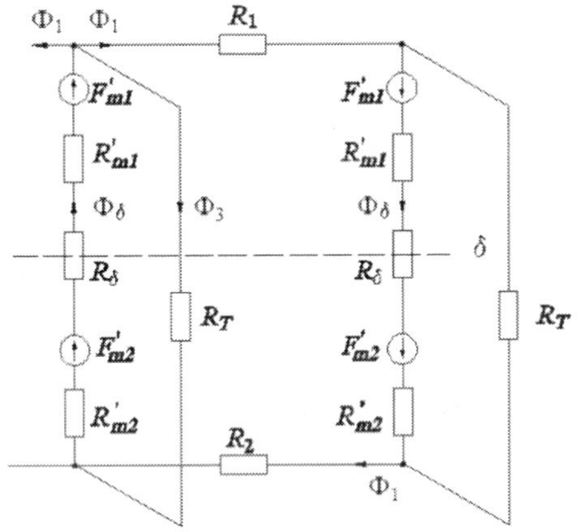

Fig. 3. Replacement scheme of the coupling magnetic circuit.

As follows from the expression, the magnetic resistances of the sections both in the transverse plane (R_1, R_2) and in the longitudinal (R_T) affect the magnitude of the magnetic flux in the air gap. Since their magnitude is determined by the geometry of the corresponding air gaps, there is some uncertainty in the estimation of the permanent magnets outer faces areas for each component of the magnetic flux. For engineering calculations the following algorithm is proposed to calculate the components of the useful flux based on the solution of two plane problems, based on the physical picture of the distribution of magnetic fluxes in the considered magnetic system:

1. The planar problem for the circuit in Figure 1,a is solved, and the magnetic flux in the air gap $F_{\delta 1}$ is calculated in neglect of the flows in the other plane. Thus, the magnetic circuit resistance of the magnetic system transverse contour is fixed.

2. The plane problem for the circuit in Figure 1,b is solved, and the magnetic flux in the air gap $F_{\delta 2}$ is calculated in neglect of the flows in another plane. This fixes the level of resistance of the magnetic circuit in the longitudinal (axial) circuit.

3. From the assumption that the distribution of the areas in the cross section of the permanent magnets for the fluxes F_1 and F_3 (Fig.3) in the real system corresponds to the relationship between the flows calculated in item 1 and item 2, the resistance of R_1, R_2 and R_T is adjusted, since their value is inversely proportional to the corresponding areas.

4. The magnetic flux in the air gap is calculated according to the replacement scheme in Figure 3, taking into account the corrected value of the resistance as per item 3.

Let's consider each of the listed items of the calculation algorithm.

2.2 Calculation of the magnetic flux $F_{\delta 1}$

The coefficient of reduction is calculated taking into account all the scattering fields along the perimeter of the permanent magnets so that the values of the magnets MMF can be used both for calculation of the fluxes of $F_{\delta 1}$ and $F_{\delta 2}$, and then of the flux of F_{δ}. This takes into account the scattering conductivity $\lambda_{\sigma p}$ along the edges of the permanent magnets between the neighboring lateral faces of each magnet (the conductivity of the spherical quadrants, in Figure 1 a are not shown) [9]. Since the distance between the lateral faces of the magnets may be greater than half the h_m, in this case, in addition to the conductivity $\lambda_{\sigma 1}$ between the lateral faces of the neighboring magnets, the scattering conductivity $\lambda_{\sigma 6}$ along the lateral faces of each magnet should be added [9]. Then the conductivity of the magnet scattering of the upper half-coupling is defined as the sum $\lambda_{\sigma B}=\lambda_{\sigma 1}+\lambda_{\sigma 6}+\lambda_{\sigma p}$. The scattering conductivity between the side faces can be found from the formula [10]:

$$\lambda_{\sigma}=\frac{k\Phi_{\sigma}}{2h_m H_c}=0{,}5k\mu_0\frac{lh_m}{\alpha_{cp}}, \qquad (9)$$

where k=1,05-1,1 – coefficient that takes into account the bulging of the force lines along the path of closure of the scattering flux; $\alpha_{cp}=(\Delta+\Delta_1)/2$.

The conductivities $\lambda_{\sigma 6}$ и $\lambda_{\sigma p}$ are found by the formulas [9]:

$$\lambda_{\sigma 6}=2\cdot 0{,}26\mu_0(l+b_1);\ \lambda_{\sigma p}=4\cdot 0{,}77\mu_0 h_m. \qquad (10)$$

Magnetic conductivity of the upper magnet:

$$\lambda_{m1}=\frac{S_{m1}B_r}{h_m H_c}. \qquad (11)$$

Reduction coefficient is $k_{\sigma 1}=1+\frac{\lambda_{\sigma B}}{\lambda_{m1}}$. The reduced MMF of the upper magnet is $F'_{m1}=\frac{h_m H_c}{k_{\sigma 1}}$. The reduced magnetic resistance of the upper magnet is:

$$R'_{m1}=\frac{h_m}{S_{m1}\mu_r\mu_0 k_{\sigma 1}}. \qquad (12)$$

Similarly, the reduced values of the MMF F'_{m2} and the resistance R'_{m2} of the lower magnet are calculated.

Since the permanent magnets are parallelepipeds, and the air gap is formed by the parallel faces (when the coupling is idling), the magnetic resistance of the air gap we find by the formula:

$$R_{\delta}=\frac{\delta}{\mu_0 l b_{cp}}, \qquad (13)$$

where $b_{cp}=(b_1+b_2)/2$.

When calculating the magnetic resistances R_1 and R_2, calculated expressions are used for the conductivity of air gaps between rectangular surfaces lying in one plane [9], corrected by an angle β corresponding to the arrangement of the magnets along the circumference.

$$R_1=\frac{1}{\frac{\mu_0}{\beta_1}l\ln\left(1+2\frac{0{,}5b_1+\sqrt{0{,}25b_1^2+0{,}5b_1\Delta_1}}{\Delta_1}\right)}$$
$$R_2=\frac{1}{\frac{\mu_0}{\beta_2}l\ln\left(1+2\frac{0{,}5b_2+\sqrt{0{,}25b_2^2+0{,}5b_2\Delta'_2}}{\Delta'_2}\right)}. \qquad (14)$$

If the holders with permanent magnets are made of electrical steel, then for an approximate calculation, the resistance of the sections with steel can be neglected in comparison with the air gap resistance. Then we can take R_1=R_2=0, which will greatly simplify the calculations. After the decision has been made about the dimensions of the device magnetic circuit, a verification calculation is performed to determine the torque characteristics, taking into account the possible saturation of the sections with steel.

Thus, the magnetic flux $F_{\delta 1}$ in the air gap is equal to:

$$\Phi_{\delta 1}=\frac{4\left(F'_{m1}+F'_{m2}\right)}{4\left(R_{\delta}+R'_{m1}+R'_{m2}\right)+R_1+R_2}. \qquad (15)$$

2.3 Calculation of the magnetic flux $F_{\delta 2}$

To calculate the magnetic flux $F_{\delta 2}$, in addition, we must find the resistance R_T. Its value depends on the ratio of the linear dimensions of the permanent magnets and the air gap size and is inverse of the corresponding conductivity $\lambda_T = g2b\mu_0$, where $g=f(m,n)$ is the geometric dimension function given graphically in [9]. In this case with $m=(2h_m+\delta)/2h_m$, $n=b_{cp}/\delta$ the magnetic flux is:

$$\Phi_{\delta 2} = \frac{F'_{m1} + F'_{m2}}{R'_{m1} + R'_{m2} + R_\delta + R_T}. \qquad (16)$$

2.4 Correction of the resistance of the replacement circuit

Below the expressions are presented for the corrected values of the magnetic circuit resistance:

$$R_{1k} = R_1 \frac{\Phi_{\delta 1} + \Phi_{\delta 2}}{\Phi_{\delta 1}}; \quad R_{2k} = R_2 \frac{\Phi_{\delta 1} + \Phi_{\delta 2}}{\Phi_{\delta 1}}$$
$$R_{Tk} = R_T \frac{\Phi_{\delta 1} + \Phi_{\delta 2}}{\Phi_{\delta 2}}. \qquad (17)$$

2.5 Calculation of the magnetic flux F_δ

We substitute the expressions for the corrected resistances (17) into expression (8) and obtain the value of the magnetic flux:

$$\Phi_\delta = \frac{4\left(F'_{m1} + F'_{m2}\right)}{4\left(R'_{m1} + R'_{m2} + R_\delta\right) + \left(1 - k_T\right)\left(R_{1k} + R_{2k}\right)}, \qquad (18)$$

where $k_T = \dfrac{R_1 + R_2}{R_{1k} + R_{2k} + 4R_{Tk}}$.

Then this value of the magnetic flux is substituted into expression (2) for calculating the torque of magnetic coupling.

3 Modeling by the finite element method

The above procedure was verified with the help of the finite element method on *2D*-models of the magnetic coupling. The difference between the values of the failure torque, calculated by the proposed method, and the torque values, obtained by the finite element method, for specific specified sizes of magnetic couplings does not exceed 10%. It should be noted that the accuracy of the calculation is largely determined by the thoroughness in determining the geometry of the magnetic flux components paths and the correct application of the calculated formulas for the conductivity of the corresponding magnetic gaps [9].

Figure 4 shows the magnetic field pattern for *2D*-analysis at the rotation angle Q=0°. Figure 5 shows the pattern of the magnetic field in the axial direction. Since the calculation of the *2D* model by the finite element method gives us the values of the magnetic fluxes $F_{\delta 1}$ and $F_{\delta 2}$, we can control and, if necessary, adjust the value of the scattering conductivities, involved in determining the reduction coefficient, and the corresponding MMF of the permanent magnets.

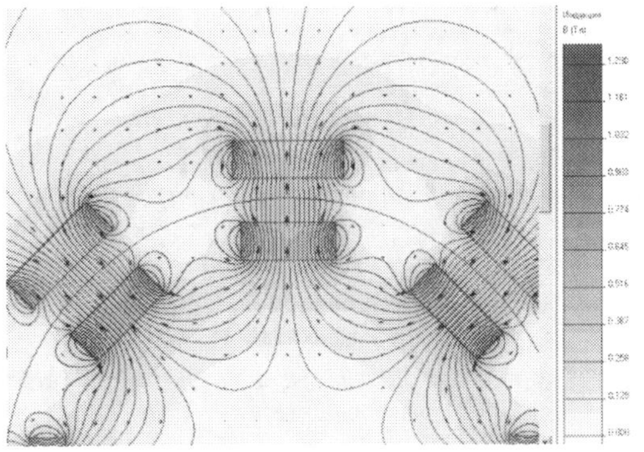

Fig. 4. Magnetic field of a plane problem for the rotation angle Q=0°.

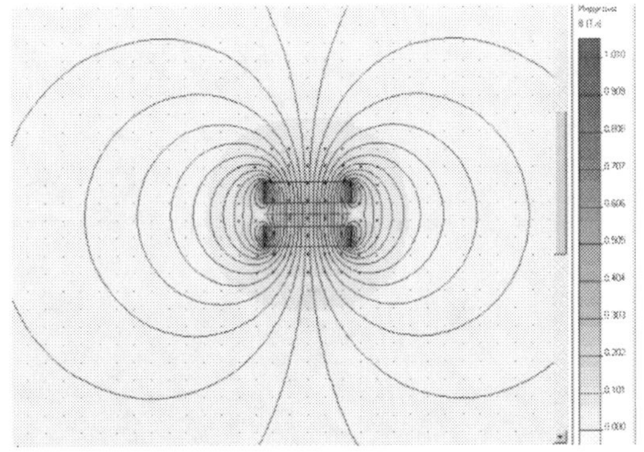

Fig. 5. Magnetic field in the axial direction.

4 Conclusion

The proposed method makes it possible to calculate the failure torque of magnetic couplings, in which the reinforcement and structural parts are made of non-magnetic materials ($\mu = 1$) and can be recommended for verification calculation of magnetic couplings at the stage of elaborating design options and choosing the number of poles pairs.

References

1. A. Ja. Krasil'nikov, Vestnik mashinostroenija [Bulletin of Machine Building], **4** (2017) (in Russian)
2. A. Ja., Krasil'nikov, A.A. Krasil'nikov, Vestnik mashinostroenija [Bulletin of Machine Building], **6** (2009) (in Russian)
3. L.Thierry, S.Mezani, Simple Analytical Expressions for the Force and Torque of Axial Magnetic Couplings. IEEE Transactions on Energy Conversion, (2012)
4. M. H. Nagrial, J. Rizk, A. Hellany, World Academy of Science, Engineering & Technology International Journal of Mechanical, Aerospace, Industrial, Mechatronic and Manufacturing Engineering, **5**(7) (2011)

5. D. MeeKer, Finite Element Method Magnetics. Available at: http://femm.berlios.de/ (Accessed 20 February 2017)
6. J. Fontchastagner, Y.Lefèvre, F.Messine, IEEE Transactions on Magnetics, **45**(3) (2009)
7. S. Mohammadi, M. Mirsalim, S. Vaez-Zadeh. IEEE Transactions on Energy Conversion, **29,** 1 (2014)
8. Je.V Lapshin, I.V. Tolstouhov, *Trudy Mezhdunarodnogo Simpoziuma "Nadezhnost' i kachestvo"* [Proc.Int.Symp."Reliability and quality"], Penza, 2, (2014) (in Russian).
9. Pjatlin Ju.M., *Postojannye magnity. Spravochnik.* [Permanent magnets. Directory], (Moscow, Energy, 1971), (in Russian)
10. V.D. Kosulin, Scientific and technical conference "Zavalishenskie readings, St. Petersburg, SUAI, (2011) (in Russian)

MATEC Web of Conferences 113, 01008 (2017) DOI: 10.1051/matecconf/201711301008
12th International Scientific-Technical Conference on Electromechanics and Robotics "Zavalishin's Readings" - 2017

Determination of transformer oil quality by the acoustic method

Anastasiya Tyuryumina*, *Andrey* Batrak *and Victor* Sekackiy

Siberian Federal University, 26 Kirenskiy st., Krasnoyarsk, 660074, Russian Federation

Abstract. Large power transformers are the most expensive and strategically important components of any power generation and transmission system. Their reliability has crucial importance for the energy system operation. Insulation breakdown can generate a serious failure of a large power transformer, which causes substantial costs to repair and financial loss due to power outage. Most power utilities are, therefore, highly motivated to assess the actual condition of the transformer insulation system. The majority of transformer condition check methods are only available for off-line periodic diagnostics. The acoustic emission (AE) method as a complex online diagnostics technique to identify evolving failures is proposed. Results of measuring the acoustic signals generated by the presence of impurities in transformer oil and determining the impurities influence on its condition as an indicator of the transformer condition is presented.

1 Introduction

The power transformers are the most important and expensive part in a power system and considered to be reliable components of electrical stations if their design life is 20-35 years (increasing to 50 years in practice with appropriate maintenance). Effective condition monitoring of these assets can protect loss of supplies to large populations of customers, while the threat of explosions and fire, causing irreparable damage, can injury the station and personnel and pose a risk to safety and the environment with potentially serious legal and regulatory implications. Therefore, the welfare of transformers remains at the forefront of utilities' asset management and obligations [1].

A survey of senior utility managers [2] showed that more than two-thirds of answers were related to the adverse stock of transformers, an issue which was one of the most important problems for utilities. The majority of power transformers currently in operation were installed in the 60's and 70's of the previous century. They might have reached a critical age with an operating time of 30 years and more. Based on obtained answers, Kurtz et al. [2] determined that the amount of problematic units is maximal when transformers reach 35-40 years (Fig. 1).

Utilities have to perform modernization or replacement of service-aged transformers, depending on each transformer technical condition and operational costs. A modernized transformer may serve 10-15 years but requires a careful maintenance, regular inspections and checks. Therefore, the investment in a new transformer may be postponed. The modernization costs are significantly less than a new unit cost [3].

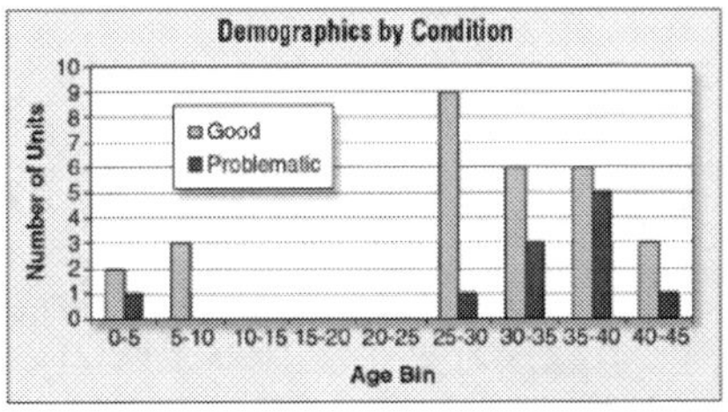

Fig. 1. Typical demographic data sorted by condition [2].

Malewski et al. [3] calculated the costs of a typical transformer. The cost of components and labor involved in modernization was 23% of a new transformer cost. At the same time the cost of condition assessment (1%) was negligible compared to the modernization cost. It was especially relevant for high power transformers because their investment cost in relation to the total cost of transmission, division and distribution devices was about 20%. In extreme cases transformer failure cost might exceed its construction cost five times [4].

Therefore, EAM is an important issue of operation management and should be applied to power transformer maintenance including advanced fault prevention and immediate diagnostics to avoid unexpected breakdowns. Some components of EAM systems in the majority of countries were discussed in [5-11].

Therefore, that the utility managers were unanimous in their assessment that improved methodologies are necessary for power transformers [2]. Currently, there has been a growing interest in the condition assessment of transformer insulation. Since the majority of transformer components are located within the main tank, the transformer oil sample can provide the information about its condition [12].

* Corresponding author: eternity17@list.ru

Estimated lifetime of an insulation system was reduced for the half if average life time temperature of transformer increased by 8°C over maximum allowable operating temperature. Practically, it means that estimated lifetime of a 50 year-old transformer is reduced to 25 years if maximum allowed operating temperature (90°C) is exceeded on average by 8°C [13].

There is variety of reasons to increase operating temperature. This study prefers to discuss water content, cellulose and gas phrase as the main reasons.

2 Effect of impurities on transformer oil condition

2.1 Water influence

The importance of moisture presence in paper-oil systems has been recognized since the 1920s [14]. Pure transformer oil consists of saturated hydrocarbons such as paraffins and naphthenes. Because of their nonpolar molecule structure, pure oil can dissolve only a very small quantity of water dispersed as water vapour according to Henry's law [15]. One drop of water is enough to saturate one liter of new oil [16].

Water causes three dangerous effects: it decreases the dielectric withstand strength, accelerates cellulose aging, and causes the emission of bubbles at high temperatures [17-20]. The knowledge about the actual moisture concentration is required to control transformer condition. Therefore, different methods and instrumentation are used to measure moisture.

Equilibrium diagram methods attempt to derive moisture in solid insulation from moisture content in oil (Du el al. [14], Fabre et al. [18], Oommen [21], Griffin et al. [22]) However, the methods fail, where aging of oil and water titration has the major impact [23].

Karl Fischer introduced the titration method in 1935. Titration basically means adding a reagent of a known concentration (titre) to an unknown substance until the concentrations are balanced. It measures moisture content related to oil weight (ppm as µg water/g oil) or as relative moisture saturation to saturation value (%).

Nowadays this method is considered to be the most reliable method and it also serves as a benchmark for other methods. Nevertheless, this method is subject to several faults, which may lead to a bad comparability of results. They are listed below:

- there is always a moisture ingress from atmosphere during sampling, transportation and sample preparation;
- cellulose binds water with chemical bonds of different strengths and releases water under the influence of heating and time;
- laboratories differently treat constraints that are not covered by standards, e.g. the solvent for oil extraction;
- sometimes direct injection and the heating method lead to different results of moisture measurement due to the influence of oil additives and ageing products [15].

In addition, the Karl Fischer titration method is not a real-time one, and requires appropriate chemical instrumentation. It is also prone to a high level of error and its results are often difficult to interpret.

A group of methods have attracted a lot of attention in recent years. They rely on the measurement and analysis of insulation dielectric properties. This diagnostics is based on the fact that paper and oil change their dielectric properties when moisture levels increase. Therefore, measuring parameters can be the dissipation factor, polarization and depolarization currents (PDC) and return voltages (RVM). Also, to evaluate water content in oil, polymer relative humidity/relative saturation sensors are frequently used [16].

However, the majority of dielectric methods do not distinguish moisture impact on an insulation aging from impact of other processes. This creates a lot of confusion over the reliability of dielectric response methods.

Water in a solid insulation is known to play a role of an ageing process catalyst. However, transformer oil in power transformers is polluted with refining remains, ageing byproducts and cellulosic particles. In aged oil, water can be adsorbed by impurities, e.g. cellulose fibres, when the saturation level is exceeded and water solubility increases [15].

2.2 Cellulose influence

Paper used as an insulation material includes 85 % of cellulose, 3-7 % of lignin and 10 % of hemicelluloses. Transformer paper ages under the influence of heat, air and moisture and eventually loses all mechanical integrity, with risk increasing dramatically due to insulation damage [24].

Fourier transform infrared (FTIR) and near-infrared (NIR) spectroscopy are used to characterize ageing of cellulose and assess paper insulation condition. They determine the differences in spectra of new and aged materials e.g. Ali et al. [25] studied aged cellulose and obtained a 0.994 correlation between the spectra and ageing time, with an error of 95 hours prediction for paper samples with up to 3000 hours of ageing.

Generally, the methods studying cellulose in transformer focus on determining its condition, although, brittle paper may break away from transformer windings and block ducts [26]. We propose to determine cellulose sediment influence on transformer oil condition with the AE method and use obtained information to evaluate transformer condition.

2.3 Influence of dissolved gases

Another detrimental effect which can occur in transformer oil is gaseous bubbles. This phenomenon results in dangerous effects, such as formation of vapour-filled cavities on insulation surface and in oil, with dramatic decrease of insulation dielectric strength; deimpregnation of the turn insulation.

The breakdown process starts with a microscopic bubble, an area of large distances between corpuscles, where ions or electrons can initiate avalanches. These microscopic bubbles originate from current impulses on an electrode. The next current impulse injects charge carriers into the bubble, leading to current amplification and finally to breakdown [27].

Applying method of dissolved gas analysis (DGA) and taking samples of mineral oil inside transformer, we can determine what gases are present and find out their concentration levels. Available gases from chromatographic analysis of insulation oil could contain concentrations of dissolved carbon monoxide (CO), carbon dioxide (CO_2), nitrogen (N_2), hydrogen (H_2), methane (CH_4), acetylene (C_2H_2), ethylene (C_2H_4), and ethane (C_2H_6) [28].

Halstead [29] argued that there was a relationship between fault temperature and the composition of gases dissolved in oil. In [30] Griffin determined key gases and what faults these gases could identify. Zhang et al [31] established diagnostics criteria for types and amounts of gases in oil. Compositions of the dissolved gases, rates of generation, and specific content ratios can be used to indicate transformer condition.

Some artificial intelligence techniques present diagnostic schemes to analyze the dissolved gas data e.g. Huang et. al. [32], Islam et al. [33], Zhang et. al [34]. Each of these techniques can identify a single impurity in transformer oil. Therefore, applying a combination of these techniques to determine three impurities is inexpedient, requires much time, incurs financial and labour costs. We propose to use the AE method to determine these impurities and their influence on transformer oil condition as an indicator of transformer condition.

3 Acoustic emission method

The AE method has been studied for 25 years [35], in particular in Poland, Brazil and Taiwan, to determine partial discharge impact [35-38]. Various physical phenomena accompany the occurrence and development of partial discharges (PDs) in insulation systems. They are chemical changes of insulation, the occurrence of the current pulse and emission of an electromagnetic wave, and a percussive elastic strain with an acoustic wave generation [35]. To avoid these effects, the AE method can be used. It is now a significant supplement for other measuring methods used in diagnostics of transformer insulation systems, current and voltage measuring transformers, power condensers and bushing insulators.

The most significant merit of the AE method is the possibility of its application in difficult operation conditions. It enables detecting the PD occurrence tracing in power equipment operation, locating the places of PD generation in insulation systems. Therefore, the AE method fills the gap in PD metrology of high-power equipment. Procedures of AE pulses' measure and registration are discussed in [39].

We used the AE method to measure acoustic signals generated by impurities (water, cellulose, gas phrase) in transformer oil and to analyze their frequency spectrum to acquire the information about transformer condition.

4 Materials and methods

As the studied sample the transformer oil of T-750 and T-1500 brands was applied, the independent variables were quantity of impurities in oil samples and generator frequency, the dependent variable was sound wave power value.

The measurement setup consisted of a transformer tank filled with electro-insulating oil and impurities, two electro-mechanical (piezoelectric) transducers, AE signal generator, a personal computer (PC) and designed software. The principal scheme of acoustic signal processing is presented in Figure 2.

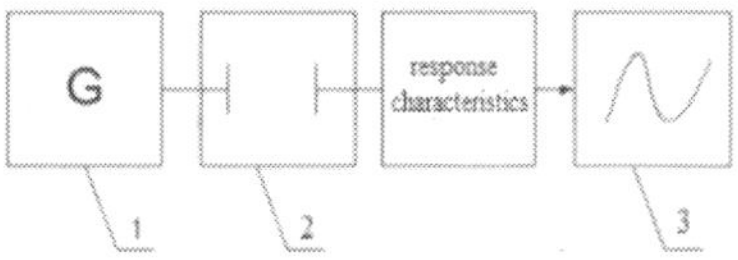

Fig. 2. Principal scheme of acoustic signal processing: 1 - generator; 2 - piezoelectric transducers, 3 - measuring system.

To measure the AE pulses, a piezoelectric transducer was used, it was placed inside the tank. Its application made the measurement of sinusoidal acoustic signals possible at amplitude characteristics in the frequency range from 1 kHz to 10 kHz with a step from 2 to 20 kHz properly.

The registration of the AE pulses was done with the original software compatible with a PC and, to the best of our knowledge, new to the field. The AE pulses registered underwent the analysis in the time, frequency and time-frequency domains and were visualized with the computer program Matlab 6.0.

First, the comparison test of a new and aged transformer oils was performed to check the AE method sensitivity. The acoustic signals of both oils were studied in 1-10 kHz frequency range at room temperature.

To determine the effect of water, cellulose, gas on transformer oil condition, we added them to different samples of new oil and measured the AE signals.

Water was added to the first transformer oil sample with a pipette. The mixture was weighed. Sound wave power was measured in 4 points, where percentage scale of water mass concentration in transformer oil was 0%, 0,224%, 0,355%, 0,0994%. This range was chosen in view of oil hygroscopicity. Measurement error did not exceed 0,000496 g [40-43].

Similarly, we added cellulose to the second transformer oil sample. Cellulose mass concentration in transformer oil being 0%, 0,06%, 0,18%, 0,31%. Gas addition was performed by air input through a slotter pipe with the speed of 3,33 ml/s in a tank with oil sample [42].

5 Results

The comparison test showed that the AE signals measured in the samples of new and aged transformer oils were significantly different. Therefore, the content of impurities in transformer oil had an impact on sound wave power (see Fig. 3). The obtained AE signals were recorded and the response characteristics of transformer oil with impurities were represented.

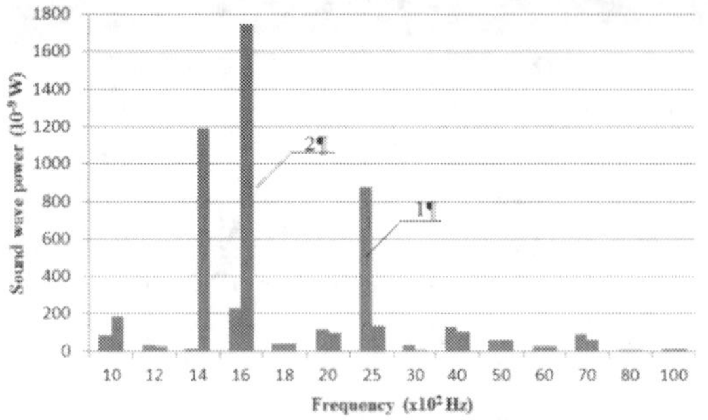

Fig. 3. Response characteristics of a new and aged transformer oils: 1 new transformer oil, 2 aged transformer oil.

The obtained sound power vs. frequency diagrams of oil samples with water, with cellulose and gas phrase were represented in Figures 4-6 respectively.

The response characteristics of model oils without impurities differed from the response characteristics of oil sample with impurities. Therefore, the presence of impurities influenced transformer oil condition.

The effect of impurities was determined applying the Fisher's test. As a result, water and cellulose influenced transformer oil quality, in contrast to gas phase. The AE method was not sensitive to determining gas phase in transformer oil in the selected the frequency range.

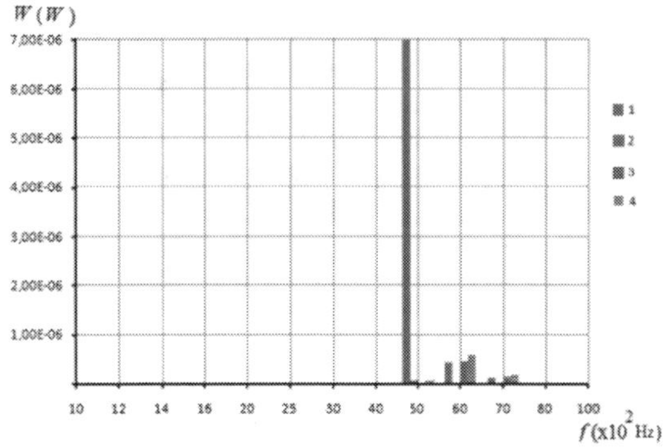

Fig. 4. Response characteristics of transformer oil with water: 1 - water 0,355%; 2 - water 0,224%; 3 - model oil (water 0%), 4 - water 0,0994%.

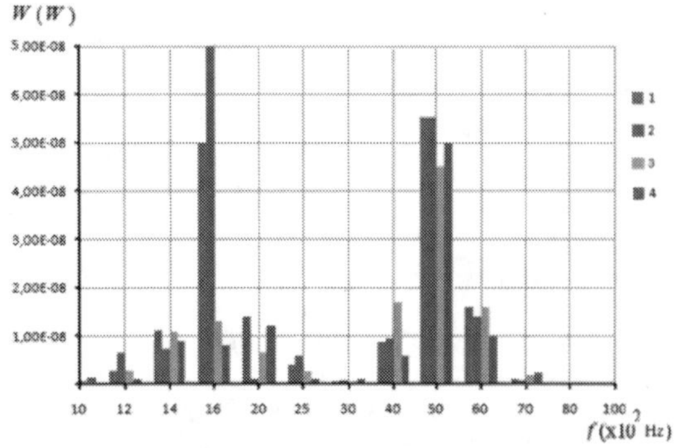

Fig. 5. Response characteristics of transformer oil with cellulose: 1 - cellulose 0,06%; 2 - cellulose 0,18%; 3 - cellulose 0,31%; 4 - model oil (cellulose 0%).

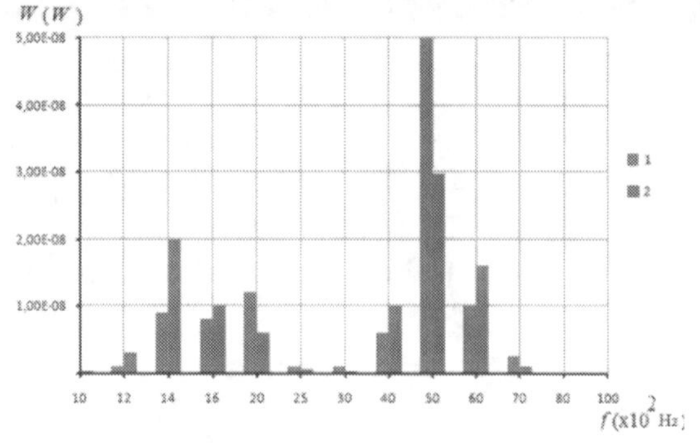

Fig. 6. Response characteristics of transformer oil with/without gas phase: 1 - gas phase presence; 2 - gas phase absence.

Applying the regression procedure to the 2 kHz, 5 kHz, 8 kHz frequencies were studied to obtain a correlation between the sound wave power and water content in oil. The correlation coefficient values were 0,96, 0,84, 0,94 for the 2 kHz, 5 kHz, 8 kHz frequencies respectively. Therefore, the dependence between the sound wave power and water in the range from 0 to 0,335% was established as linear which had type:

$$Y = b_0 + b_1 X \quad (1)$$

Determining the regression coefficients, for the 2 kHz, 5 kHz, 8 kHz frequencies the following equations were obtained:

$$Y_{2\,kHz} = 9{,}94 \cdot 10^{-9} - 2{,}76 \cdot 10^{-8} \cdot X \quad (2)$$

$$Y_{5\,kHz} = -4 \cdot 10^{-8} + 1{,}22 \cdot 10^{-6} \cdot X \quad (3)$$

$$Y_{8\,kHz} = 8{,}54 \cdot 10^{-10} + 4{,}1 \cdot 10^{-9} \cdot X \quad (4)$$

Figure 7 represented modal and experimental dependences between sound wave power and water concentration for the 2 kHz frequency [44].

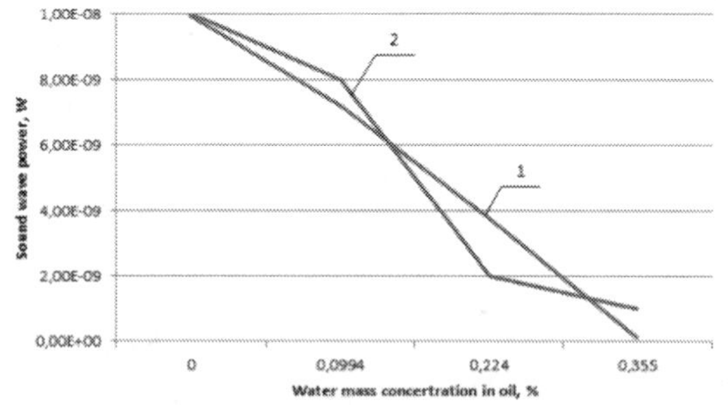

Fig. 7. Comparison of model and experimental curves of water content for 2 kHz frequency: 1 - model dependence; 2 - experimental dependence.

Similarly, the 1 kHz, 1,6 kHz, 5 kHz frequencies were studied to obtain a correlation between the sound wave power and cellulose. The correlation coefficients were 0,03, 0,02 and 0,53. Therefore, this dependence was not linear and had more sophisticated character.

Our future work may include determination of more sophisticated model of the correlation between the sound wave power and cellulose content in transformer oil and establishing a gas effect on transformer oil condition.

Compared to other methods, the AE method had a number of advantages. For example, its possibility to determine the content of several impurities applying the same equipment in field, e.g. in a substation or in a power station. Therefore, this method is available for complex online diagnostics. Compared to other methods it is simpler, cheaper, less time-consuming as we do not need to transport oil samples to an equipped laboratory, use sophisticated reagents, and skills.

6 Conclusion

Water, cellulose and gas phase were determined as the main reasons of transformer oil degradation. Various methods used to determine the impact of the impurities were presented in the paper. The AE method was highlighted as the most perspective method for online complex diagnostics of transformer oil condition. A detailed explanation of the experiments with the AE method was presented.

Taking into account the fact that transformer oil contains about 70% of the available diagnostics information, the obtained results can be important for the utilities which aim to reduce maintenance costs, lengthen transformer's life, increase operation safety, minimize accidental and severe failures of power transformers, and for energy consumers who aim at reliable energy supply for the low cost.

References

1. S.M. Strachan, S. Rudd, S.D.J. McArthur, M.D. Judd, S. Meijer, E. Gulski, IEEE Transactions on Dielectrics and Electrical Insulation, **1** (2008)
2. C. Kurtz, G. Ford, M. Vainberg, M. Lebow, B. Ward, Transmission & Distribution World, July (2005)
3. R. Malewski, J. Subocz, M. Szrot, J. Plowucha, R. Zaleski, CIGRE Session 2008, Paris (2008)
4. T. Boczar, P. Fracz, Physics and chemistry of solid state, **3** (2006)
5. A. Tyuryumina, V. Sekackiy, A. Batrak, The modern materials, technics and technologies, **3**(3) (2015)
6. A. Tyuryumina, A. Batrak, *Issues of modern science theory and practice: records of V International theoretical and practical conference* (2016)
7. A. Tyuryumina, A. Batrak, V. Sekackiy, The young scientist, **8** (2016)
8. A. Tyuryumina, A. Batrak, V. Sekackiy, The young scientist, **28** (2016)
9. A. Tyuryumina, *Svobodniy Avenue – 2016: proceedings of the scientific conference* (2016)
10. A. Tyuryumina, *Svobodniy Avenue – 2016: proceedings of the scientific conference* (2016)
11. A. Tyuryumina, V. Sekackiy, A. Batrak, *The complex systems in extreme conditions: proceedings of XVIII All-Russian symposium with international participation (*2016)
12. P. Verma, Y.R. Sood, J. Singh, *Proceedings of the 16th International Symposium on High Voltage Engineering (*2009)
13. C. Sumereder, M. Muhr, B. Körbler, *17th International Conference on Electricity Distribution*
14. Y. Du, M. Zahn, B.C. Lesieutre, A.V. Mamishev, S.R. Lindgren, Feature Article, **1** (1999)
15. M. Koch, S. Tenbohlen, J. Blennow, I. Hoehlein, *15th ISH* (2007)
16. T. Gradnik, M. Konkan-Gradnik, N. Petric, N. Muc, *IEEE International Conference on Dielectric Liquids* (2011)
17. F.M. Clark, Transactions of electrical engineering, **61** (1942)
18. J. Fabre, A. Pichon, *International Conference on Large High Voltage Electric System (CIGRE)* (1960)
19. H.P. Moser, Special print of Scientia Electrica, (EHV-Weidmann Lim., St., Johnbury, Vermont, USA, Section C, 1979)
20. W.A. Fessler, W.J. McNutt, T.O. Rouse, EPRI report, **EL-5384** (1987)
21. T.V. Oommen, *Electrical/Electronics Insulation Conference* (1983)
22. P.G. Griffin, C.M. Bruce, J.D. Christie, *Minutes of the Fifty Fifth Annual International Conference of Doble Clients* (1988)
23. M. Koch, *XIVth International Symposium on High Voltage Engineering* (2005)
24. A.M. Emsley, G.C. Stevens, Cellulose, **1** (1994)
25. M. Ali, A.M. Emsley, H. Herman, R.J. Heywood, Polymer, **42** (2001)
26. C. Sumereder, M. Muhr, B. Körbler, *17th International Conference on Electricity Distribution* (2003)
27. M. Butcher, *Mechanisms of charge conduction and breakdown in liquid dielectrics*, (PhD dissertation, Graduate Faculty of Texas Tech University, 2005)
28. Chia-Hung Lin, Chien-Hsien Wu, Ping-Zan Huang, Expert Systems with Applications, **36** (2009)
29. W.D. Halstead, Journal of the Institute of Petroleum, **59** (1973)
30. P.J. Griffin, STP998. American Society for Testing and Materials, **1** (1988)
31. Y. Zhang, X. Ding, Y. Liu, P.J. Griffin, IEEE Transactions on Power Delivery, **11(4)** (1996)
32. Y.-C. Huang, H.-T. Yang, C.-L. Huang, IEEE Transactions on Power Delivery, **12(2)** (1997)
33. S.M. Islam, T. Wu, G. Ledwich, IEEE Transactions on Dielectrics and Electrical Insulation, **7(2)** (2000)
34. T. Boczar, Physics and chemistry of solid state, **3** (2007)

35. T. Boczar, P. Fracz, Physics and chemistry of solid state, **3** (2006)
36. D. Park, S. Choi, G. Kil, *The 7th WSEAS International Conference on Power Systems* (2007)
37. O.G. Santos Filho, S.L. Zaghetto, G.O. Pereira, 17th World conference on nondestructive testing, Shanghai, China (2008)
38. D. Wotzka, T. Boczar, D. Zmarzly, Acta Physica Polonica, A **3** (2009)
39. T. Boczar, Journal of Acoustic Emission, **17**(3-4) (1999)
40. A. Batrak, A. Tyuryumina, A. Nikitina, Heavy equipment industry, **3** (2013)
41. A. Batrak, A. Tyuryumina, A. Nikitina, *XIII International youth scientific conference Intellect and science* (2013)
42. A. Batrak, A. Tyuryumina, A. Nikitina, *XIV All-Russian youth scientific conference with international participation Intellect and science* (2014)
43. A. Batrak, T. Chupac, A. Tyuryumina, A. Nikitina, Heavy equipment industry, **4-5** (2014)
44. A. Batrak, A. Tyuryumina, A. Nikitina, *V International youth scientific conference Youth and XXI century* (2015)

MATEC Web of Conferences 113, 01009 (2017) DOI: 10.1051/matecconf/201711301009
12th International Scientific-Technical Conference on Electromechanics and Robotics "Zavalishin's Readings" - 2017

Genetic algorithm optimization for pitch angle control of variable speed wind turbines

Mikhail Burakov* and *Vladislav* Shishlakov

State University of Aerospace Instrumentation, B. Morskaya St. 67, 190000 Saint-Petersburg, Russia

Abstract. In the article, the problem of optimization of the control law for rotating the blades of a wind generator with the purpose of increasing its output power is considered. The dynamics of a wind generator can be described by a nonlinear mathematical model, in which the input effect is the variable wind speed. This model is used to develop a fuzzy regulator whose parameters are tuned using a genetic algorithm. The computational experiments carried out show the practical importance of the obtained control law to ensure the effective use of the wind generator.

1 Introduction

Wind energy is a promising direction for the development of renewable energy sources, especially in combination with solar energy in the form of autonomous combined energy complexes [1].

Wind power plant (VEU) is a device for converting kinetic energy of wind flow into mechanical energy of rotor rotation and conversion of this energy into electric energy. The principle of operation of wind turbines: the wind flow rotates the blades and activates the rotor on which they are mounted, the rotor and blades drive the main shaft, and as a result of the generator's rotation, electricity is produced at the output [2]. According to Betz's law, a wind turbine can never have an efficiency factor greater than 59.3% [3]. In other words, the energy coming from the movement of the wind cannot be completely transformed into useful energy. The wind must move through the turbine, maintaining the optimum speed of rotation of the wind wheel. The aerodynamic power of the rotor can be adjusted by means of the angle of rotation of the blades of the wind wheel. The dynamics of the wind turbine is described by a nonlinear mathematical model, therefore it is difficult to construct an analytical dependence determining the law of the variation of the angle of inclination of the blades from the wind speed or the rotational speed of the turbine shaft.

In this paper, we consider a variant of genetic optimization of the control law for wind turbines based on experiments with its simulation model.

2 Mathematical model of wind generator

Consider the mathematical description of the wind turbine as a control object [4, 5].

The kinetic energy E, which the air stream has, depends on its mass and velocity can be determined by the formula:

$$E = \frac{mv^2}{2}, \qquad (1)$$

where m is the mass of the air stream; V - wind speed, m/s.

The air flow in the wind turbine is shown in Figure 1. The second mass air flow rate m is equal to the product of the second air flow per density ρ of the air flow.

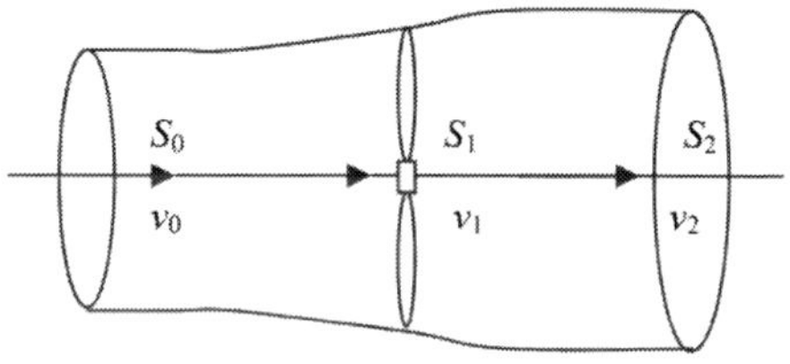

Fig. 1. Flow of air masses in a wind turbine.

According to the law of conservation of energy, the mass of the air stream of jet m_0 flowing through section S_0 is equal to the mass m_1 of this stream flowing through section S_1 and the mass of this stream flowing through section S_2, provided that the air flow is steady:

$$m_0 = m_1 = m_2 = const,$$

$$m = \rho S_0 v_0 = \rho S_1 v_1 = \rho S_2 v_2.$$

Where $v_2 < v_1 < v_0$, S_1 is the area swept by the wind turbine, S_0 and S_2 are the cross-sectional areas of the wind flow passing through the wind turbine, respectively, to the wind wheel and behind it, ρ is the air density.

This expression is called the equation of continuity of the air stream, it establishes the relationship between the jet section and wind speed.

The power of the moving air is equal to the derivative of the kinetic energy (1):

* Corresponding author: bmv@sknt.ru

$$P_{\text{wind}} = \frac{dE}{dt} = \frac{1}{2}mv = \frac{1}{2}\rho S v^3. \quad (2)$$

For the circular section, relation (2) can be written in the form:

$$P_{\text{wind}} = \frac{1}{2}\rho\pi R^3 v^3. \quad (3)$$

Aerodynamic power of the rotor:

$$P = \tau_{\text{aero}}\omega, \quad (4)$$

where τ_{aero} is the aerodynamic moment of rotation applied to the rotor by the wind and ω is the angular velocity of the rotor.

The power factor of the windmill (wind power factor) is defined as the ratio of the aerodynamic power of the rotor to the wind power:

$$C_{\text{p}} = \frac{P}{P_{\text{wind}}}.$$

The power factor $C_p(\lambda, \beta)$ is a function that depends on the two most important turbine parameters λ and β, where β is the rotation angle of the blades, λ is the coefficient characterizing the screw operation or turbine speed.

Coefficient C_p - one of the main parameters characterizing the efficiency of wind turbines, it determines the average generation of electricity at a particular facility.

Another important parameter of the wind turbine is the tip speed ratio λ, defined as the ratio of the circumferential velocity of the blade ends to the effective value of wind speed:

$$\lambda = \frac{\omega R}{v}, \quad (5)$$

where R is the radius of the circle (*m*), swept by the end elements of the blades; ω - angular frequency (*rad/sec*); V - wind speed (*m/sec*).

The value of λ is in the range 0.2 – 10. For large wind turbines, $\lambda > 1$, for wind turbines with a large number of blades $\lambda \approx 3$, for a wind turbine with three blades and a high rotational speed $\lambda \approx 6 - 10$.

Rotor torque:

$$\tau_c = k_{\text{opt}}\omega^2, \quad (6)$$

where the coefficient k_{opt} [4]:

$$k_{\text{opt}} = \frac{1}{2}\rho S R^3 \frac{C_{\text{p}}^{\max}}{\lambda_*^3},$$

where λ * is the value of λ in which the maximum power factor Cp is reached.

Angular acceleration of the rotor

$$\dot{\omega} = \frac{1}{J}(\tau_{\text{aero}} - \tau_c), \quad (7)$$

where J is the moment of inertia of the rotor, and the aerodynamic moment of rotation is described by the formula:

$$\tau_{\text{aero}} = \frac{1}{2}\rho S R C_{\text{q}}(\lambda,\beta)v^2,$$

where

$$C_{\text{q}}(\lambda,\beta) = \frac{C_{\text{p}}(\lambda,\beta)}{\lambda}.$$

Then

$$\tau_{\text{aero}} = \frac{1}{2}\rho S R \frac{C_{\text{p}}(\lambda,\beta)}{\lambda}v^2, \quad (8)$$

Then, substituting (6) and (8) in (7), we obtain an expression for the angular velocity of the generator

$$\dot{\omega} = \frac{1}{2J}\rho S R^3 \omega^2 \left(\frac{C_{\text{p}}(\lambda,\beta)}{\lambda^3} - \frac{C_{\text{p}}^{\max}}{\lambda_*^3} \right), \quad (9)$$

In this way,

$$\dot{\omega} < 0, \text{ if } C_{\text{p}}(\lambda,\beta) < \frac{C_p^{\max}}{\lambda_*^3}\lambda^3; \sin^{-1}\theta$$

$$\dot{\omega} > 0, \text{ if } C_{\text{p}}(\lambda,\beta) > \frac{C_p^{\max}}{\lambda_*^3}\lambda^3.$$

The dependence of the wind turbine power factor Cp on λ and β is expressed in the form [4, 5]:

$$C_{\text{p}}(\lambda,\beta) = C_1\left(\frac{C_2}{\lambda_i} - C_3\beta - C_4 \right) e^{\frac{-C_5}{\lambda_i}} + C_6\lambda, \quad (10)$$

Then

$$\frac{1}{\lambda_i} = \frac{1}{\lambda + 0.08\beta} - \frac{0.035}{\beta^3 + 1}.$$

where the coefficients are: $C_1 = 0.5176$, $C_2 = 116$, $C_3 = 0.4$, $C_4 = 5$, $C_5 = 21$ and $C_6 = 0.0068$.

Using (5), (6), (8) and (9), we can represent the dynamic model of the wind turbine in the form of the structure shown in Figure 2.

Figure 3 shows the dependence of the power factor of the turbine on the speed and the angle of inclination of the blades, found by formula (10).

As Figure 3, the maximum value of $C_p = 0.48$ is reached for $\beta = 0$ and $\lambda = 8.1$

3 Principles of control system organization

When controlling the wind turbine, the slope of the blades β acts as a control signal (Fig. 4). A qualitative description of the operation modes of the wind turbine as a function of the wind speed is given in Figure 5.

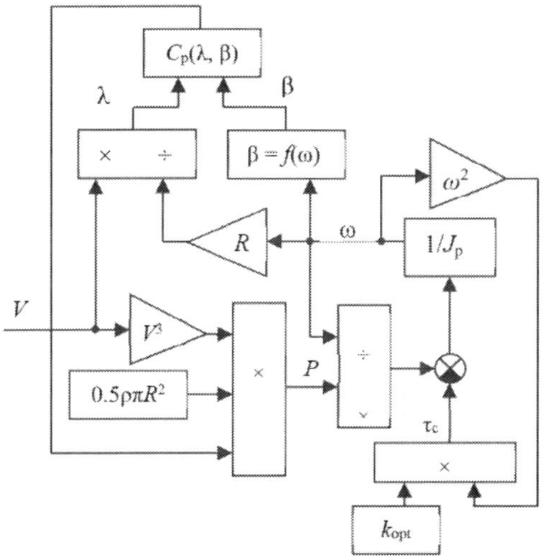

Fig. 2. Block diagram of a wind turbine as a control plant.

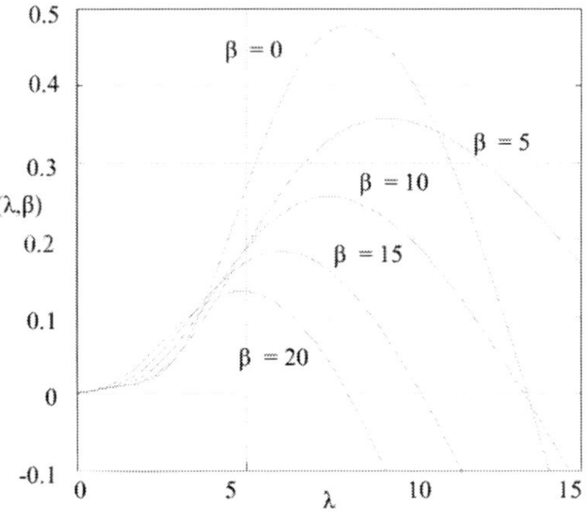

Fig. 3. Dependence of the power factor on λ and β.

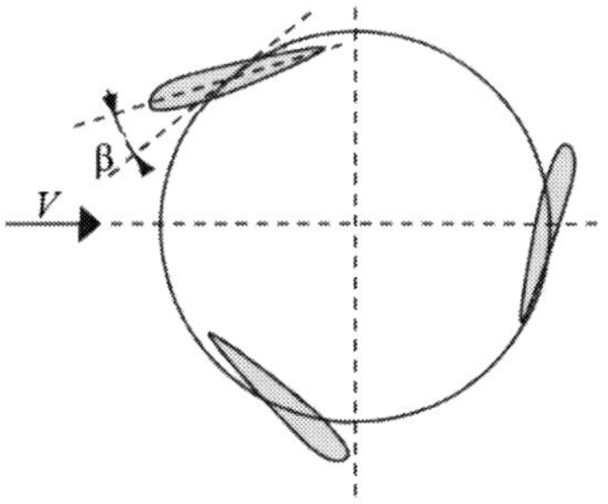

Fig. 4. Wind turbine with a vertical arrangement of blades.

Figure 5 depicts four regions:

I. Low wind speed (0 – 5 m/sec) – the generator is not connected to the network.

II. The average wind speed (5 – 15 m/sec), the generator is connected, but does not produce rated power.

III. High wind speed (15 – 25 m/sec) – the generator is connected to the network and generates a nominal power.

IV. Critical wind speed ($\geq$ 25 m/sec) – the generator is disconnected and the turbine is stopped.

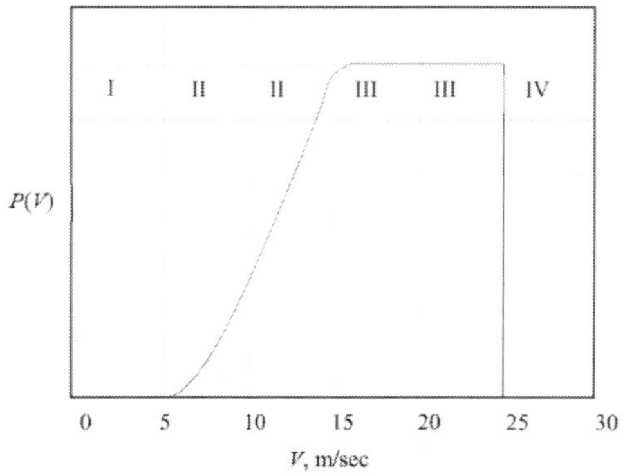

Fig. 5. Example of wind turbine operation mode.

The change in the inclination of the blades makes it possible to control the angular velocity of rotation of the turbine. The following management strategies are possible [4-6]:

1. Control of the blade inclination as a function of wind speed $\beta = f(V)$. This approach is easy to implement, but it is difficult to provide high quality, because accurate measurement of wind speed is complex (Fig. 6).

The dynamics of the servo drive is usually described by an aperiodic link:

$$\beta = \frac{\beta^*}{1 + T_b s},$$

where T_b is the servo time constant.

2. Control by the error of angular velocity (Figure 7).

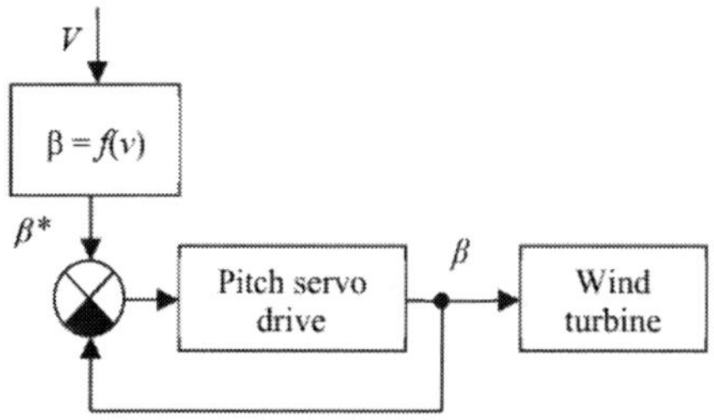

Fig. 6. Wind turbine control by wind speed.

4 Genetic algorithm optimization of the control law

As shown in Figure 3 and 5, as well as the above mathematical description, the control law of the wind turbine must be nonlinear, therefore, a fuzzy logic control (FLC) can be an adequate tool for organizing the control loop of the wind turbine [7, 8].

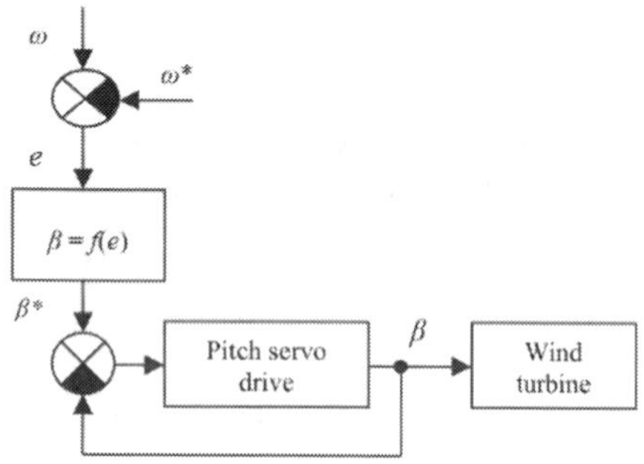

Fig. 7. Control of the wind turbine in angular velocity.

The main problem of applying FLC is the need to formalize the control law in the form of fuzzy rules that use linguistic variables to describe the inputs and outputs of the regulator.

Using triangular membership functions, it is possible to describe the control law of FLC in the form of odd-symmetric piecewise linear functions [9-10].

In accordance with Figure 6 and 7, the FLC for control of the wind turbine must realize a nonlinear static mapping: $\beta = f(V)$ or $\beta = f(\omega)$. Thus, it is possible to use P-type FLC (Fig. 8 and 9).

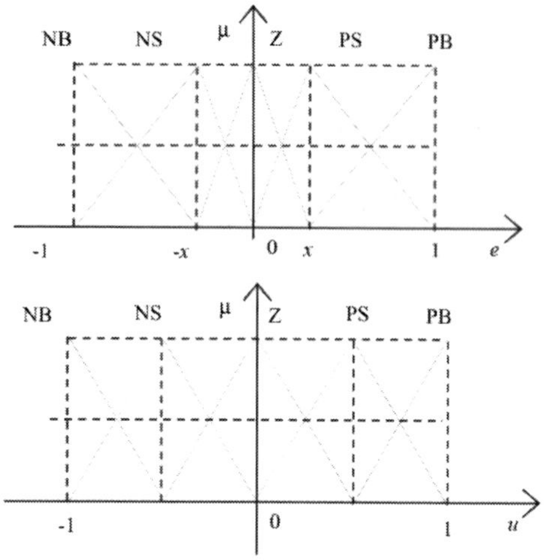

Fig. 8. Linguistic description of FLC's input and output.

When controlling the wind turbine it is impossible to heuristically describe the dependence of the inclination of the blades on the wind speed or wind speed of the wind turbine shaft. Here it is required to solve the optimization problem on the basis of experiments with the simulation model of the object. The genetic algorithm (GA) can serve as a tool for solving the problem here.

The principles of genetic adjustment of FLC are described in [9-11]. The use of GA suggests coding the parameters of the problem using chromosomes, the constituent parts of which (genes) correspond to separate parameters. The totality of chromosomes forms a population evolving in time. The goal of evolution is to improve the fitness of chromosomes, which describes the quality of the solution of the task.

The general scheme of the evolutionary optimization of FLC is shown in Figure 10.

In the problem under consideration, FLC has one input and one output, so the chromosome has a length of $2N$, where N is the number of terms used to describe wind speed. A general view of the chromosome is shown in Figure 11.

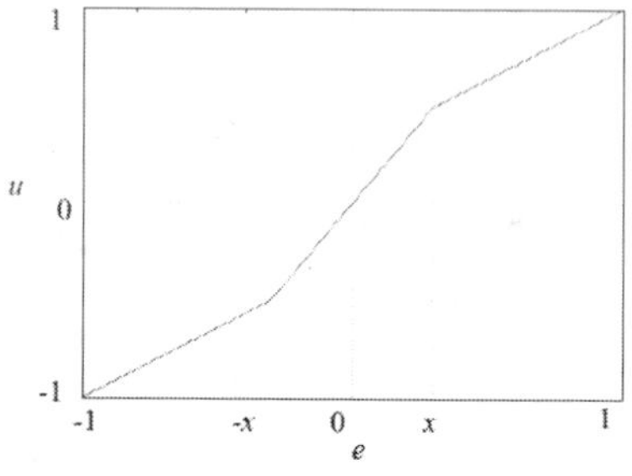

Fig. 9. Control curve of FLC.

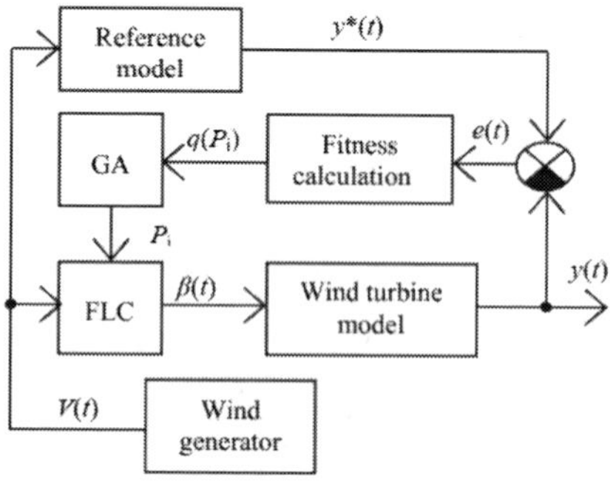

Fig. 10. Evolutionary optimization of FLC.

V_1	V_2	...	V_n	β_1	β_2	...	β_N

Fig. 11. Coding parameters of FLC.

The optimization algorithm cyclically starts a wind generator that generates a test action $V(t)$. At each start-up, the controller receives the parameters P_j, which corresponds to the chromosome. At the end of the transient process, the parameter set receives the fitness rating $q(P_i)$, which is formed by comparing the output of the object $y(t)$ and the reference process $y^*(t)$. Then the population is transformed under the influence of genetic selection, crossing and mutation operators, and new testing is taking place. The criterion for the end of the process is usually the long absence of improvements or the exhaustion of the number of given iterations.

Depending on which control option is selected - wind turbine control in wind speed or wind turbine control with respect to the angular velocity - choose the option $y(t) = P(t)$ or $y(t) = \omega(t)$. Accordingly, we consider $y^*(t) = P_R$ or $y^*(t) = \omega_R$, where P_R and ω_R are the rated power and rotational speed of the wind turbine.

Then, in controlling the wind speed, the following chromosome estimate can be used:

$$q(P_i) = \sum_{j=1}^{n} \left| P_R - P_j \right|,$$

and when controlling the angular velocity, estimate:

$$q(P_i) = \sum_{j=1}^{n} \left| \omega_R - \omega_j \right|,$$

where n is the number of times considered.

5 Simulation result

In the simulation, a wind turbine with the following parameters was considered: R = 3.1 m; J = 25.3 kg·m2; ω = 25 rad/sec; V_H = 15 m/sec. The result of genetic optimization is shown in Figure 12.

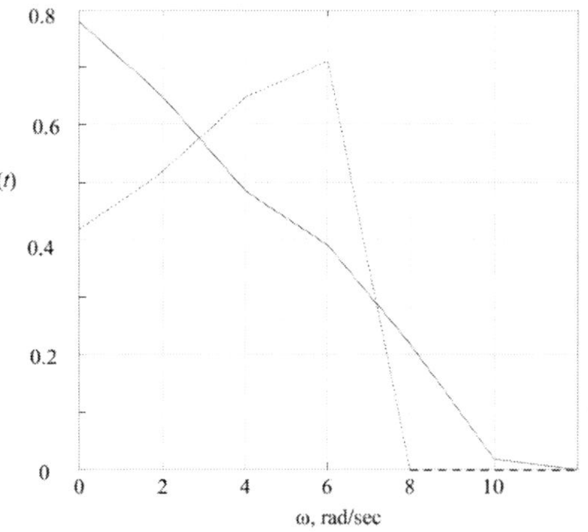

Fig. 12. The control law the inclination of the blades according to the frequency of rotation (continuous) and the wind speed (dotted line).

As Figure 12, at high wind speeds, the optimal blade angle should be close to zero. At low and medium speeds (up to 10 m/sec), the control law on the frequency of rotation is close to linear. The speed control is preferred due to strong fluctuations of the wind force.

Let us consider an example of a simulation of a wind turbine control system. Let the wind speed change in accordance with Figure 13.

Figure 14 shows the dynamics of output power variation during simulation. At low wind speeds, the control of the pitch of the blades has a significant effect on the output power.

Figure 15 shows the dynamics of the angular velocity of the wind wheel, Figure 16 – a change in the coefficient C_p, Figure 17 – a change in the speed of the wind wheel. In the graphs, the solid line corresponds to the control of the inclination of the blades by means of FLC, and the dotted lines correspond to the constant inclination of the blades β = 0 rad.

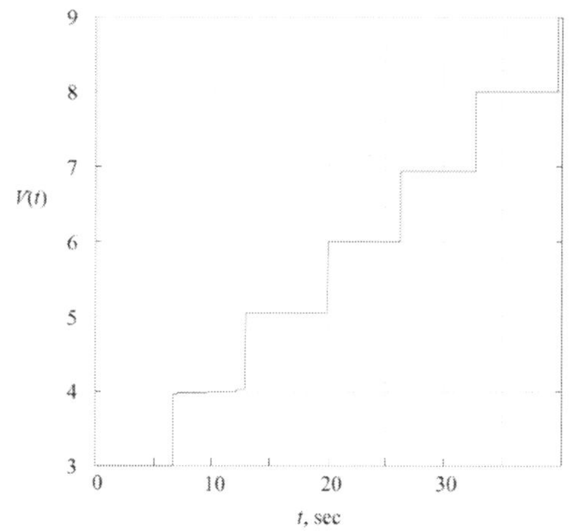

Fig. 13. Change in wind speed during simulation.

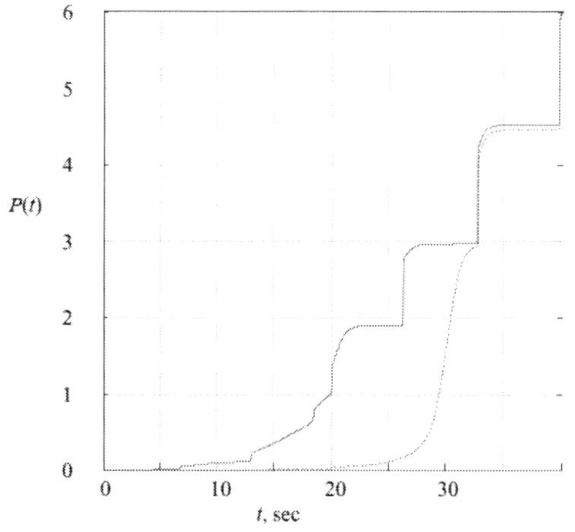

Fig. 14. Change in output power.

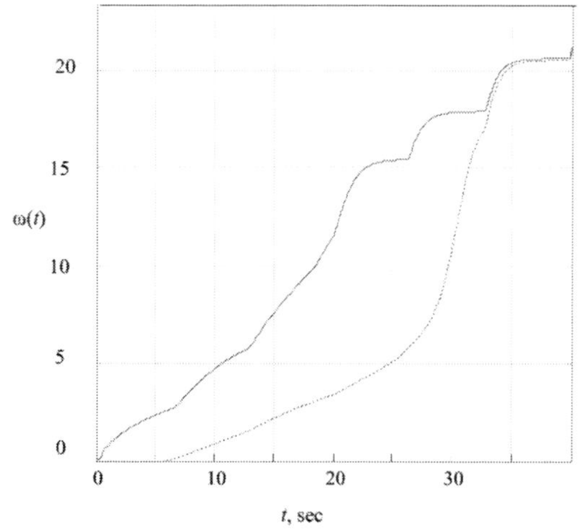

Fig. 15. Variation of the angular velocity of the wind wheel.

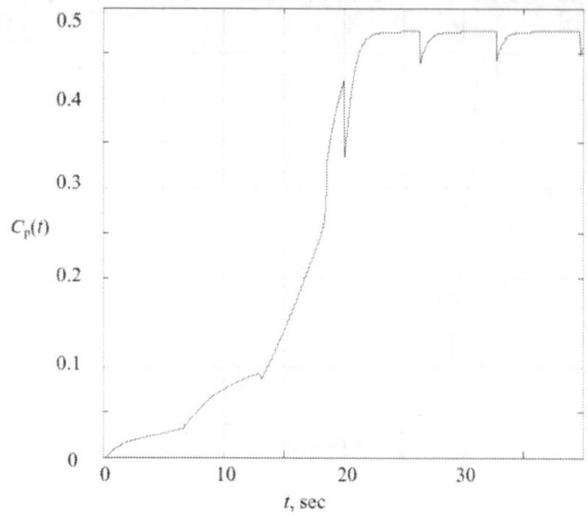

Fig. 16. Changing the power coefficient.

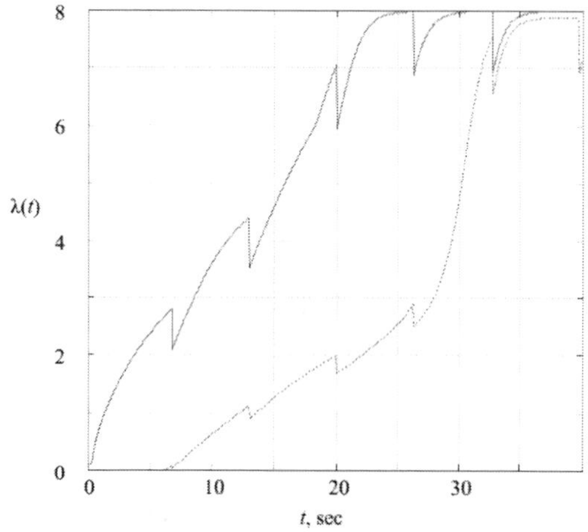

Fig. 17. Changing the tip speed ratio.

6 Conclusion

The approach to the synthesis of the wind turbine control law is based on the genetic optimization of the P-type FLC parameters as a result of experiments with a nonlinear computer model that takes into account all the features of the real object. The computational experiments carried out have shown that a change in the angle of inclination of the blades of the wind wheel can significantly increase the output power at low and average wind speeds. The obtained result can be useful in designing autonomous wind power plants, as well as combined renewable energy sources of low power.

References

1. L.I. Chubraeva, An.L. Ronzhin, A.V. Shyshlakov, Al.L. Ronzhin, V.F. Shyshlakov, SPIIRAS Proc. **2**(33), 207-226 (2014)
2. Burton T., Sharpe D., Jenkins N., Bossanyi E., *Wind Energy Handbook* (John Wiley & sons, Chichester, UK, 2001)
3. Hau E., *Wind turbines: fundamentals, technologies, application, economics*. (Springer Verlag, 2006)
4. Laks J., Pao L., Wright, *Control of Wind Turbines: Past, Present and Future*, ACC June 2009
5. L.Y. Pao, K.E. Johnson, Proc. Amer. Ctrl. Conf., June 2009
6. V. Ramakrishnan, IJESC, **1**(2) (2009)
7. Passino K. M., Yurkovich S. *Fuzzy control* (Addison Wesley Longman, Inc. 1998)
8. O. Mengi, Turk J Elec Eng & Comp Sci, **20**(2) (2012)
9. M.V. Burakov, V.G. Kurbanov, JEAS, **11**(9) (2016)
10. M.V. Burakov, V.G. Kurbanov, JEAS, **12**(4) (2017)
11. M.V. Burakov, A.S. Konovalov, Kluwer International Series in Engineering and Computer Science, **664** (2002)

Low-speed electric drive of the solar battery pivot device

Dmitry Yakimovsky, *Dmitrii* Polozhentcev and *Daniel* Dzhukich*

JSC "Command devices research institute", St. Petersburg, Russia

Abstract. The problem of designing a control system for monitoring angular velocities and accelerations has been considered with regard to a low-speed actuating electric drive application. The block diagram for the electric drive, which weight-dimension characteristics are subject to strict requirements, is presented herein. The results of experimental development of the actuating electric drive are discussed below.

1 Introduction

Most spacecraft have appendages, like solar arrays or antennas, to be deployed, and in some cases these appendages need to be continuously rotated or repointed during the mission [1]. Solar arrays are rotated to trace the source of solar radiation corresponding to the spacecraft's rotation in order to generate power at optimal levels [2]. Actuators are widely used for this purpose, and although currently there exist various types of state-of-the-art actuators based on smart material technology (piezoelectric actuators, shape memory alloys, etc.), conventional technologies, namely electric motors (stepper motors, permanent magnet synchronous motors, etc.) in combination with gearboxes, remain the most preferred ones with a lot of research dedicated to them [1, 3-6].

Modern actuating electric drive (AED) of the solar battery pivot device (SBPD) of spacecraft is designed for slow operational rotation velocity of the output shaft (OS), which may be as low as tenths, hundredths or even thousandths of a degree per second. Other features of the SBPD AED include restrictions for OS rotation acceleration and the high dry friction torque occurring in the contact brush device transmitting power from the solar battery to the spacecraft power supply system. Tough restrictions on weight and dimensions of onboard drives, the requirement to withstand the effects of outer space and ensure long term service of spacecraft significantly affect the design and structure of the electric drive. Extensive research has been conducted to achieve optimal operational results, overcome various restrictions and solve arising issues. For example, research [7] concludes that a system with fuzzy PID control eliminates "flat top" position tracking and "dead zone" speed tracking, which are generated by traditional PID, and thereby decreases the effect of friction on the performance of the servo system. Other researches advocate the need for combined control systems. Paper [8] demonstrates that the use of a combined position and speed controller results in better output speed accuracy, lower current consumption and lower temperature.

The purpose of this paper is to create a SBPD AED based on the brushless motor with a single low-resolution (16–18 bits) induction angle sensor. A drive creating angular rotation velocity of the OS within the range of ± (0; 0.0039…0.5) °/s, with discreteness of NMT 0.000244 °/s, is considered as an example. The OS rotation acceleration value must fall within the range of 0.0012 to 0.01 $°/s^2$, and inaccuracy of the resulting steady state angular velocity may not exceed 10 % for the whole range. Shaft dry friction torque may reach 0.5 N·m, while external disturbing torque varies within the range from -0.3 to 0.3 N·m. To realize speed control, angle loop PD regulator was applied which was supplemented with two components: gradually increasing speed for acceleration restriction and an additional non-linear link ensuring almost constant overcoming of dry friction. This approach proved suitable for achieving the given parameters of the system.

2 Design and modeling

2.1 Rotation velocity control

The velocity control system relies on availability of a velocity indicator. Rotation velocity can be calculated from the angle sensor readings. The risk of error in steady state can be minimized by using a PI-controller within the control loop.

However, if an encoder angle sensor is used in the feedback loop, a different approach to angular rotation velocity control appears to be more practical: namely integration of velocity assigning signal followed by operation in tracker mode with control by angular position of the OS. In this case, PD-controller may be used for angle control [9, 10].

Block diagram of such SBPD AED operating in tracker mode based on the linearly varying angle is shown in Figure 1.

* Corresponding author: info@niikp.spb.ru

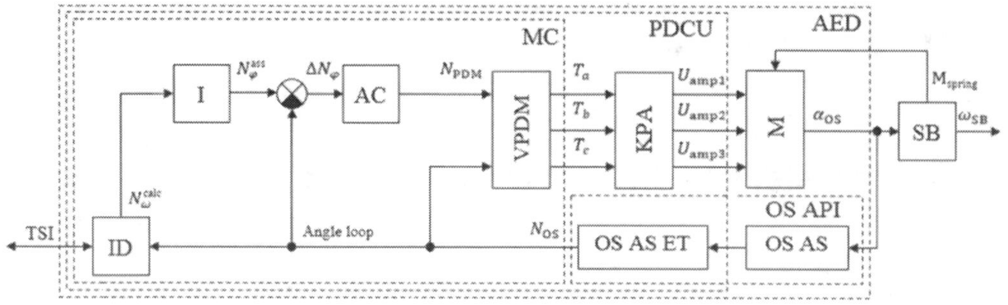

Fig. 1. Block diagram of AED operating in tracker mode based on angle readings.

Where I is velocity integrator; AC is angle controller; VPDM is vector pulse-duration modulator; KPA is key power amplifier; M is motor; SB is solar battery; OS AS is OS angle sensor; OS AS ET is electronic transducer of the output shaft angle sensor; ID is interface device; MC is microcontroller; PDCU is pivot device control unit; AED is actuating electric drive; N_{ω} is code of the assigned angular rotation velocity of the OS; $N_{\varphi}^{assigned\,(ass)}$ is calculated code of OS rotation; ΔN_{φ} is angle loop error signal; $N_{pulse-duration\ modulator\ (PDM)}$ is VPDM code input value; T_a; T_b; T_c is duration of VPDM signals controlling KPA; $U_{amplitude(amp)1}$; U_{amp2}; U_{amp3} is amplitude values of motor feed linear voltages; $\alpha_{output\ shaft\ (OS)}$ is OS angular position data; $\omega_{solar\ battery\ (SB)}$ is SB angular velocity values; M_{spring} is spring torque created by SB; N_{OS} is code of measured rotation angle of OS; TSI is trunk serial interface.

Velocity integration coefficient is calculated by the following formula:

$$C_{\omega} = \frac{C_{slope\ (sl)}}{\delta}, \qquad (1)$$

where C_{sl} is "code-velocity" characteristics slope, δ is "angle-code" conversion discreteness of angle sensor.

"Code-velocity" characteristics slope is defined as:

$$C_{\omega} = \frac{\omega_{ass.max}}{2^{n}-1}, \qquad (2)$$

where $\omega_{ass.max}$ is maximum assigned velocity, n is number of used bits of register that stores the assigned velocity value (not considering the sign bit).

The outputs of mathematical model of AED operation at the assigned velocity of 0.1 °/s are shown in Figure 2. It can be seen that the required accuracy of the SBPD OS angular rotation velocity control is attained. However, the acceleration diagram shows multifold exceeding of the OS rotation acceleration in relation to the assigned value.

2.2 Angular acceleration control

The easiest way of solving this problem could have been to limit the current amplitude in the actuating motor winding, and thus curtail the resulting torque (i.e. acceleration). However, as friction torque is a variable (i.e. its exact value is not known) that changes with angular position of the OS, it would be impossible to fine-tune such a current regulator.

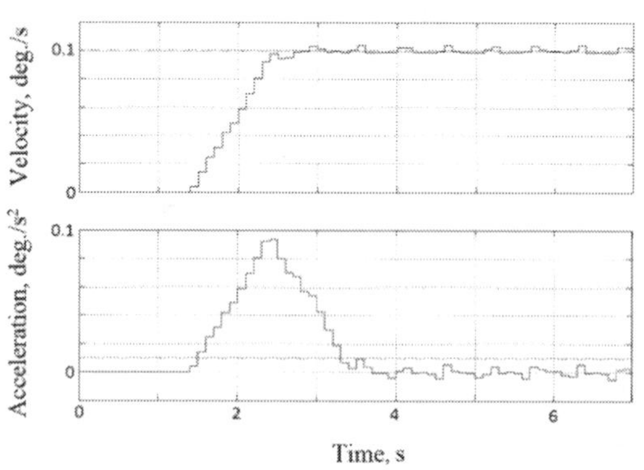

Fig. 2. Outputs of model of AED operation at the assigned velocity of 0.1 °/s.

To solve the problem, a velocity code generator (CG_{ω}) has been integrated into the AED structure, as shown in Figure 3. On receipt of N_{ω}, the velocity code generator generates a linearly increasing (decreasing) code for angular velocity $N_{\omega}^{calculated\,(calc)}$ with limitation at the level corresponding to the code N_{ω}. In order to keep the OS angular rotation acceleration (ε_{OS}) within the range of 0.0012 to 0.01 °/s², the rate of increasing (decreasing) of calculated code of the motor shaft angular velocity is defined by the following difference equation:

$$N_{\omega}^{calc}(n) = k_{generator\ integration\ (GI)} \cdot T_s \cdot N_{\omega}(n) + +N_{\omega}^{calc}(n-1), \qquad (3)$$

where n is the reference number of the AED operation cycle; T_s is sampling period; k_{GI} is integration factor of generator:

$$k_{GI} = \varepsilon_{OS} \cdot \frac{1}{\omega_{OS}}, \qquad (4)$$

where $\omega_{OS} = N_{\omega}(n) \cdot D_{OS}$.

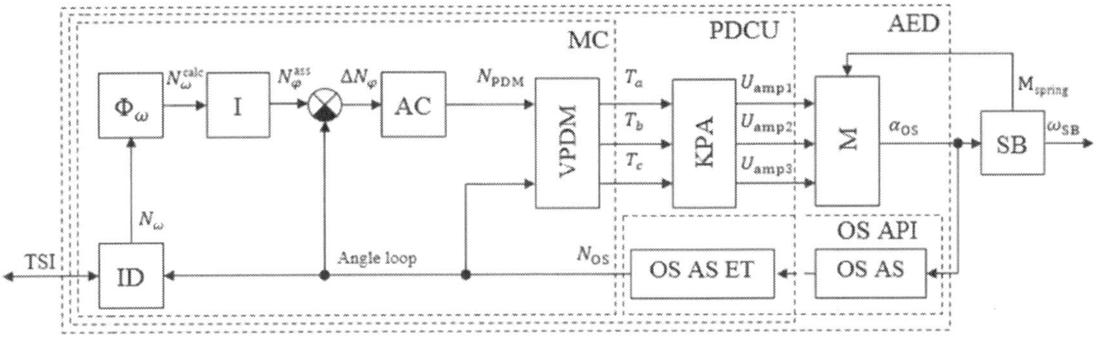

Fig. 3. Block diagram of AED with velocity code generator operating in tracker mode controlled by angle position variations.

In order to keep OS acceleration within the specified range from 0.0012 to 0.01 °/s², mean OS angular acceleration was selected on the basis of the modelling results:

$$\begin{cases} \varepsilon_{OS} = 0.005, at\ N_\omega > N_\omega^{calc} \\ \varepsilon_{OS} = -0.005\ at\ N_\omega < N_\omega^{calc} \\ \varepsilon_{OS} = 0, at\ N_\omega = N_\omega^{calc} \end{cases}. \quad (5)$$

Then

$$\begin{cases} k_{GI} = \frac{\varepsilon_{OS}}{N_\omega(n)\cdot D_{OS}}, at\ N_\omega > N_\omega^{calc} \\ k_{GI} = \frac{-\varepsilon_{OS}}{N_\omega(n)\cdot D_{OS}}, at\ N_\omega < N_\omega^{calc} \\ k_{GI} = 0, at\ N_\omega = N_\omega^{calc} \end{cases}. \quad (6)$$

Outputs of mathematical model of AED operation at the assigned velocity of 0.1 °/s set by the generator are shown in Figure 4. It can be seen that the required accuracy of control is attained for both angular velocity and angular acceleration of SBPD OS. The time lag between issuing the angular velocity command and start of movement is more than 5 seconds, which is due to the high dry friction torque.

2.3 Overcoming dry friction torque

In the considered SBPD, dry friction torque value may be up to 0.5 N·m (the value may vary between instruments and depends on OS angular position), which makes ≈ 30 % of rated torque of actuating motor. Such friction torque value may cause a significant delay of OS response to the assigned velocity code. To solve the problem, a non-linear link (NL) within the control loop generates a code $N_{\text{non-linear link (NL)}}$ to increase the rated drive slope within the range of small angular misalignments by adjusting the calculated code N_{PDM} based on the code sign of calculated angular velocity. Thus, block diagram of the considered AED is modified as shown in Figure 5, where NL is non-linear link, N_{output} is angle controller code output value. NL is described by the formula:

$$N_{\text{NL}} = K_{\text{NL}} \cdot sign(N_\omega^{\text{calc}}), \quad (7)$$

where K_{NL} is the value determined at AED adjustment.

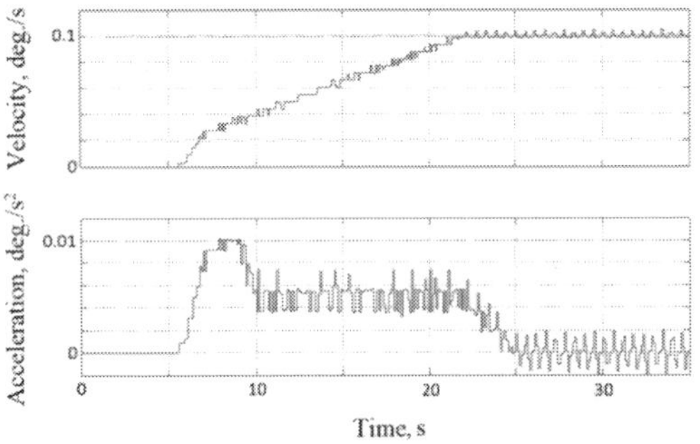

Fig. 4. Output of model of AED operation at the assigned velocity of 0.1 °/s set by the generator.

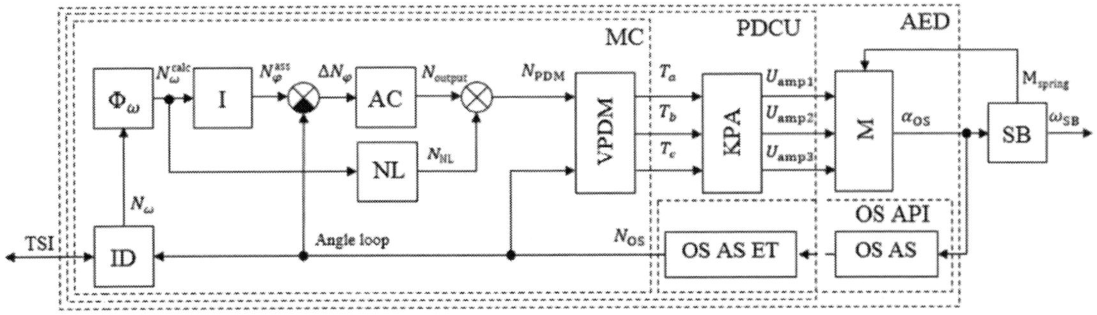

Fig. 5. Block diagram of AED with velocity code generator and non-linear link operating intracker mode based on the varying angle.

Such value shall not exceed dry friction torque value, but shall compensate its greater part.

Outputs of mathematical model of AED operation with NL and generator of assigned velocity of 0.1 °/s are shown in Figure 6. It can be seen that the OS pickup time has significantly reduced, and readjustment based on acceleration has decreased.

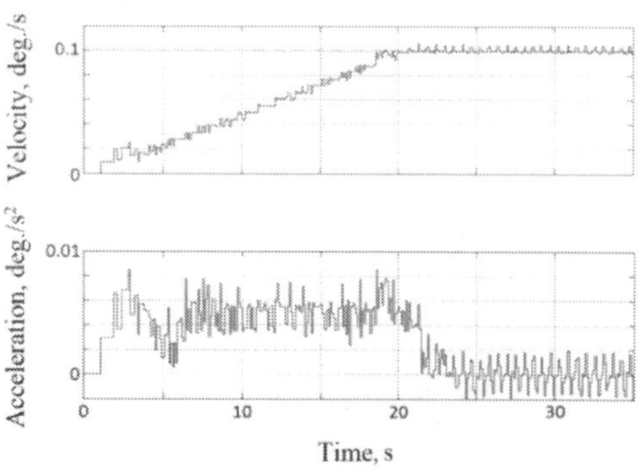

Fig. 6. Outputs of model of AED operation with NL and generator of assigned velocity of 0.1 °/s.

3 Validation

Figures 7–8 show the results of experimental operation of SBPD AED with inertial load of the assigned rotation velocities of -0.1 and -0.0039 °/s, respectively. It can be seen that both acceleration and velocity are controlled by the control system with relatively high accuracy. Inaccuracy of OS angular rotation velocity in the steady state was 1.3 %, and 2.56 % for Figures 7–8, respectively.

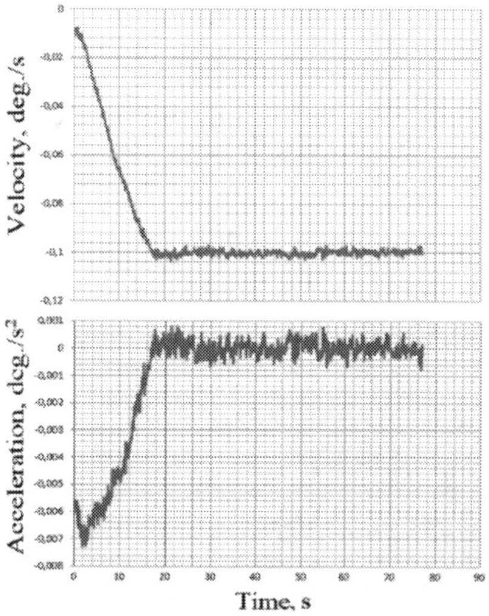

Fig. 7. Real SBPD AED operation at the assigned rotation velocity of -0.1 °/s.

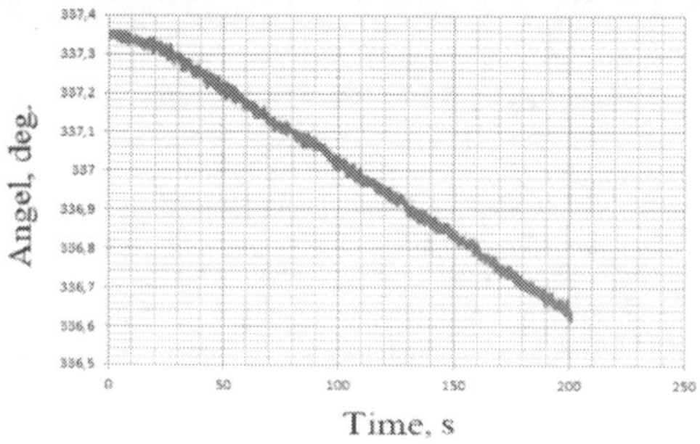

Fig. 8. Real SBPD AED operation at the assigned rotation velocity of -0.0039 °/s.

4 Conclusion

Thus, the results of the SBPD AED modelling and prototype tests demonstrated the effectiveness of the adopted technical solutions. The proposed design of actuating drive ensures accurate control of rotation velocity with controlled acceleration value in the conditions of a large unstable dry friction torque.

References

1. C. Allegranza, L. Gaillard, R. Le Letty, S. Patti, L. Scolamiero, M. Toso, ACTUATOR 2014, 14th International Conference on New Actuators, 283-288 (2014)
2. X. Li, W. Cheng, W. Han, Int. J. Hybrid IT, **7**(2) 123-138 (2014)
3. Z.Y. Shu, J.L. Liu, X.X. Zhang, Small and Spec. El. Mach., **41** (2013)
4. A. Kumar, M. Valarmathi, Int. J. Adv. Res. in Electrical, Electronics and Inst. Eng., **2**(4), 1197–1202 (2013)
5. Z. Marinković, D. Marinković, G. Petrović, P. Milić, Teh. V., **19**(4), 717-725 (2012)
6. A.S. Arora, G. Singh, IRES 6th Int. Conf., 107-115 (2015)
7. B. You, M. Qiao, J. Xu, D. Liu, Math. Pr. Eng. **2012** (2012)
8. I. Todić, M. Miloš, M. Pavišić, Teh. V., **20**(5), 853-860 (2013)
9. R.M. Trakhtenberg. *Pulse astatic electric drive systems with discrete control* (1982)
10. Ch. Phillips, R. Harbor, *Feedback control systems* (2001)

Probability estimation of ignitions formation in the electrical network

Sergey Solyonyj*, *Vladislav* Shishlakov and *Oksana* Solenaya

SUAI, Department of Electromechanics and Robotics, 190000 Saint-Petersburg, Bolshaya Morskaya str., 67, Russia

Abstract. Their fire safety is one of the major requirements, which is imposed to electric networks. However extensiveness of methods of her assessment and a different of approaches to definition of factors of danger leads to complication of the choice of a technique and process of an assessment. The work purpose is development of a uniform method of mathematical modelling and an assessment of safe operation of electric networks.

1 Introduction

Within the Federal target program «Fire Safety in the Russian Federation for the Period till 2017» № 1481 from 12.30.2012, carrying out researches in the field of development of mathematical models and methods by definition and forecasting of probability of emergence of the fires, effective ways of their prevention and mitigation of consequences is provided. It also agrees, to the Federal law of the Russian Federation № 117-FL «About introduction of amendments to the Federal law № 123-FL «Technical regulations about requirements of fire safety» from 7.10.2012 [1], development of the new technical solutions and organizational actions, use operating and creation of new normative documents which application will allow to ensure fire safety of the industrial and household electrified facilities is provided.

The analysis of statistics of the fires in the Russian Federation because of damages of electric conducting from 2006 for 2016 [2] demonstrates that if not to develop new and not to improve the known methods, ways of ensuring safe operation of systems of low-voltage power supply, then in them to confidential probability 0,95 there will be about 42000 fires from the electrotechnical reasons [3].

Therefore, the work directed to creation new and improvement of the known mathematical models, methods of an assessment and forecasting of safe operation of systems of low-voltage power supply, and also development of new technical and organizational actions for improvement of a complex of means of protection, low-voltage electric network up to 1 kV, is an urgent scientific and technical task which solution will allow not to allow cases of emergence of ignitions of combustible material (isolation, etc.) in systems of low-voltage power supply and will ensure their fire safety at the level of the existing normative documents [4].

Due to the adoption of laws [1, 5] special relevance is acquired development and deployment in domestic practice of evidence-based recommendations about ensuring fire safety of systems of low-voltage power supply, use of methods by means of which the opportunity of an assessment of the existing level of fire danger is given and to compare him to the value $1\ 10^{-6}$, which is legislatively established state standard [4].

2 Problem definition

We will assume that ignition of combustible material on an electrical network because of damage of electrical wiring happens in case of coincidence in space and time of the following three processes: there was a damage of an electrical network which is energized; the system of protection refused actuating; there was a combustible material near the place of damage of an electrical network [6-8].

We will provide state change of an electrical network in the form of uniform Markov process of $\alpha(t)$ with two discrete states and the continuous time eventually. We will designate through «0» it a status of an electrical network in which damages of its insulation aren't watched, and through «1» – duration of time in which there is an electrical network after damage of its insulation (protection reaction time).

We will designate parameters of process $\alpha(t)$ through λ_1 and μ_1 where λ_1 characterizes intensity or speed of state change of an electrical network, frequency of transitions from a status «0» in a status «1», and μ_1 – is characterized by the frequency of transitions from a status «1» in a status "0".

We will designate through $\beta(t)$ uniform Markov accidental process with two discrete states – «0» and «1» and the continuous time which characterizes state change of system of protection: «0» – up state of system of protection (the waiting mode) and «1» – the damaged status of system (comes to light only as a result of diagnostics). Parameters of process $\beta(t)$ following: λ_1 – the speed of change of operable periods (working state

* Corresponding author: ssv555ssv@yandex.ru

of system of protection) on no serviceable; μ_2 – the speed of change of no serviceable periods during which the system of protection was in the refused status, on operable. Diagnostics of system of protection is carried out through each Θ_1 of year.

Existence or absence of combustible material in close proximity to accidental insulation failure of an electrical network it is representable similar uniform markov process of $\xi(t)$ with two discrete states: «0» – near an electrical network which is energized there is no combustible material (dust, paper, liquid flammable materials, etc.) and «1» – near a spacer of an electrical network there is a combustible material.

We will designate parameters of process $\xi(t)$ through λ_3 and μ_3 where λ_3 – the frequency of appearance of combustible material in close proximity to an electrical network, and characterizes by μ_3 the speed of change of availability of combustible material with its complete or partial elimination which will provide a fire safety of a securable object.

Availability of combustible material near a spacer of conductors of an electrical network or its customers comes to light and eliminated as a result of conducting preventive checks with Θ_2 time interval.

Ignition of combustible material on an electrical network will happen at the time of coincidence in space and time of three accidental processes of $\alpha(t)$, $\beta(t)$ and $\xi(t)$ is able 1 that is when: $\alpha(t)=1$, $\beta(t)=1$, $\xi(t)=1$.

We will assume that in initial time point of $\alpha(t)=0$, $\beta(t)=0$, $\xi(t)=0$ and parameters of processes $\alpha(t)$, $\beta(t)$, $\xi(t)$, that is their characteristics $\lambda_1;\mu_1$, $\lambda_2;\mu_2$, $\lambda_3;\mu_3$ are known.

The task consists in that knowing parameters of processes $\lambda_1;\mu_1$, $\lambda_2;\mu_2$, $\lambda_3;\mu_3$ to define: average time before the first ignition of τ_1, dispersion of this time of D_1, probability of finding of electric network in each of possible conditions of $P_i(t)$, probability of ignition of combustible material in electric network $F(t)$.

3 Assessment of safe operation

We will consider set of processes of $\alpha(t)$, $\beta(t)$ and $\xi(t)$ as one regular uniform Markov process of $æ(t)$ with eight discrete states and continuous time (Fig. 1: ζ_k – duration of finding of process of $æ(t)$ in each of eight possible states $k=1\div8$; $\alpha(t)$, $\beta(t)$, and $\xi(t)$ – uniform casual Markov processes with two discrete states and continuous time; $\alpha_i^{(0)}$, $\alpha_j^{(1)}$, $\beta_i^{(0)}$, $\beta_j^{(1)}$, $\xi_i^{(0)}$, $\xi_j^{(1)}$ – time intervals between events and their duration $i,j=1\div n$; Θ_1, Θ_2 – time intervals between diagnostics of system of protection and verifications of combustible material near emergence of a dangerous electric source in electric network; τ_0^1 – time before the first ignition of combustible material in electric network).

At any moment process of $æ(t)$ can be in one of eight possible discrete states: $e_1(0,0,0)$, $e_2(1,0,0)$, $e_3(0,1,0)$, $e_4(0,0,1)$, $e_5(1,0,1)$, $e_6(0,1,1)$, $e_7(1,1,0)$, $e_8(1,1,1)$.

The condition of $e_8(1,1,1)$ absorbing. Having got to this state, the system can't leave it. The being able $e_8(1,1,1)$ occurs ignition of combustible material.

We will designate through ζ_k, $k=1\div8$, time of stay of process of $æ(t)$ is able e_k. For regular uniform Markov process of $æ(t)$ with discrete number of states and continuous time random variables of ζ_k don't contradict an exponential function of distribution of probabilities:

$$F_\zeta(t)=P\{\zeta<t\}=1-e^{-\lambda_{ii}t} \quad (1)$$

We will designate through $P_{ii}(\Delta t)$ probability that process of $æ(t)$ for a small period of Δt will remain to the being able e_i, $i=1\div8$, and through $P_{ij}(\Delta t)$ – probability that the system in time Δt will pass from e_i a state into a condition of e_j, $j=1\div8$.

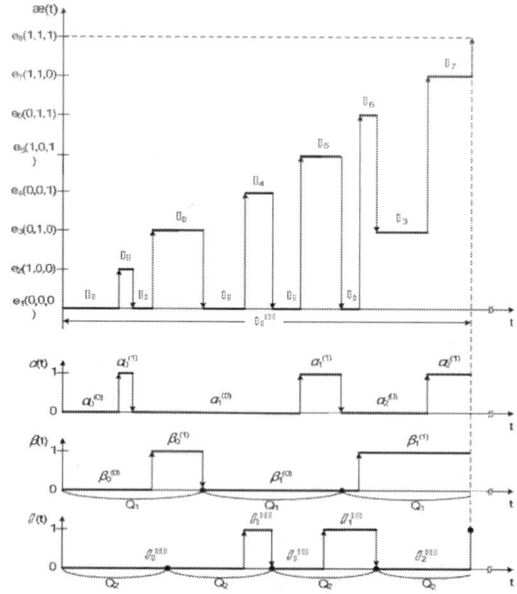

Fig. 1. Possible realization of uniform casual Markov process with eight discrete states and continuous time.

The probability of transitions of $P_{ii}(\Delta t)$ and $P_{ij}(\Delta t)$ decides on the help of formulas:

$$P_{ii}(\Delta t)=P[æ(t+\Delta t)=e_i/æ(t)=e_i]=P\{e_i\to e_i\}=1-\lambda_{ii}\Delta t+0(\Delta t), \quad (2)$$

$$P_{ij}(\Delta t)=P[æ(t+\Delta t)=e_j/æ(t)=e_i]=P\{e_i\to e_j\}=1-\lambda_{ij}\Delta t+0(\Delta t). \quad (3)$$

In a formula (2) the size $\lambda_{ii}\Delta t$ considers to within the second order of a trifle that during Δt there will be no transition of system from a condition of e_i in a condition of e_j, that is process will remain to the being able e_i.

Follows from a formula (3) that the size $\lambda_{ij}\Delta t$ considers to within members of the second order of a trifle that during Δt there will be a transition of system from a condition of e_i in a condition of e_j.

Using a matrix of probabilities of transitions $P(\Delta t)$ [6] we find a matrix A:

$$A=\lim_{\Delta t\to 0}\frac{Q(\Delta t)-I}{\Delta t}, \quad (4)$$

where I – a single matrix.

The matrix of $Q(\Delta t)$ is from $P(\Delta t)$ matrix by an exception of the absorbing state – the last line and the last column.

Relation limit $0(\Delta t)$ to Δt at Δt aspiring to «0» is equal «0»:

$$\lim_{\Delta t\to 0}\frac{0(\Delta t)}{\Delta t}=0\,, \tag{5}$$

Using matrix of probabilities of transitions $P(\Delta t)$, (4) and (5), we will receive a matrix A ($c_1=\lambda_1+\lambda_2+\lambda_3$, $c_2=\mu_1+\lambda_2+\lambda_3$, $c_3=\lambda_1+\mu_3+\lambda_3$, $c_4=\lambda_1+\lambda_2+\mu_3$, $c_5=\mu_1+\lambda_2+\mu_3$, $c_6=\lambda_1+\mu_2+\mu_3$, $c_7=\mu_1+\mu_2+\lambda_3$).

$$A=\begin{pmatrix} -c_1 & \lambda_1 & \lambda_2 & \lambda_3 & 0 & 0 & 0\\ \mu_1 & -c_2 & 0 & 0 & \lambda_3 & 0 & \lambda_2\\ \mu_2 & 0 & -c_3 & 0 & 0 & \lambda_3 & \lambda_1\\ \mu_3 & 0 & 0 & -c_4 & \lambda_1 & \lambda_2 & 0\\ 0 & \mu_3 & 0 & \mu_1 & -c_5 & 0 & 0\\ 0 & 0 & \mu_3 & \mu_2 & 0 & -c_6 & 0\\ 0 & \mu_2 & \mu_1 & 0 & 0 & 0 & -c_7 \end{pmatrix}. \tag{6}$$

We will find a matrix of N as follows:

$$N=\lim_{\Delta t\to 0}\frac{I-Q(\Delta t)}{\Delta t}\,, \tag{7}$$

Using matrix of probabilities of transitions $P(\Delta t)$, (5) and (7), we find:

$$N=\begin{pmatrix} c_1 & -\lambda_1 & -\lambda_2 & -\lambda_3 & 0 & 0 & 0\\ -\mu_1 & c_2 & 0 & 0 & -\lambda_3 & 0 & -\lambda_2\\ -\mu_2 & 0 & c_3 & 0 & 0 & -\lambda_3 & -\lambda_1\\ -\mu_3 & 0 & 0 & c_4 & -\lambda_1 & -\lambda_2 & 0\\ 0 & -\mu_3 & 0 & -\mu_1 & c_5 & 0 & 0\\ 0 & 0 & -\mu_3 & -\mu_2 & 0 & c_6 & 0\\ 0 & -\mu_2 & -\mu_1 & 0 & 0 & 0 & c_7 \end{pmatrix}, \tag{8}$$

The average time of τ_1 before emergence of ignition of combustible material in electric network, dispersion of D_1 and function of probability $P'(t)$ of finding of system during t time in each of i of possible states, $i=1\div r$, it is possible to find using systems of the equations:

$$\tau=N^{-1}\xi, \tag{9}$$

$$D=(2N-I)\tau-\tau^2, \tag{10}$$

$$P'(t)=P(t)A, \tag{11}$$

where ξ – a vector column which all elements are equal to 1; τ – vector column $\tau=[\tau_i]^7_{i=1}$; τ^2 – $\tau^2=[\tau^2_i]^7_{i=1}$ vector column; $P'(t)$ – vector line $P'(t)=[\,P_i'(t)]^7_{i=1}$; $P(t)$ – vector line $P(t)=[\,P_i(t)]^7_{i=1}$.

Using a matrix (8) and system of the algebraic equations (9), we will write down system of the equations for finding of τ_1 in a look:

$$\begin{pmatrix}\tau_1\\ \tau_2\\ \tau_3\\ \tau_4\\ \tau_5\\ \tau_6\\ \tau_7\end{pmatrix}=\begin{pmatrix} c_1 & -\lambda_1 & -\lambda_2 & -\lambda_3 & 0 & 0 & 0\\ -\mu_1 & c_2 & 0 & 0 & -\lambda_3 & 0 & -\lambda_2\\ -\mu_2 & 0 & c_3 & 0 & 0 & -\lambda_3 & -\lambda_1\\ -\mu_3 & 0 & 0 & c_4 & -\lambda_1 & -\lambda_2 & 0\\ 0 & -\mu_3 & 0 & -\mu_1 & c_5 & 0 & 0\\ 0 & 0 & -\mu_3 & -\mu_2 & 0 & c_6 & 0\\ 0 & -\mu_2 & -\mu_1 & 0 & 0 & 0 & c_7 \end{pmatrix}^{-1}\cdot\begin{pmatrix}1\\1\\1\\1\\1\\1\\1\end{pmatrix}. \tag{12}$$

Using a matrix (8) and system of the equations (10) we find dispersion of D_1 from system of the equations of a look:

$$\begin{pmatrix}D_1\\ D_2\\ D_3\\ D_4\\ D_5\\ D_6\\ D_7\end{pmatrix}=\left[2\begin{pmatrix} c_1 & -\lambda_1 & -\lambda_2 & -\lambda_3 & 0 & 0 & 0\\ -\mu_1 & c_2 & 0 & 0 & -\lambda_3 & 0 & -\lambda_2\\ -\mu_2 & 0 & c_3 & 0 & 0 & -\lambda_3 & -\lambda_1\\ -\mu_3 & 0 & 0 & c_4 & -\lambda_1 & -\lambda_2 & 0\\ 0 & -\mu_3 & 0 & -\mu_1 & c_5 & 0 & 0\\ 0 & 0 & -\mu_3 & -\mu_2 & 0 & c_6 & 0\\ 0 & -\mu_2 & -\mu_1 & 0 & 0 & 0 & c_7 \end{pmatrix}^{-1} - \begin{pmatrix} 1&0&0&0&0&0&0\\ 0&1&0&0&0&0&0\\ 0&0&1&0&0&0&0\\ 0&0&0&1&0&0&0\\ 0&0&0&0&1&0&0\\ 0&0&0&0&0&1&0\\ 0&0&0&0&0&0&1 \end{pmatrix}\right]\begin{pmatrix}\tau_1\\ \tau_2\\ \tau_3\\ \tau_4\\ \tau_5\\ \tau_6\\ \tau_7\end{pmatrix}-\begin{pmatrix}\tau_1^2\\ \tau_2^2\\ \tau_3^2\\ \tau_4^2\\ \tau_5^2\\ \tau_6^2\\ \tau_7^2\end{pmatrix} \tag{13}$$

Using (6) and (11) we will find probability of finding of the source-protection-Wednesday system in each of possible states by means of the system of the linear differential equations which is written down in a matrix look:

$$\begin{bmatrix}P_1'(t)\\ P_2'(t)\\ P_3'(t)\\ P_4'(t)\\ P_5'(t)\\ P_6'(t)\\ P_7'(t)\end{bmatrix}^T=\begin{bmatrix}P_1(t)\\ P_2(t)\\ P_3(t)\\ P_4(t)\\ P_5(t)\\ P_6(t)\\ P_7(t)\end{bmatrix}^T\begin{bmatrix} -c_1 & \lambda_1 & \lambda_2 & \lambda_3 & 0 & 0 & 0\\ \mu_1 & -c_2 & 0 & 0 & \lambda_3 & 0 & \lambda_2\\ \mu_2 & 0 & -c_3 & 0 & 0 & \lambda_3 & \lambda_1\\ \mu_3 & 0 & 0 & -c_4 & \lambda_1 & \lambda_2 & 0\\ 0 & \mu_3 & 0 & \mu_1 & -c_5 & 0 & 0\\ 0 & 0 & \mu_3 & \mu_2 & 0 & -c_6 & 0\\ 0 & \mu_2 & \mu_1 & 0 & 0 & 0 & -c_7 \end{bmatrix}. \tag{14}$$

The system of the equations (14) decides under the following entry conditions: $P_1(0)=1$, $P_2(0)=1$, $P_3(0)=1$, $P_4(0)=1$, $P_5(0)=1$, $P_6(0)=1$, $P_7(0)=1$.

For a case when:

$$\lambda_1\leq 100\mu_1,\ \lambda_2\leq 100\mu_2,\ \lambda_3\leq 100\mu_3 \tag{15}$$

From system of the equations (12) we find:

$$\tau_1=\frac{\mu_1\mu_2\mu_3}{\lambda_1\lambda_2\lambda_3(\mu_1+\mu_2+\mu_3)}\,, \tag{16}$$

In case to check operability of system of protection through time interval Θ_1 (checks are absolutely reliable), then we will determine the parameter of process μ_2 as follows:

$$\mu_2=\frac{1}{\Theta_1-\frac{1}{\lambda_2}(1-e^{-\lambda_2\Theta_1})}, \tag{17}$$

In case $\lambda_2\Theta_1<0{,}1$ then:

$$\mu_2=\frac{2}{\lambda_2\Theta_1^2}. \tag{18}$$

If availability of combustible material in electric network to control each Θ_2 of year and to consider that checks are absolutely reliable, then:

$$\mu_3=\frac{1}{\Theta_2-\frac{1}{\lambda_3}(1-e^{-\lambda_3\Theta_2})}. \tag{19}$$

In case $\lambda_3\Theta_2<0{,}1$ then:

$$\mu_3=\frac{2}{\lambda_3\Theta_2^2}. \tag{20}$$

We will substitute in a formula (16) of μ_2 and μ_3 value, received by means of formulas (18) and (20), then:

$$\tau_1=\frac{\mu_1\cdot\frac{2}{\lambda_2\Theta_1^2}\cdot\frac{2}{\lambda_3\Theta_2^2}}{\lambda_1\lambda_2\lambda_3(\mu_1+\frac{2}{\lambda_2\Theta_1^2}+\frac{2}{\lambda_3\Theta_2^2})}. \tag{21}$$

In case $\mu_1<<\mu_2$ and $\mu_1<<\mu_3$, a formula (21) can be presented in the form:

$$\tau_1=\frac{1}{0{,}25\lambda_1\lambda_2^2\lambda_3^2\Theta_1^2\Theta_2^2}. \tag{22}$$

When performing a condition:

$$\tau_1=\sqrt{D_1}. \tag{23}$$

The probability of ignition of combustible material in electric network during time of t can be defined by a formula:

$$F(t)=1-e^{-H_1t}, \tag{24}$$

where $H_1=\frac{1}{\tau_1}$.

If the condition (23) isn't satisfied, then $F(t)$ is defined as follows:

$$F(t)=1-\sum_{i=1}^{7}P_i(t), \tag{25}$$

where $P_i(t)$ are defined from the solution of system of the linear differential equations (14).

5 Conclusion

Ignition of combustible material on an electrical network can occur every time in case of coincidence in space and time of the next three accidental events: there was an accident on an electrical network which is energized; the system of protection which ensures fire safety of an electrical network refused actuating; there was a combustible material near the place of possible accident of an electrical network. Process of formation of ignition of combustible material on an electrical network because of its damage owing to appearance of dangerous electrical sources and a failure to actuating of system of protection can be described adequately by means of set of three uniform accidental Markov processes of $\alpha(t)$, $\beta(t)$ and $\xi(t)$ with the discrete number of statuses and the continuous time. The proposed system is based on equations (12) and (13), by which it is possible to define an average period of τ_1 before the first ignition of combustible material on an electrical network and dispersion of this time of D_1. The system of the linear differential equations (14) which allows to estimate probability of ignition of combustible material on an electrical network during t time from the frequency of appearance of damages to it, reliability of system of protection and periods of its diagnostics of Θ_1, and also from the frequency of appearance of combustible material near possible sources of initiation of ignition of combustible material and intervals of time between its detection of Θ_2, is received.

References

1. Federal Law of the Russian Federation from 22.07.2008 № 123-FL «Technical Regulation on the Requirements of Fire Safety» with Change comply with the Russian Federation Federal Law № 117-FL from 10.07.2012.
2. The Official Website of the Ministry of the Russian Federation for Civil Defense, Emergencies and Elimination of Consequences of Natural Disasters]. Available at: http://www.mchs.gov.ru/activities/stats/Pozhari/2016_god (accessed 30.05.2017).
3. A.P. Kovalev, O. Ya. Solenaya, Electricity, **11**, 36-45 (2014)
4. State Standard 12.1.004-91. Fire Safety. General Requirements. Moscow, Standartov Publ. (1992)
5. Federal Law of the Russian Federation from 21.12.1994. № 69-FL «About Fire Safety» with Change Comply with the Russian Federation Federal Law № 234-FZ from 13.07.2015.
6. A.P. Kovalev, O.Ya. Solenaya, Industrial power, **11**, 43–46 (2013)
7. K. Verfondern, X. Yan, T. Nishihara, H.-J. Allelein, International Journal of Hydrogen Energy, **42**(11), 7551-7559 (2017)
8. W. Chen, K. Zhou, S. Yang, C. Wu, Renewable and Sustainable Energy Reviews, **75**, 98-105 (2017)

Electrotechnical complex of drill rig with adjustable electric drives

Oleg Nikulin[1,*] and *Vitaliy* Shabanov[2]

[1]Management Company Tatburneft, LLC Almetievsk, 51 Fahretdin Street, 423450, Russia
[2]Ufa State Petroleum Technological University (USPTU), Electrotechnical and Electrical Equipment of Enterprises, 1 Kosmonavtov Street, 450044, Russia

Abstract. The paper considers the electrotechnical complex of a drill rig with adjustable electric drives of the main mechanisms. A computer model has been developed for the electrical complex of the drill rig, which allows studying processes in case of interruptions in the power supply system, changes in the technological parameters of drilling, etc. The article presents the simulation results for short circuits in the power supply system of the drilling rig and for drilling a well.

1 Introduction

Drill rig power supply is implemented by electric networks from 6 to 10 kV. In most cases drill rig power supply is implemented mainly from weak electric networks of commensurate capacity or autonomous power supply systems [1]. At the same time, the requirements to control systems of drill rig are increasing [2, 3], the operating conditions of electric drives and relay protection become more complicated and increase the relevance of developing computer models of electrotechnical complex of drill rig (ETC DR).

Computer models of ETC DR use the experience of developing models of relay protection systems [4], electric motors [5, 6] and power supply systems [7]. Modelling of ETC DR is widely used for research and analysis of the mutual influence of the electric power source and the main electric drives of the drilling rig, both in production drilling and in the construction of geological exploration wells due feeding the drilling rig from autonomous diesel power plants, including for studying the quality of electricity in various modes work of the main electric drives. Such studies, implemented in [8], allow establishing that the use of a frequency-controlled asynchronous electric motor makes it possible to improve the electromagnetic compatibility of the electric power source and the main drive of the drilling rig while they work together and reduce the cost of energy supply. Electric drive of drill winches works in a short-time mode with frequent starts. This determines the important role of research and optimization of launching modes. Studies of the ETC DR in the start-up modes with the use of mathematical modelling in the MATLAB/Simulink software allow obtaining the dependence of the characteristics of an uncontrolled and frequency-controlled asynchronous electric motor upon its starting from time and on the parameters of the power supply network [8]. Comparison of the obtained dependences has shown that the use of a frequency-controlled asynchronous electric drive allows reducing the average starting current of the electric motor by 80% and reducing the voltage deviation in the electric network by 60%. Comparison of the maximum values of active and full powers during the start-up of an uncontrolled and frequency-controlled asynchronous electric motor has shown that the use of a frequency converter allows choosing a diesel generator by power 1.5 times less than in the case of an unregulated electric drive. This makes it possible to more fully use the diesel engine in terms of power and, thereby, reduce the consumption of energy resources during the construction of the well.

In work [9], using mathematical modelling in the MATLAB/Simulink software, the authors studied the effect of defects and malfunctions of asynchronous drilling rig engines on the electric network. On the basis of results of the research, oscillograms of the change in the active and reactive power of the induction motor consumed from the network were obtained in the presence of rotor malfunctions and a significant effect of the faulty motor on the power quality indicators and the power consumed by the drilling rig was revealed. It is shown that the presence of rotor defects in the motor leads to modulation of the amplitude of the supply voltage, which leads to significant variations in the speed of the electric motors of the drilling mechanisms and increases the active, reactive and full power consumed by them. All these factors reduce the mechanical drilling speed, increasing the time, energy and economic costs of the well construction.

Electrotechnical complex of DR as a research object is inextricably linked with the technological process of drilling, the kinematics of drilling mechanisms and their loads, which should also be taken into account in modelling. It is important to take into account the

* Corresponding author: oleg309@yandex.ru

properties of the drill string while modelling the electromechanical transient processes in the ETC DR. So in [10] the structural model of the drill string is considered, as a consecutive connection of elementary links. A study of this structure on dynamic properties was performed using a package of mathematical programs MatLab. The analysis of transient processes for various operating modes of the drill string is implemented. In [11], studies of electromechanical processes in the ETC DR were implemented taking into account the properties of the drill string for the DR with an autonomous power supply from the diesel power station during geological prospecting drilling.

This review shows the relevance of the development of computer models of ETC DR taking into account the power supply system and the actual technological parameters of the drilling process. Consider the development of the computer model ETC DR on the example of drilling rig BU-2000 / 125EP, a single-line power supply scheme which is shown in Figure 1.

Computer models of the electrotechnical complex of drill rig use experience modelling systems relay protection [4], electric motors [5, 6] and electricity [7]. The paper considers the development of a computer model. This development is based on the example of electrotechnical complexes of drill rig, the drill rig's power supply single-line diagram, which is shown in Figure 1.

2 Methods used for experiments

To create and research the electrotechnical complex of drill rig we will use the MATLAB system. This system is based on algorithms of vector computation, providing high speed and accuracy of calculations, as well as the possibility of visual control of the processes occurring in the model. Merging the SIMULINK package and the vast library of electrical units of the package SimPowerSystem in the system MATLAB provides the ability to construct simulation and functional models of complex electrical systems, including the electrotechnical complexes of drill rig [8].

The model contains the following blocks and subsystems (Fig.2): source voltage -Three-PhaseSource, voltage 6000 V, frequency 50 Hz; «Airline-PowerLine» is represented by a block of Three-PhaseSeries RLC Branchjob. In the task window is introduced the system settings of the external power supply; three-phase AC switches HVVS1...HVVS4. Control of the switches is effected by signaling to the com port. A single level of the control signal causes the closure of the keys, and the zero level is the command to open the keys. In addition, the disconnection of the circuit in each phase is performed when the current reaches this zero level [8].

To measure amperage and voltage meters M1 ... M4 were used; three-phase transformers - T_TD (6kV/600V, 1100 kVA for power supply of the top drive system (TD)), T_DW (6kV/400V 1000 kVA for electric power draw-works (DW)) and T_MP (6kV/400V, 1600 kVA power for electric drive drill pump (MP)); rectifier R_TD and invertor I_TD of the upper drive system; DC rectifiers of electric motors (EM) of the drill winches and drill pumps R_MP и R_DW; of electric drives chokes at the input of rectifiers MP and DW L_MP and L_DW; electric motors AM_TD, DC_DW, DC_MP top drive systems, drawwork and drill pumps.

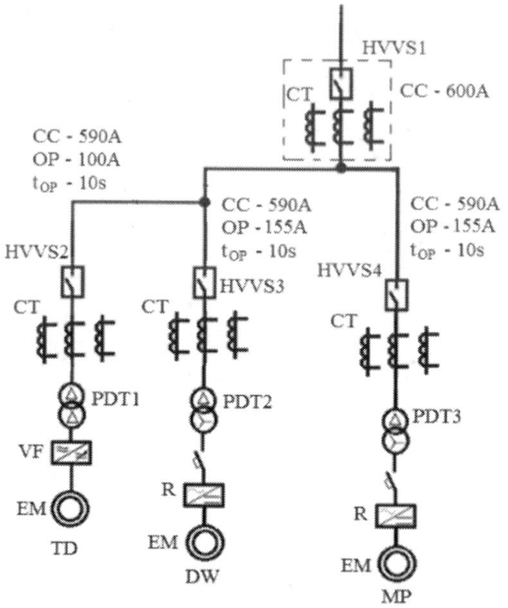

Fig. 1. The drill rig's power supply single-line diagram HVVS1...HVVS4 – high-voltage vacuum switches; PDT1...PDT3 – power dry transformers; CT – current transformers; VF – frequency converter; EM – electric motor; R – rectifier; CC – current cutoff; OP – overcurrent protection; TD – top drive; DW – drawwork; MP – mud pump.

The active resistances and inductances of the windings of the transformers, as well as the magnetizing circuits are given in relative units.

In the parameters of the electric motors, the values of the active resistances and inductances obtained in the identification process [6] of electric drives on drill rig (the data from the management systems of frequency converters and DC/DC converter) are entered. Recommendations were considered, a stationary reference frame [9] was used for electric motors for stator voltages that are unbalanced or intermittent.

The inverter uses sinusoidal modulation [10], a signal of the specified type is a sinusoid with a frequency of 50 Hz, the carrier frequency is a triangular signal with a frequency of 1650 Hz. This corresponds to the frequency modulation factor mf=33 (50 Hz · 33 = 1650) [11]. Power-up sensor "Chirp Signal" is an original development and generates a three-phase signal from the initial to the final frequency. The parameters which are set under the law U/f= the window mode. The formation of the increase and decrease of the signal is provided according to the specified law. This depends on the actual frequency value and the reference at the input from the "Reference" block. System of pulse-phase control of DC / DC converters of a control motors with feedback on the speed and current. Data on electric motor loading from the mat-file using blocks, due to the formation of load blocks that are input to the motor Tm.

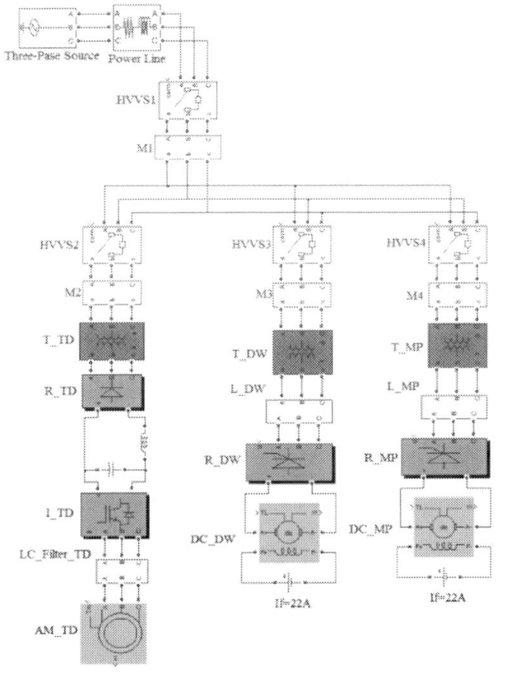

Fig. 2. Computer model of the electrotechnical complex of drill rig.

3 The main results

Figure 3 shows the relay protection model. We use the From block for control of currents, it takes data from the meter M. The Matlab system calculates the amplitude values of currents and voltages. Therefore, an RMS block is used to derive an effective current value that calculates the RMS value. To ensure the shutter response time MT3 we used block Discrete On/Off Delay.

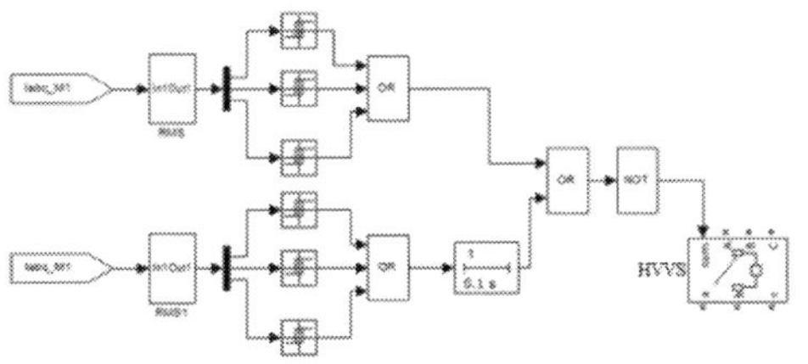

Fig. 3. Relay protection model.

Simulation of short circuits behind the transformer T_MP that feed the electric drive Mud Pump is a three-phase shorting "Three-Phase Fault" (Fig. 4). At the time 6s, an interfacial short circuit is produced. The simulation results are shown in Figure 5.

The protection relay tripping currents (TO in Figure 1) were not selected correctly and do not provide a trip for the circuit breaker in case of a short circuit, this was revealed by simulation. Therefore, the currents were reduced to 200A, which ensured the disconnection of the circuit breaker in the short-circuit behind the transformers. The results of the three-phase short-circuit simulation showed the necessity to correct the relay protection operation currents. As a consequence, this was done at the existing drill rigs.

The model of an electrotechnical complex of the drilling rig allows the modeling and investigation of the electrotechnical parameters at development drilling.

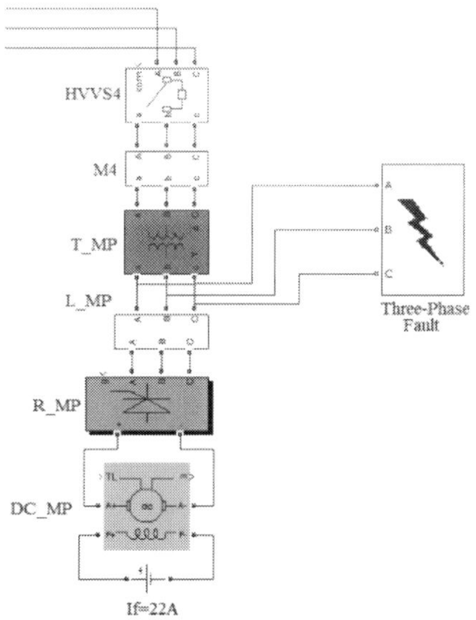

Fig. 4. Model simulating a short circuit.

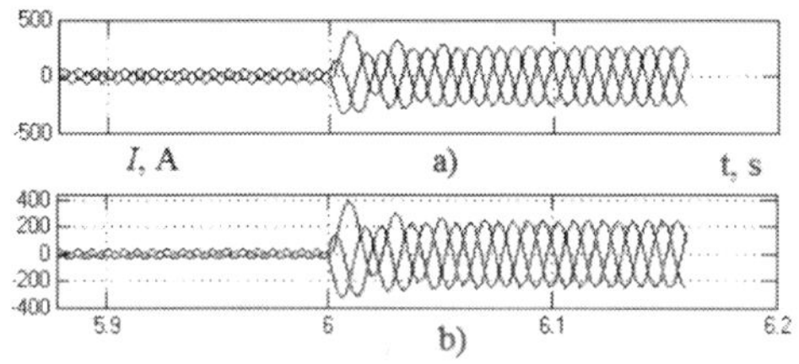

Fig. 5. The results of a simulation of short circuit behind the transformer T_MP: a) the currents in the power center for a breaker BB1- behind the switch HVVS1; b) the currents behind the breaker HVVS4.

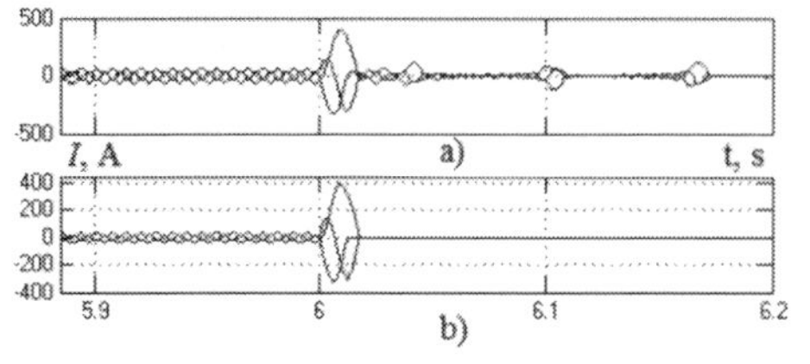

Fig. 6. Simulation results of short circuit behind the transformer T_MP with an adjusted operating current, relay protection of the switch HVVS4: a) the currents in the power center for a breaker HVVS1; b) the currents for circuit-breaker HVVS4.

Figures 7-8 show the results of the simulation of the electrical complex of the drill rig during the operation of well No 32409. Interval of depth of drilling of one boring candle from 603 to 628 m.

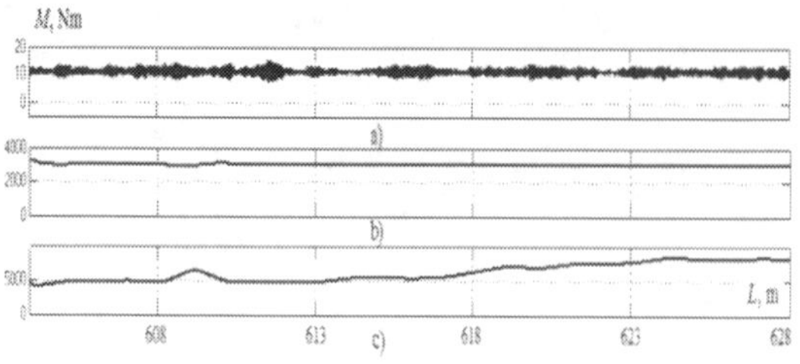

Fig. 7. Results of modeling the air gap torque of the electric motor.

a) the air gap torque of the electric motor of the top drive system; b) the air gap torque of the electric motor of the draw-work; c) the air gap torque of the electric motor of the mud pump

Data on electric motor loading from the mat-file using blocks, due to the formation of load blocks that are input to the motor Tm. Then data are unloaded for a certain interval from the control system of parameters of drilling. The load on the electric motor of the draw-work is kept constant, this provides a predetermined load on the bore bit.

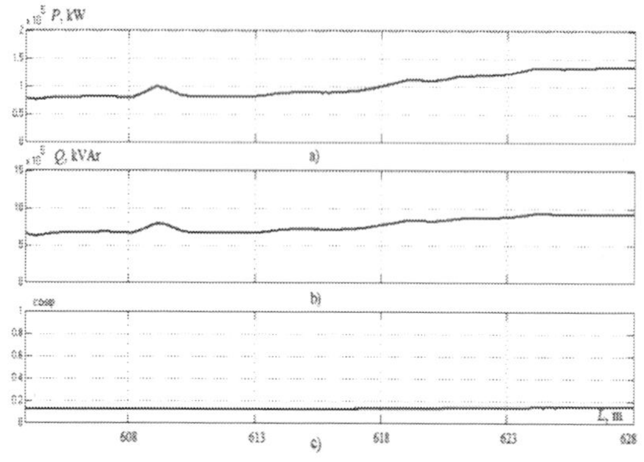

Fig. 8. Results of simulation of active (a) power and reactive (b) power and power factor (c) in the center of the power of the drill rig.

Figure 8 shows that in shallow drilling the electric network power factor is low.

When the well deepens the load of the drilling pump, the draw-work, the system of the top drive and the rotor increases, that changes all the parameters of the electrotechnical complex, including the power factor and the quality of electrical energy [12, 13]. This is confirmed by the results of the simulation (Figures 7, 8): they show that when drilling a well N32409 at an interval of 603-628 m with an increase in the load of the electric motors of the upper drive system, the mud pump and the draw-work, the power factor is increases.

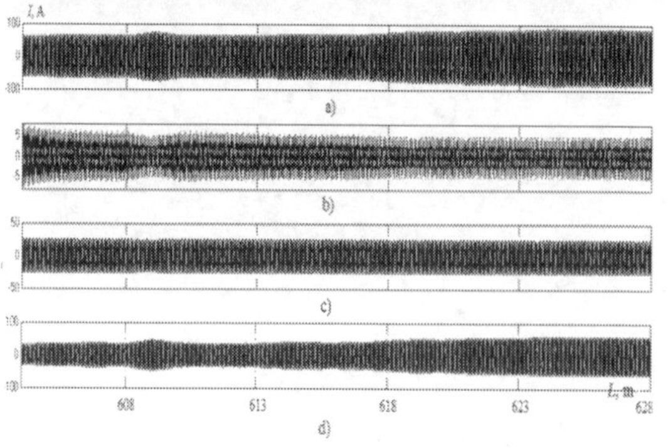

Fig. 9. Results of current simulation: a) the current in the power center; b) the current of the upper drive system; c) the current of the draw-work; d) the current drilling pump.

4 Conclusions

The developed model of the electrotechnical complex of the drill rig allows intestigation of the processes in the drill rig system in the event of emergency conditions, and selelcting the relay protection operation currents. It also allows researching processes using actual technological parameters in electric drives during the drilling of the well.

References

1. N. Y. Shevyreva, *Quality Improvement of the Electric Power at Power Supply of Drill Rigs with Variable Frequency Drives* (Ph.D. thesis in Engineering Science, 2016)
2. B. I. Abramov, O. I. Kozhakov, V. A. Shilenkov, V. K. Vasilyev, D. A. Ram, V. E. Kirillov, *Proceedings of the VIII International (XIX All-Russian) Conference on The Automatic Electric Drive*, 2 (2014)
3. B. Lotfy, K. Salakhkhoor, Oil and Gas, **3** (2013)
4. I.G. Kuliš, M. Skok, *23rd International Conference on Electricity Distribution Lyon* (2015)
5. A. Smolarczyk, R. Kowalik, E. Bartosiewicz, *18th Power Systems Computation Conference* (2014)
6. X. Zhan, G. Zeng, J. Liu, Q. Wang, S. Ou, International Journal of Advanced Computer Science and Applications, **6** (2015)
7. M.Y. Fedorova, Southern Ural state university Publishing the newsletter, **15** (2011)
8. Y.V. Shevyriv, D.A. Morgayov Bulletin of the Yuno-Ural State University. Series: Power. South Ural State University (National Research University), **15** (2015)
9. M.I. Grigoryev, S.V. Bukreev, Professional education and society. College of Modern Technologies. Hero of the Soviet Union Panova, **2** (2014).

10. I.Y. Bykov, S.F. Zaikin, B.A. Perminov, V.B. Perminov, S.V. Poletaev, Z.Kh. Yagubov, Scientific and methodical electronic journal. Interregional Center for Innovative Technologies in Education, **20** (2017)
11. S.V. Bukreev, Advances in modern science, **10** (2016)
12. I.V. Chernykh, *Simulation of Electrotechnical Devices in MATLAB, Simulink and SimPowerSystems*. (DMK Press, 2008)
13. P.C. Krause, O. Wasynczuk, S.D. Sudhoff, IEEE Press (2002)
14. N. Mohan, T.M. Undeland, W.P. Robbins, Power Electronics: Converters, Applications, and Design, John Wiley & Sons, Inc., New York **8.4.1.** (1995)
15. Power System Blockset User's Guide. The Mathworks Inc (2000)
16. V.A. Shabanov, O.V. Nikulin, Rogtec Russian Oil and Gas developments, **44** (2016)
17. O.V. Nikulin, A.A. Zhilenkov, S.G. Chernyi, Automation in the automation in industry, **3** (2017)

Active element influence on the motor's torque

Antonina Dolgih*, *Vladimir* Martemyanov and *Valeriy* Borikov

Tomsk Polytechnic University, Institute of Non-Destructive Testing, 634050, Tomsk, Russia

Abstract. The paper presents the numerical and experimental studies of the influence of the torque motor active element on the motor's torque. The tape active element is a novel type of a motor's stator organization, where the conventional winding is replaced by a tape winding. The force (torque) dependence over the rotor pole position using COMSOL is given; the tape winding resistance and the turns number are defined. The relative motor's characteristics are investigated and the maximum torque over the certain poles pair number is obtained. The application of the proposed active element in brushless DC motor is considered. The results show the possibility of the further synthesis of the torque motor.

1 Introduction

The advantages of torque motors allow using them in the direct drive development. This is confirmed by the fact that, at present, many companies develop and improve such motors. The history of the torque motors development and the problems of their improvement in details are described in the works of Alexander Mikerov [1, 2]. Theoretical principles of the widespread brushless permanent magnet motors are given in the majority of known sources, for example [3-5]. Nowadays, in order to improve the torque motors characteristics, both magnetic circuits and windings are being investigated [6, 7].

The particular attention is paid to the thermal mode of the motor operation, since a high temperature can destroy the winding and degrade the permanent magnets properties. The increased temperature during the operation is caused by the large current flowing through the windings. The motor overheating reduction can be achieved by its intensive cooling. The fan application connected to the rotor is not rational due to the low rotation speed. Therefore, in some cases, liquid cooling is used. For example, RIB torque motors [8] are slotted, permanent magnet direct drive motors with an internal rotor. The primary part is a fully cast stator with external liquid cooling. On the outer stator surface there are ring canals. TMB motors [9] also contain a canal structure with a cage for water cooling. However, this cooling method is difficult to use when the torque motor represents a part of a system located on the moving objects. In this situation, the torque motor, whose stator active element is a multilayer winding from a conductive tape with side cuts, is preferable [10].

The benefits of this active element include the novel design and technological approach to the motor stator manufacture; the possibility of high current load with a good heat removal; the formation of required functional dependence of the motor's torque over the angular movement.

The active element presents a set of serially connected plates, through which the current flows in the diagonal direction. The current transverse components interact with the magnetic field of the rotor poles and create a torque. The length of each plate a is equal to the pole division τ of the motor magnetic circuit. Diagonal character of current, flowing on each plate, is provided by the side cuts, located from the each tape side with a pole division pitch [11].

The purpose of this work is to show the influence of the active element characteristics on the motor's torque.

2 The active element characteristics

2.1 Torque dependence

The character of current flowing through the active element is distributed. The force F acts on the rotor pole and depends on the flux density B, the tape thickness Δ and the double integral from the distributed current density [12]:

$$F = B \cdot \Delta \cdot \iint j(x, y)dxdy. \qquad (1)$$

This force can be determined numerically using COMSOL software [11, 12].

Generally, the force dependence $F(x)$ over the relative position of plate and rotor pole has the form shown in Figure 1. Accordingly, the torque dependence T over the rotor angular position α will also be displayed in the same way.

The character of the resulting dependence $M(\alpha)$ in many practical cases is unacceptable. For the required dependence, it is proposed to apply the additional side cuts along with the main ones. An example of this method is shown in Figure 2.

* Corresponding author: ivanovatonya@tpu.ru

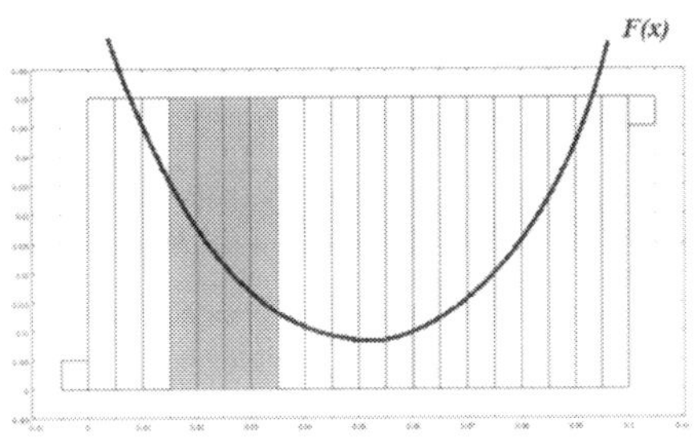

Fig. 1. The force dependence over the rotor pole position.

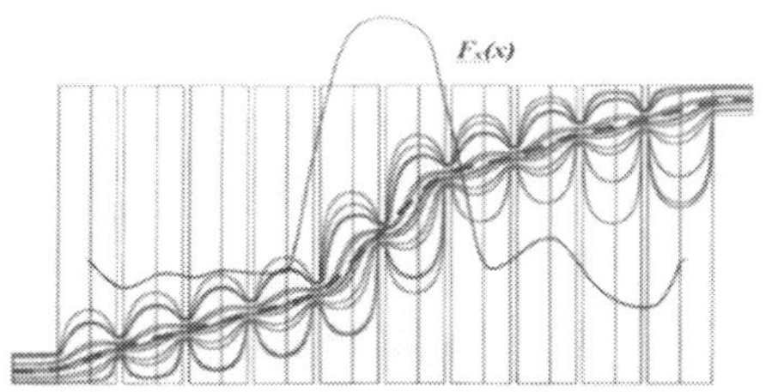

Fig. 2. The required force dependence using the additional side cuts.

The particular interest represents "the triangular" torque dependence over the angular position. This dependence can be realized with only two additional cuts on each plate (Fig. 3a). Figure 3b shows the dependence of the double integral $D(x)$ over the plate length. The double integral is taken from the expression (1).

2.2 Plate resistance

During the numerical simulation of the proposed active element, it was found that the resistance of a single plate depends on the ratio of its length a and width b. Varying this ratio $\beta = a/b$, a certain minimum resistance value is observed (Fig. 4). The parameter $\beta_e = b_e/b$ is the relative electrode width, where b_e is the tape section size, remaining after the cut.

Besides, the plate resistance is defined experimentally using thin aluminum tape with the following geometrical parameters: b=0,05 m; β_e=0,2; Δ=0,0007 m. The results are shown in Figure 5.

2.3 Winding turns number

Let us suppose that the number of winding turns N is small, and the plates geometry in all layers is almost similar. Then the each plate resistance R will be permanent. Also, suppose that the magnetic flux density is uniform.

As shown earlier, the force acting on one plate is defined as:

$$F = \Delta \cdot B \cdot D(I_0, x_c). \qquad (2)$$

Further, for simplicity, the double integral $D(I_0,x_c)$ over the area, bounded by the magnetic flux density action and with the center determined by the coordinate x_c, will be denoted by D. The value of this integral is determined by passing through the plate the current I_0.

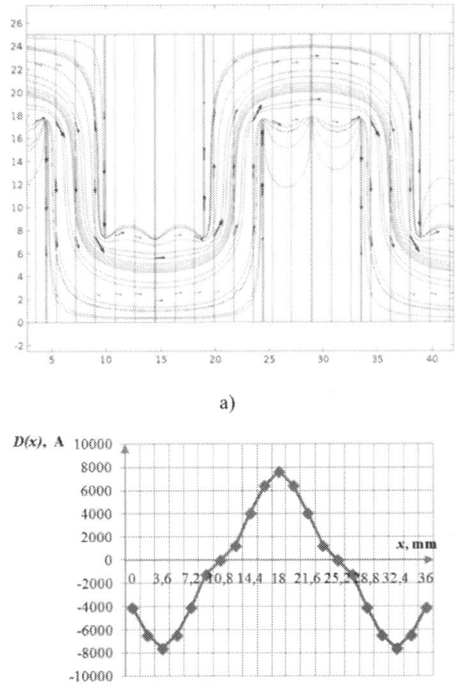

Fig. 3. "The triangular" torque dependence.

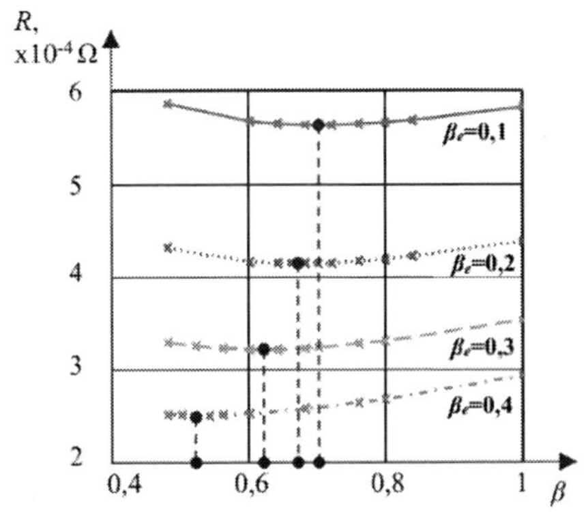

Fig. 4. The dependence of the plate resistance over the ratio of its length and width.

The current flowing through the tape winding depends on the source voltage U, the plate active resistance R, the number of poles pairs p, the number of winding turns N:

$$I = \frac{U}{2p \cdot N \cdot R}. \qquad (3)$$

Since the force is determined by the double integral D, which depends linearly on the current, let us write the expression for the force in the following form:

$$F = \Delta \cdot B \cdot \frac{U}{2p \cdot N \cdot R \cdot I_0} \cdot D. \quad (4)$$

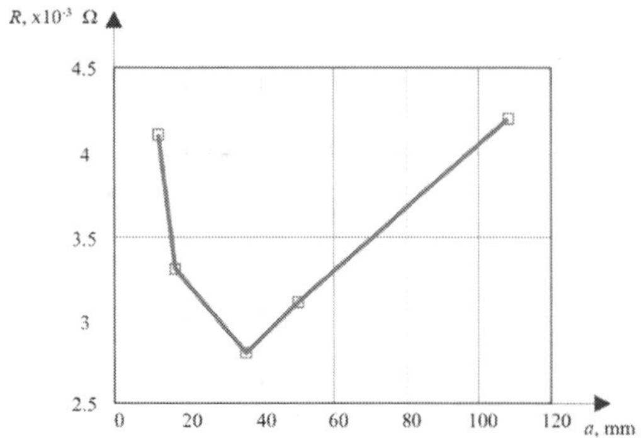

Fig. 5. The experimental dependence of the plate resistance over the ratio of its length and width.

The motor's torque is equal to the sum of the torques, created by each plate of all winding layers. The winding average radius is defined as:

$$r_{av} = r_0 + \frac{N \cdot \Delta}{2}, \quad (5)$$

where r_0 is the tubular base radius on which the tape is wound.

In this case, the torque is determined as follows:

$$T = 2p \cdot N \cdot F \cdot r_{av} = \Delta \cdot B \cdot \frac{U}{R \cdot I_0} \cdot D \cdot \left(r_0 + \frac{N \cdot \Delta}{2} \right). \quad (6)$$

Let us estimate the influence of the number of winding turns on the developed torque value. On the one hand, the number of plates increasing should lead to the torque growth. On the other hand, the total tape winding resistance will increase by the same amount, and consequently the current will decrease. In this case, the number of winding turns does not affect the torque value.

The torque change due to the average winding radius increase is insignificant. But at the same time, a smaller number of winding turns will decrease magnetic circuit air gap and lead to the growth of the flux density B. When defining the minimum value of N, it is necessary to proceed from the possibilities of both the power source and the winding heat resistance. The number of turns must be minimal and is determined by a permitted current:

$$N \geq \frac{U}{2p \cdot R \cdot I_{perm}}. \quad (7)$$

2.4 Relative characteristics

Let us introduce the concept of the relative torque to determine the number of magnetic circuit poles pairs and, correspondingly, the winding parameters at which the maximum torque is observed.

$$k_M = \frac{M_i}{M_1} = \frac{B_i}{B_1} \cdot \frac{R_1}{R_i} \cdot \frac{D_i}{D_1}. \quad (8)$$

In this case, M_1, B_1, R_1, D_1 characterize the motor with one poles pair p=1; M_i, B_i, R_i, D_i characterize the motor with poles pairs number p=i.

Firstly, we will only take into account the parameters characterizing the winding. There are the plates resistance and the double integrals of the distributed currents. For their determination the winding numerical models with the average radius range r_{av}=0,025÷0,075 m are created in the COMSOL software. The tape width b = 0,05 m and 0,025 m; the tape thickness $\Delta = 10^{-4}$ m; the tape material is a copper.

The coefficient of pole overlap is α_p=0,5. The pole center position is at the point x_c=0,25·a_i, where the plate length a is equal to the pole division (a=τ). The graphs of the resistances ratio R_1/R_i over the pole pairs are shown in Figure 6. The graphs correspond to the case r_{av}=0,027 m and b =0,025 m.

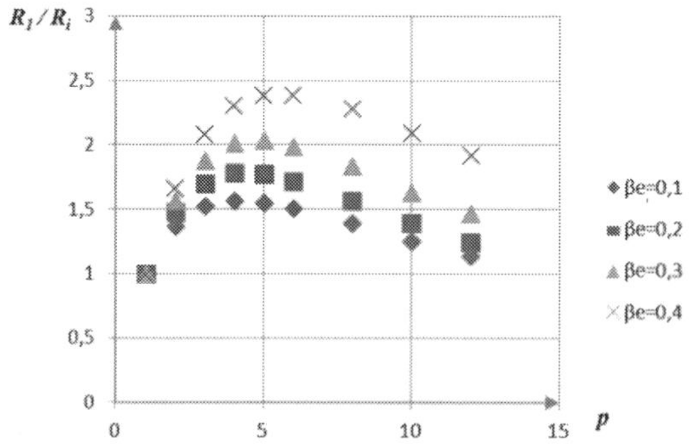

Fig. 6. The character of relative resistances changing.

For other considered variants of tape winding, the characters of relative resistances changing are preserved. Their maximum values are in the range of the poles pairs number p= 4÷6. Figure 7 shows the dependencies of the calculated double integrals values for the tape winding with the parameters given above.

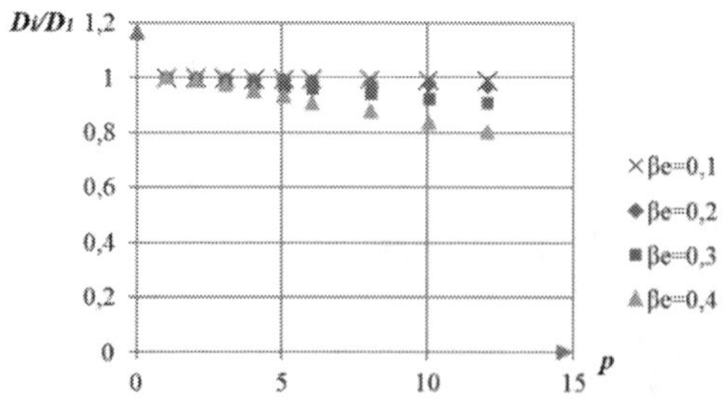

Fig. 7. Relative double integral values changing.

As seen from the graphs (Fig. 7), the double integral value D varies insignificantly with the change in the number of pole pairs, especially for small β_e. Therefore, we can conclude that the main effect on the motor's torque growth exerts the active resistance of the winding plates.

Since the aim of this work is not to define the magnetic circuit optimum parameters, a simple parametric model of the motor was constructed in T-flex CAD with the possibility of changing the poles pairs number and the air gap size (Fig. 8).

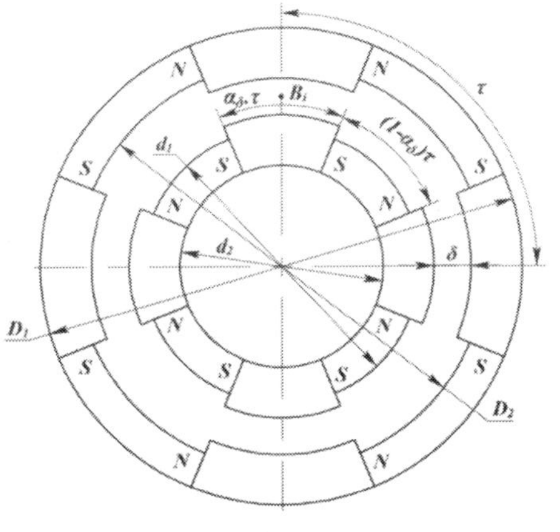

Fig. 8. The model of torque motor magnetic circuit.

Figure 9 shows the flux density changing in the air gap of the investigated magnetic circuit. The air gap δ varied from 3 to 7 mm.

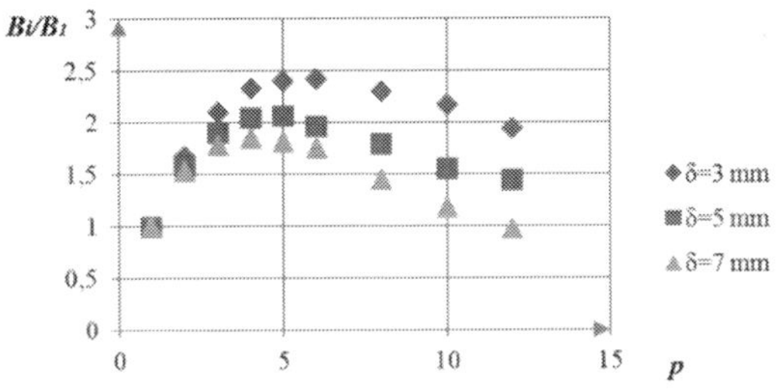

Fig. 9. The flux density changing in the air gap over the different poles pairs number.

Analyzing the obtained results, the maximum torque for the tape winding model of the selected geometric parameters will be observed with the poles pairs number in the range from $p = 4$ to $p = 6$. The relative torque dependence $k_T(p)$ is shown in Figure 10.

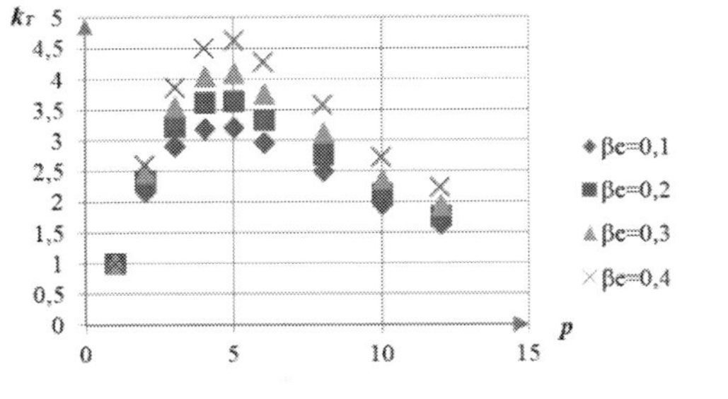

Fig. 10. The relative torque over the different poles pairs number.

The power consumed by the motor is determined by the expression:

$$P = \frac{U^2}{2p \cdot R \cdot N}. \tag{9}$$

Similarly to the previous calculations, let us introduce a concept of the relative power consumption:

$$k_p = \frac{P_i}{P_1} = \frac{R_1}{p \cdot R_i} = \frac{1}{p} \cdot \left(\frac{R_1}{R_i} \right). \tag{10}$$

The results of calculations for tape winding with the given earlier parameters are shown by the graphs in Figure 11.

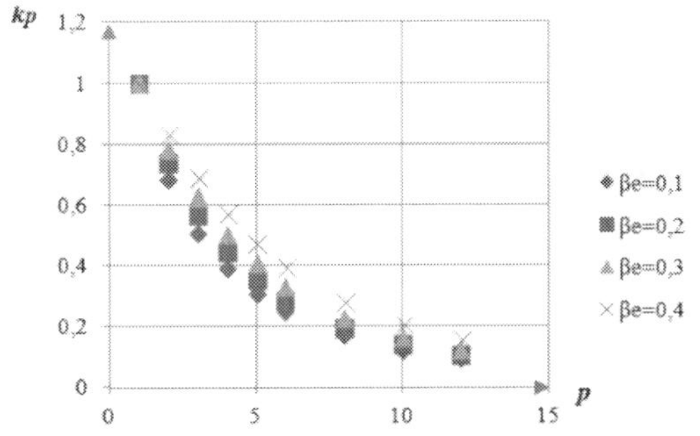

Fig. 11. The relative power consumption over the different poles pairs number.

In practical work, the static Q-factor [13], as the torque motor characteristic, is of interest

$$Q = \frac{M}{m \cdot P^{0,5}}. \tag{11}$$

The relative static Q-factor $k_Q(p)$ was estimated, taking into account the assumption that the motor mass m increases insignificantly with the increase in the number of poles pairs. The results of the calculations are reflected by the graphs in Figure 12.

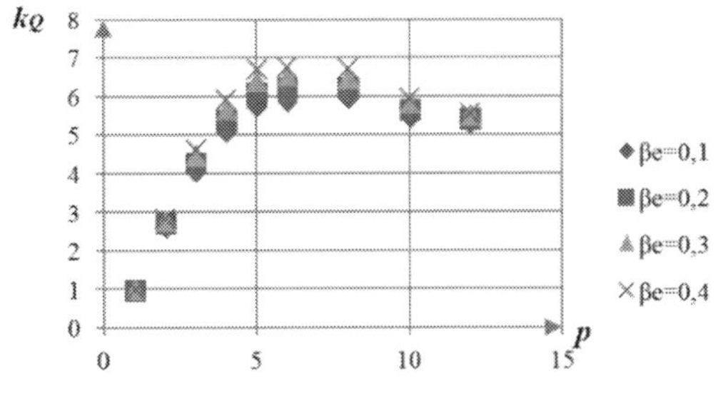

Fig. 12. The relative static Q-factor.

The numerical simulation results show that the torque motor optimum parameters will be observed with the poles pairs numbers from $p = 4$ to $p = 6$, with the given geometric parameters of the tape winding.

3 Results and discussion

As a result of this work, the force (torque) dependence and the variants of the required torque dependence implementation are shown. The resistance of the tape winding single plate has a minimum value at the certain ratio of the plate length and width. The number of winding turns does not influence the motor's torque. The analysis of the relative motor's characteristics is carried out. The maximum torque is observed at a certain number of poles pairs. The obtained results allow making the synthesis of the proposed torque motor.

Also, it is important to consider the application of the proposed active element in electromechanical devices. The two-phase brushless DC (BLDC) motor can be realized using two tape windings. The cuts of the second tape winding should be shifted relative to the first tape winding by half of the pole division [14].

In this variant, the torque combined characteristic will not have zero points. If the torque dependencies of both windings are identical and represent a combination of isosceles triangles, the resulting motor's torque will be unchanged for all rotor positions. Ideally, the torque ripple will be absent (Fig. 13).

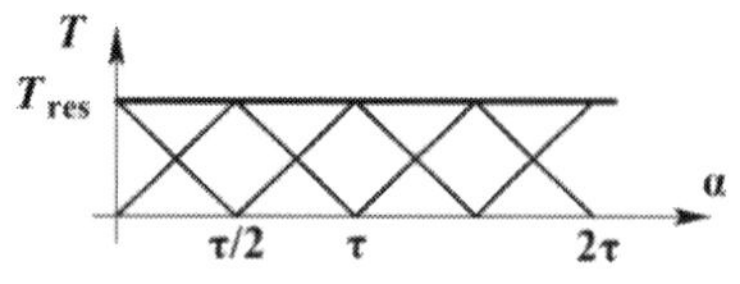

Fig. 13. The two-phase BLDC motor torque.

The second tape winding is coaxial with the first and is under the influence of the same rotor magnetic flux. There are two variants of the second tape winding location: it can be located side by side in the axial direction, performed during simultaneous winding of the two tapes; or it can be placed concentrically, covering the first winding. The design of such motor is shown in Figure 14.

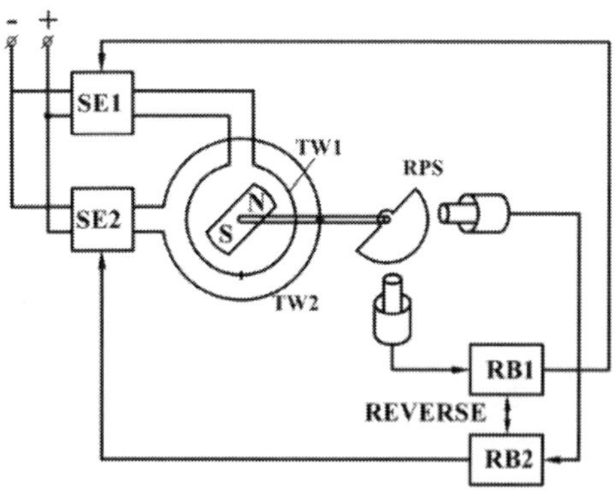

Fig. 14. The BLDC motor design.

The following symbols are used: TW - the tape winding, SE - the switch element, RB - the reverse block, RPS - the rotor position sensor. RPS can be implemented on the inductive sensor and its rotor is made as a disk of ferromagnetic material of the appropriate shape. RPS switches the signal between the tape windings. If it is necessary to change the direction of rotation, as well as the sign of the torque, the tape winding current should flow in an opposite direction. This can be done by the reverse block of the BR, where the signal from RPS is reversed.

4 Conclusion

The results of the work show the influence of the characteristics of the proposed active element on the motor's torque and allow continuing the research. At present time, the numerical simulations of torque pulsations with the different current values flowing through the set of two tape windings are being realized.

References

1. G. Dubrovskiy, A. Mikerov, V. Dzhankhotov, J. Pyrhonen *EPE-ECCE Europe*, 6910754 (2014)
2. A. G. Mikerov, J.J. Pyrhönen, J.J. Vauterin *EUROCON*, 4400568 (2007)
3. D. Hanselman, *Brushless Permanent Magnet Motor Design. Second Edition.* (Magna Physics Publishing, United States of America, 2006)
4. A. Hughes B. Drury, *Electric Motors and Drives. Fundamentals, Types and Applications. 4th Edition.* (Elsevier-Newnes, Oxford, 2013)
5. P. Madaan, *Brushless DC Motors. Part 1. Construction and Operating Principles.* (Cypress Semiconductor, India, 2013)
6. Z.Q. Zhu and D. Howe, IEEE Transactions on Energy Conversion **15**(407) (2000)
7. L. Zhu, S.Z. Jiang, Z.Q. Zhu, C.C. Chan, IEEE Transactions on Magnetics **45**(4), 2023 (2009)
8. INA – Drives & Mechatronics GmbH & Co. KG, *IDAM Direct Drives* (2014)
9. ETEL, *Torque Motors*. (2017)
10. V. Martemjanov, A. Dolgih (Ivanova), RF2441310, (2012)
11. V. Martemjanov, A. Dolgih (Ivanova), SIBCON, 6693632 (2013)
12. A. Dolgih, V. Martemyanov, MATEC Web of Conf. 48, 01004 (2016)
13. A.V. Demagin, Electric machines for direct drive of instrument systems. (Scientific-Production. Association "Azimut", Saint-Petersburg, 1991) (In Russian)
14. V. Martemjanov, A. Dolgih (Ivanova), RF2454776, (2012)

Control features of the electromechanical system with end-effector considering the regulated torque

Andrei Malchikov*, *Andrei* Yatsun, *Petr* Bezmen and *Oleg* Tarasov
Southwest State University, Department of Mechanics, Mechatronics and Robotics, 305040 Kursk, Russia

Abstract. The article describes the features of digital control of an electromechanical system consisting of an electric drive, a load and a feedback system. The emphasis is on implementing a mode in which the electric drive torque can be adjusted. These results can be applied in service and industrial robotics.

1. Introduction

Effective functioning of the most of robotic systems is possible only under the following conditions: the controlled consumption of the power source, the optimal motion laws and controlled force-moment parameters of the mechanism actuating links.

High energetic efficiency of a robot is achieved by using modern productive electric drives with mechanical transmissions with a high coefficient of efficiency and optimal circuit solutions [1, 2]. One of the modern approaches to provide the efficiency of the drive is monitoring power consumption and heat losses during the execution of the required motion laws. This aim can be achieved by using a double loop control system, in which torque and motor current control are implemented in addition to end-effector position control. This method is especially relevant for mobile systems, in which the cruising range of the robot is determined by the limited battery capacity.

The example of such a system is an assistive powered exoskeleton (Fig. 1) that enables a person with lower-limb disabilities to perform upright gait, climb stairs and perform simple body movements [3-8]. It is also important for exoskeleton system to ensure torque limitation to guarantee smooth motion of the drives, user safety and long life of the mechanical transmissions.

This paper compares two approaches to design a control system the end-effector position of the electromechanical system. The first approach is based on a classic control system for the rotation angle of the end-effector without the motor torque and winding current feedback. The second one is the implementation of a double loop control system that takes into account the real-time data of the current sensors and provides the required torque limitation.

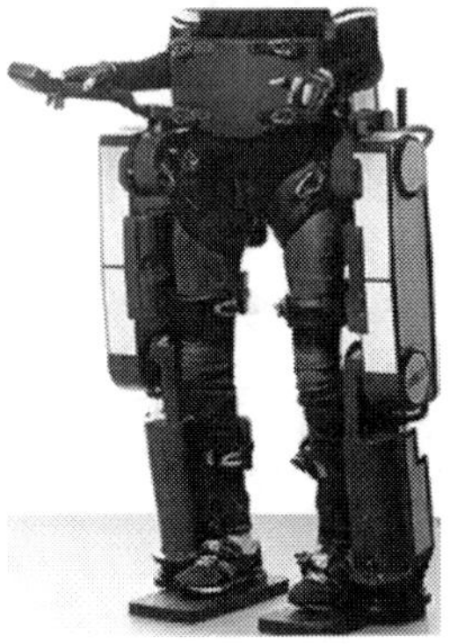

Fig. 1. Assistive powered exoskeleton ExoLite.

The authors present both the results of a numerical experiment on the mathematical model of the mechatronic drive, and the results of full-scale tests on a laboratory bench. A comparative analysis of the obtained results is made and conclusions are drawn.

2. Analytical model of the system

Figure 2 shows the analytical model of the experimental electromechanical system consisting of a mass 1 fixed to the link 2. The link 2 is driven by an electric drive consisting of a DC-motor 4 and a reducer 3.

The device works in the following way. A control signal with the desired angle φ^* and current I^* values and real-time data from the feedback sensor 8 (I' value) and the feedback sensor 9 (φ' value) comes in the inputs of the control system 6, consisting of a controller unit and the comparator unit.

The controller unit generates an output signal u^*, which is fed to the power amplifier 7 and after the control voltage u goes to the DC-motor 4.

* Corresponding author: teormeh@inbox.ru

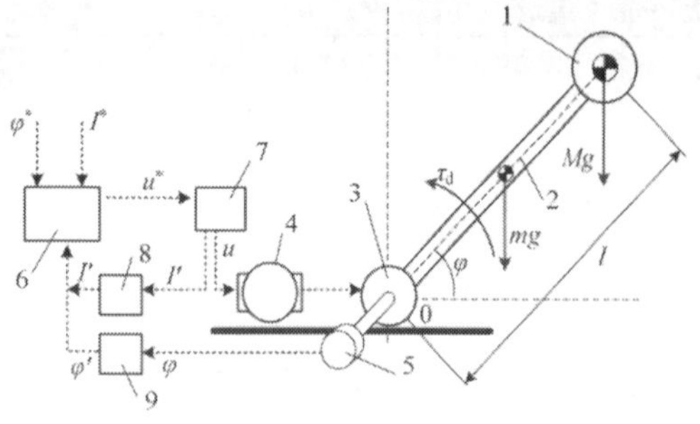

Fig. 2. The analytical model of the experimental electromechanical system.

3. Mathematical model of the mechatronical system

During investigation of the system mathematical model the following assumptions were made: the links of the system are absolutely rigid, the gaps are neglected in the reducer 3, the saturation of the magnetic flux in the motor simulation is not considered, dry friction forces act according to the Coulomb model and the viscous component is represented by a linear model. Then the equations describing the processes taking place in the mechatronic system can be written in the form:

$$\begin{cases} J\ddot{\varphi} = \tau_d - \tau_\mu - \tau_b - Mgl\cos\varphi - \dfrac{mgl\cos\varphi}{2} \\ L\dfrac{dI}{dt} + RI + c_E(\dot{\varphi}\cdot i_r) = u \end{cases}, \quad (1)$$

In this case, the torques acting on the system are described by the following equations:

- the electric drive torque is equal to:

$$\tau_d = c_M I \cdot i_r \eta, \quad (2)$$

- the viscous resistance moment is equal to:

$$\tau_b = \beta\dot{\varphi}, \quad (3)$$

- the dry friction force moment equation is equal to:

$$\tau_\mu = \begin{cases} -\tau_\mu^* \operatorname{sign}(\dot{\varphi}), \text{ if } \dot{\varphi} \neq 0; \\ -\sum_{j=0}^{n} \tau_j, \text{ if } \dot{\varphi} = 0 \text{ and } \left|\sum_{j=0}^{n} \tau_j\right| \leq \tau_\mu^*; \\ -\tau_\mu^* \operatorname{sign}\left(\sum_{j=0}^{n} \tau_j\right), \text{ if } \dot{\varphi} = 0 \left|\sum_{j=0}^{n} \tau_j\right| > \tau_\mu^*. \end{cases} \quad (4)$$

The following denotations are used in above-mentioned formulas: M – the mass of the body fixed on the link, m – the link mass, l – the length of the link, J – the inertia equivalent moment of the system:

$$J = \frac{ml^2}{3} + Ml^2 \quad (5)$$

I – the motor winding current; R – the motor winding active resistance; i_r - the total reduction ratio of the reducer; η - the efficiency coefficient of mechanical transmission; c_M - the motor moment coefficient; c_E - the motor anti-electromotive force coefficient; β - the viscous resistance coefficient; τ_μ^* - the dry friction limiting moment, τ_j - the sum of all external force moments, excluding moments of frictional forces.

4. Control system description

In the simplest case, the drive must be equipped with a closed-loop control system to ensure the required position of the link. The functional diagram of the closed-loop control system is shown in Figure 3.

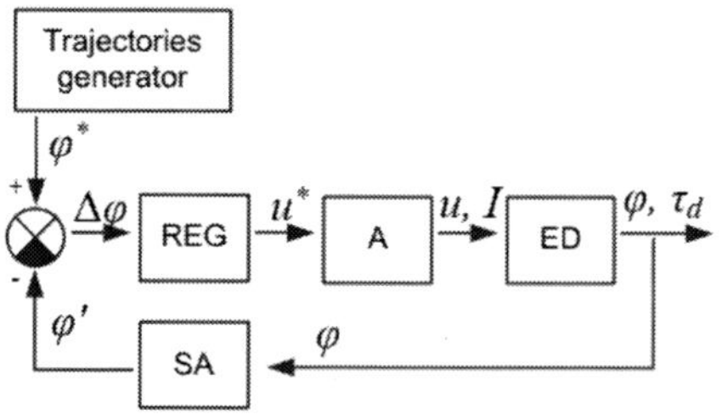

Fig. 3. Block diagram of the closed-loop control system.

The following denotations are used in Figure 3: REG. – the regulator, A – the power amplifier, ED – the electric drive, SA – the angle sensor.

In this case, the power supplied to the drive (u, I) is determined by the error ($\Delta\varphi$) value – the difference between the measured angle (φ') of the actuator and the required angle (φ^*).

In this case, P-, PI- or PID- controllers and other types of controllers, such as fuzzy logic controllers can be used.

In this paper we consider a PD-controller, for which the u^* signal value is defined as:

$$u^* = k_p e + k_d \dot{e}, \quad (6)$$

where k_p, k_d – the controller amplification coefficients.

$$e = \varphi^* - \varphi' \quad (7)$$

$$\dot{e} = \dot{\varphi}^* - \dot{\varphi}' \quad (8)$$

In mechatronic systems, P-regulators and PID-controllers are widely used to create a position control system [2-5], but in practice PD- controllers are also used, which are much easier to configure using, among other methods, manual tuning methods [10, 11]. We note that the evaluation of the integral component in the PID controller requires significant computational power of the control system [10].

With this control approach the motor current can be determined by the load on the final link. Thus, in the event of jamming or other unforeseen situations, the engine may fail or the mechanical transmission can be damaged. Usually modern controllers have adjustable protection against overheating or current limits. However, the values of these current limits cannot be changed during the movement of the device, which is especially important for multi-link systems, in which the load moment on the reducer can be determined by the system configuration and cannot be predetermined. To ensure the protection of the mechanical part of the drive, torque monitoring and the possibility to adjust the limits depending on the configuration of the mechanism are required. This is possible only when current data is taken into account directly by the control system in real-time mode.

To ensure the protective properties of the system and to provide the required parameters of energy efficiency, the authors propose a double loop control system (Fig. 4).

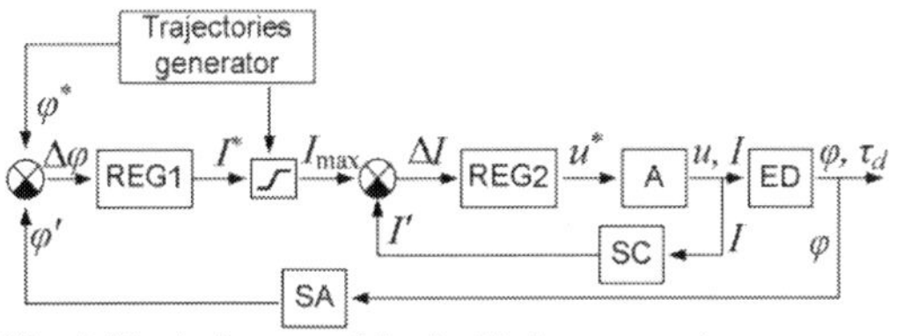

Fig. 4. Block diagram of the double loop control system.

Some denotations in Figure 4 are similar to those in Figure 3. In this case the power supplied to the drive (u, I) is not only determined by the angle error ($\Delta\varphi$), but also determined by the current error value (ΔI) obtained by comparing the current generated by the local feedback and the current value received from the sensor, denoted on the diagram as C.

Thus, the operation of the controller is described by the equation of the control signal at the amplifier input:

$$u^* = k_p^I e^I, \quad (9)$$

where k_p^I – the proportional coefficient of the current regulator (REG2 – see Fig. 4). The mistake is defined as:

$$e^I = I^* - I' \quad (10)$$

The required current value is limited by the current value I_{max} obtained from the control action generator.

$$I^* = \begin{cases} I_{max} \text{ if } (I^* \geq I_{max}) \\ I^* \text{ if } (I^* < I_{max}) \end{cases} \quad (11)$$

The initial value I^* is determined by the loop of the angle control, as follows:

$$I^* = k_p^\varphi e^\varphi + k_d^\varphi \dot{e}^\varphi \quad (12)$$

where k_p, k_d – the gain coefficients of the position controller (REG1 – see Fig. 3);

$$e^\varphi = \varphi^* - \varphi' \quad (13)$$

$$\dot{e}^\varphi = \dot{\varphi}^* - \dot{\varphi}^* \quad (14)$$

5. Results of mathematical modelling

Further we study the end-point control features. To do that we numerically integrate the system of equations (1) describing the motion of the link (the actuating link). In this case, the electric drive torque value is determined by the equation (2), and the moments of friction and viscous resistance forces by equations (3) and (4). To simulate an angular position control system, we use equations (5) and for a system with current limitation – equations (8), (10), and (11). A program was developed for numerical calculation in the MathCad software to obtain the solution.

The primary objective of modeling is to study system's operation features in the presence of a motor current limit and the result of controlling the torque on the actuating link. We obtain graphs of the system behavior for different values I_{max} for the following types of control signals (Fig. 5):

- stepped signal,
- rectangular signal.

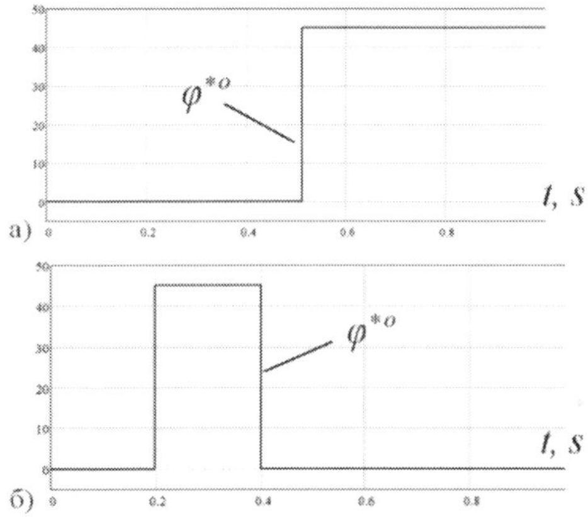

Fig. 5. Types of control signal used for simulation: a – stepped signal, b – rectangular signal.

The electromechanical system parameters used in the simulation were taken in accordance with the parameters of the experimental bench, which was created for debugging the mathematical model and conducting full-scale tests on the control system (Table 1).

We consider the modeling results for stepped control signal with an amplitude of 45 degrees (Fig. 6), and a tuned angular position control system (coefficients $k_p = 3.4$, $k_d = 4$).

Table 1. The mathematical model parameters.

Parameter name	Value
The link mass, m	0.005 kg
The mass of the body fixed on the link, M	0.024 kg
The length of the link, l	0.11 m
Operating supply voltage, U_{nom}	12 V
The motor armature active resistance, R	2.4 Ohm
The motor moment coefficient, C_M	0.0055
The motor anti-electromotive force coefficient, CE	0.0055
The total reduction ratio of the reducer, ir	30
The efficiency coefficient of mechanical transmission, η	0.85
The accuracy of the sensor, taking into account the reducer	0.4°
The sampling rate of the feedback sensor	500 Hz
The dry friction limiting moment, τ_{μ}^{*}	0.01 Nm
The viscous resistance coefficient, β	0.0001 Nm·s

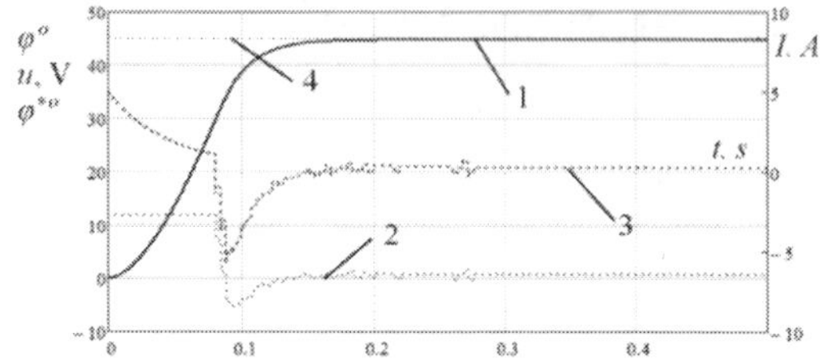

Fig. 6. Results of the angular position control system modeling at stepped control signal: 1 – angular position of the link, φ; 2 – control voltage, u; 3 – current, I; 4 – required angular position φ* of the link (stepped control signal).

We can see that the current value is 5 A at the electric motor start-up and reaches 6 A when the voltage is reversed.

Further, we show an example that demonstrates the results of rectangular control signal influence. In this case, the signal changes instantaneously by a significant amount (Fig. 5, b). Figure 7 shows that the required angular position changes from 60° to 0° at 0.06 sec.

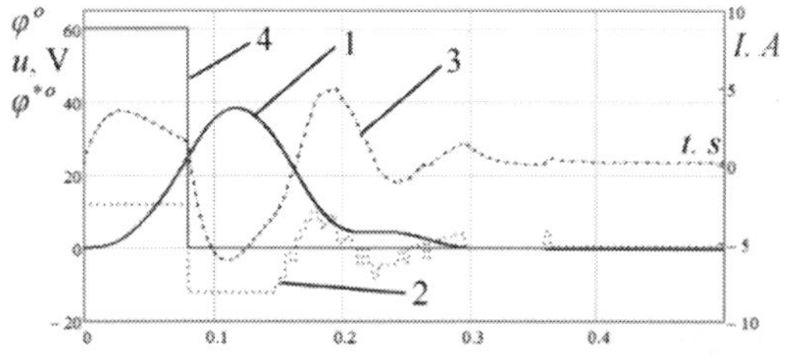

Fig. 7. Results of the angular position control system modeling at rectangular control signal: 1 – angular position of the link, φ; 2 – control voltage, u; 3 – current, I; 4 – required angular position φ^* of the link (rectangular control signal).

Figure 8 shows the character of the variation of the link angular position and the motor current when current feedback is used without limitations.

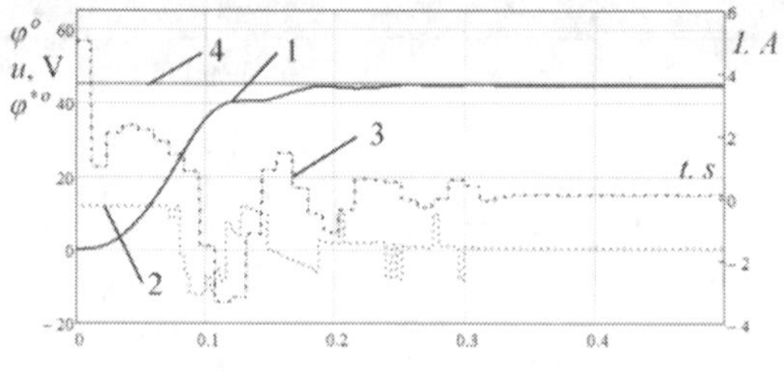

Fig. 8. Results of the control system modeling with using of current limitation: 1 – angular position of the link, φ; 2 – control voltage, u; 3 – current, I; 4 – required angular position φ^* of the link (stepped control signal).

In this case, the motor current can reach 7 A, which can lead to damage of the motor insulation or the mechanical transmission elements. A similar situation can also arise when an external force appears, which can result in significant current and forces in the reducer.

The control system of the link position with the torque control makes it possible to reduce the motor current in the transient processes while providing the required operating mode of the electric drive.

Addition of the current control loop increased the transient time to 0.3 sec as compared to single-loop control system (Fig. 9). An oscillatory character of the link's motion also occurred due to an overshoot in the feedback loop.

Further, we consider the modeling results for a stepped control signal and different values of the limiting current.

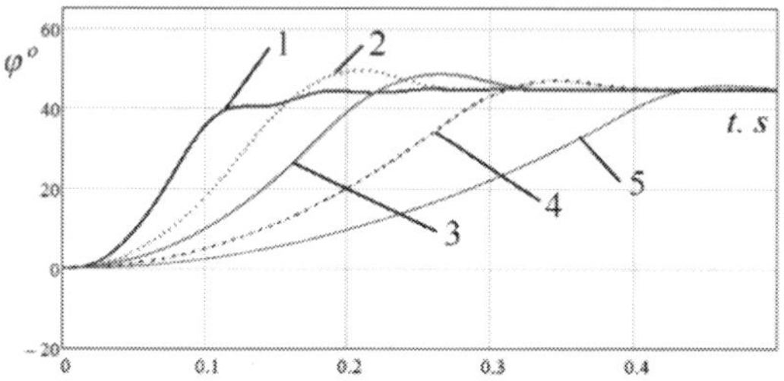

Fig. 9. Control system simulation results for a stepped control signal with current limitation: 1 - limitation 5A, 2 - limitation 1.5A, 3 - limitation 1A, 4 - limitation 0.7A, 5 - limitation 0.55A.

We can see from the numerical simulation results that limiting the maximum current preserves the system's operability but changes the parameters of the transient process. When a current limitation of 0.55 A is introduced under stepped control signal the transient time can be more than 0.5 sec and the overshoot value also changes. The current limitation does not allow the system to stop instantly – a motor rundown can be seen in the graphs. To ensure the required parameters of the transient process it is necessary to set the controller coefficients for each limiting value separately.

The simulation results for a rectangular control signal and different values of the limiting current are shown in Figure 10.

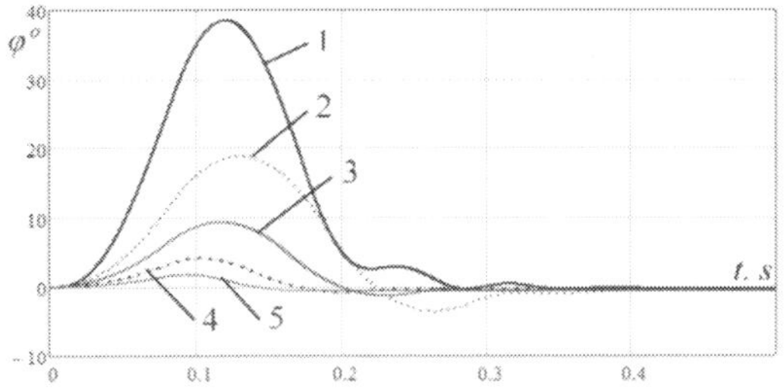

Fig. 10. Control system simulation results at the rectangular control signal with current limitation: 1 - limitation 5A, 2 - limitation 1.5A, 3 - limitation 1A, 4 - limitation 0.7A, 5 - limitation 0.55A

Modeling the system under a rectangular control signal (the pulse duration is 0.1 sec) made it possible to establish that the transient process time is 0.4 sec without current limitation. The transient process takes approximately the same time with current limitations of 0.7 and 0.5 A. However, the link does not execute the required amplitude of motion when the motor current is limited and overshoot and oscillation can also arise.

Conclusion

As shown by the simulation results motor current limitation makes it possible to avoid exceeding the torque during acceleration and braking of the motor, which then reduces the load on the electric motor and reducer of the electric drive.

However, the introduction of the input power limitation significantly changes the nature of the actuating link's movement. Current limitation increases the duration of the transient process, while an instantaneous change of the control action allows system rundown, which can lead to undesired overshoot of the control system. In order to guarantee the required system parameters, additional tuning for a specific value of the maximum current is required.

Supported from grant of the President of the Russian Federation for young scientists MK-2701.2017.8

References

1. S. Jatsun, S. Savin, B. Lushnikov, A. Yatsun, *ITM Web of Conferences*, **6** (2016).
2. S. Jatsun, S. Savin, A. Yatsun, A. Malchikov, *Advances in intelligent systems and computing*, 371, 165-172 (2015)
3. S.F. Jatsun, L. Yu. Vorochaeva, A.S.Yatsun, S.I. Savin, *International Conference on CLAWAR*, 175-182 (2015)
4. S.F. Jatsun, S.I. Savin, A.S. Yatsun, R.N. Turlapov, *26th DAAAM Intern. Symp. on Intelligent Manufacturing and Automation*, 107-112 (2015)
5. S. Jatsun, A. Malchikov, A. Yatsun, *22nd International Conference on Vibroengineering*, 8 (2016)
6. S.F. Jatsun, P.A. Bezmen, A.S. Yatsun, *22nd International Conference on Vibroengineering*, 8, 403-408 (2016)
7. S.F. Jatsun, S.I. Savin, P.A. Bezmen, *International Conference on Pure Mathematics*, 146-149 (2015)
8. S.F. Jatsun, S.I. Savin, P.A. Bezmen, *International Conference on Pure Mathematics*, 83-87 (2015)
9. L.Yu. Vorochaeva, A.S. Yatsun, S.F. Jatsun, SPIIRAS Proceedings, **3**(52), 70-94 (2017)
10. R. Izerman, *Digital Control Systems* (1984)
11. V.V. Denisenko, Sensors and systems, **6**, 35-38 (2009)
12. K.J. Astrom, T. Hagglund, *The Instrumentation, Systems, and Automation Society* (2006)
13. M. Hassan, H. Kadone, K. Suzuki, Y. Sankai, Sensors. **14**(1) (2016)
14. K. Kiguchi, T. Takakazu, F. Toshio, Fuzzy Systems: IEEE Transactions, **12** (4) (2004)
15. N. Aphiratsakun, P. Manukid, International Journal of Advanced Robotics Systems, **6**(4) (2009)
16. R. Steger, S. Kim, H. Kazerooni, *IEEE International Conference on Robotics and Automation* (2006)
17. J. Pratt, B. Krupp, C. Morse, Industrial Robot: an International Journal, **29**(3) (2002)
18. J.F. Veneman, R. Ekkelenkamp, R. Kruidhof, F.C. van der Helm, H. van der Kooij, The International Journal of Robotics Research, **25**(3) (2006)

MATEC Web of Conferences 113, 02002 (2017) DOI: 10.1051/matecconf/201711302002
12th International Scientific-Technical Conference on Electromechanics and Robotics "Zavalishin's Readings" - 2017

Improving energy efficiency of a robotic system based on multiple analytical solutions for inverse kinematics

Roman Iakovlev[1], *Vladislav* Sakuta[1], *Alexander* Denisov[1] and *Ryhor* Prakapovich[2]

[1]SPIIRAS, Laboratory of Autonomous Robotic, 199178 Saint-Petersburg, 14-th line of V.I., 39, Russia
[2]UIIP NASB, Laboratory of Robotic Systems, 220012 Minsk, Surganava str., 6, Belarus

Abstract. The paper presents the results of a study devoted to the problem of improving the energy efficiency of mechanical motion of anthropomorphic robotic systems. Achieving higher energy efficiency is largely due to the implementation of improvements directly in the algorithms that ensure the movement of a robotic system. For this purpose, several existing analytical methods for solving the inverse kinematics problem for robotic walking platforms were analyzed. According to the survey, key areas where modification can improve the energy efficiency of mechanical motion in various RS are identified. The paper discusses the algorithm developed to optimize the solutions of the inverse kinematics problem in terms of energy consumption.

1 Introduction

Despite the rapid pace of development of robotics, many problems related to the energy efficiency of robotic walking platforms remain unresolved mainly because of the extreme limitations of energy resources in such systems. The main problem in the design, modeling and implementation of solutions for kinematics for robotic walking platforms is the need to develop motion patterns with reduced power consumption to ensure sufficient time of full-fledged functioning, since most of the consumed energy is spent to provide mechanical motion.

At the same time, the solution of the inverse kinematics problem is a very difficult task in terms of computational and algorithmic complexity. Thus, the implementation of additional heuristics optimizing the energy consumption is often not feasible in practice. Nevertheless, for a number of robotic systems (RS) where the usage of analytical methods for inverse kinematics problem (IKP) solutions is possible, the introduction of such algorithms is advisable, since the implementations of analytical methods have a relatively small computational complexity.

In order to solve the IKP, a number of specialized methods simplifying the obtaining of analytical expressions have been developed [1, 2]. The essence of these methods consists in determining the rotation angles of joints from nonlinear equations for individual elements of the matrix equation. Because of the non-linearity, these methods for complex robotic systems lead to multiple solutions, but as a rule, researchers' attention is not focused on the method of choosing a particular solution from the set, and the proper one is chosen empirically.

Even though analytical methods require significant preliminary calculations for their use on each particular RS, such methods are very reliable and can provide solutions with high accuracy. Therefore, analytical methods are widely used in modern projects.

In [3], a partial solution of the kinematics problems for Darwin-OP [4] is discussed. Darwin-OP is an anthropomorphic RS with 20 degrees of freedom (DoF). Each Darwin-OP's leg has 6 DoF, and the solution of the IKP for them is of the utmost interest. /The authors chose the analytical approach as it was possible to obtain a solution of the IKP through analytical expressions because of the structural features of the system. Nevertheless, despite the fact that the problem of multiple solutions was identified in this work, the approach to its solution was not demonstrated. Thus, additional opportunities to increase the energy efficiency of the obtained solutions were missed.

In [5], a comparative analysis of modern solutions of the IKP applying to industrial RS with 6 DoF is presented. One of the approaches used is the analytical method and the problem of multiple solutions was also considered, namely an algorithm clearing the list of suitable solutions in accordance with the constraints imposed by the structural features of the RS was added. Nevertheless, the quality of the remaining decisions was not evaluated, and a random selection of the so-called "true solution" was carried out. In this case, the probability of choosing the most qualitative solution is inversely proportional to the number of decisions remaining after the limitations was taken into account.

The paper [6] discusses solving the IKP for the anthropomorphic RS NAO [7] by means of an analytical method. The method described in [1] was implemented for NAO's hands and head. Some changes in the solution are implemented for the legs of this RS. Each leg has 6 DoF. In order to obtain a less computationally complex solution, each limb viewed as a system consisting of two

* Corresponding author: iakovlev.r@mail.ru

simpler components. The 3 upper joints of each leg correspond only for the orientation of robot, and the three lower ones for the shift along the chosen orientation, which allowed obtaining IKP solutions in a simplified form. This approach is a heuristic, focused on simplifying the calculations while solving the IKP with analytical method, but at the same time the problem of multiple solutions was formally eliminated. However, eliminating the possibility of choosing a "better" solution leads to losses in terms of ensuring high energy efficiency of the IKP solutions.

This paper discusses the causes of multiple solutions of the IKP when analytical methods are used. The developed algorithm for improving the energy efficiency of mechanical motion, which is based on the modification of the process of choosing the most suitable solution, is also described.

2 Background

Consider the main/basic principles of the classical approach to solving the IKP. In order to describe the body's orientation in space, it is necessary to correlate a coordinate system with the body. One way to describe the coordinate system associated with the body is to express its three axes in the global coordinates system, and write them down as columns of a 3x3 matrix in the order [X, Y, Z]. Such a matrix is called the rotation matrix. Thus, the position of a body can be described by a vector, and its orientation by the rotation matrix.

Assume that the coordinate system *B* is connected with some body. In space, a global coordinate system *A* is defined, and vectors X_B, Y_B, Z_B define the principal directions of the coordinate system *B*. These vectors expressed in coordinates of system *A*, are written as ${}^AX_B, {}^AY_B, {}^AZ_B$. Then the rotation matrix describing the system *B* relatively to the system *A* is denoted as ${}^AR_B$ and has the following form:

$$ {}^AR_B = \left[{}^AX_B, {}^AY_B, {}^AZ_B \right] = \begin{bmatrix} r_{11} & r_{12} & r_{13} \\ r_{21} & r_{22} & r_{23} \\ r_{31} & r_{32} & r_{33} \end{bmatrix}. \quad (1) $$

In order to completely localize the joint of a certain robotic system, it is necessary to know its position and orientation. These two parameters can be combined in one called a frame of reference, which allows to completely describe one coordinate system relatively to the other.

2.1 Joints description in classical representation

Each link of a typical robotic system in terms of kinematics can be represented as a rigid body. It determines the relationship between the axes of adjacent joints of the system. The axis of the joint *i* is determined by the direction of the vector around which the *i*-th link rotates relative to the *(i – 1)*-th link.

The kinematic function of the link is to maintain a constant relationship between the axes of adjacent joints, which can be described by the following parameters: the length of the link *a* and the link's twist angle α. These parameters completely determine the relative position of the two axes in space.

For any two axes of the joints in three-dimensional space exists a well-defined measure of the distance between them, which is measured along a straight-line perpendicular to both axes and is denoted as a_i. This perpendicular always exists and it is unique except for the case when the axes are parallel. For parallel axes, there are many common perpendiculars of equal length.

The second parameter to be determined is the twist angle α_i of the link, which can be calculated as follows: take a plane for which the common perpendicular for the two axes is the normal, project the axes numbered *i-1* and *i* on this plane, then measure the angle between axes projections.

2.2 Description of the links connections in classical representations

Adjacent links of the robotic system have one common joint axis. One of the two parameters that make it possible to completely determine the way in which the links are connected to each other is called the shift of the link. This parameter characterizes the distance along the common axis from one link to the other. The shift along the axis of the i-th joint is denoted as d_i. The second parameter called the angle of the joint deflection and denoted as Q_i, is the amount of rotation of one link relative to its neighbor around the common joint axis.

Shift of the link corresponds to the distance along the axis of the *i*-th joint from the point at which this axis intersects the perpendicular α_{i-1} to the point of intersection with the perpendicular α_i. Angle of the joint deflection is the angle between the extensions of the perpendiculars α_{i-1} and α_i, measured around the joint *i*.

The position of the *i*-th link in the coordinate system of the link *i-1* can be expressed by composing the transformation matrix ${}^{i-1}_{i}T$, which general form looks as follows:

$$ {}^{i-1}_{i}T = $$

$$ = \begin{bmatrix} cQ & -sQ & 0 & a_{i-1} \\ sQ_i c\alpha_{i-1} & cQ_i c\alpha_{i-1} & -s\alpha_{i-1} & -s\alpha_{i-1} d_i \\ sQ_i s\alpha_{i-1} & cQ_i s\alpha_{i-1} & c\alpha_{i-1} & c\alpha_{i-1} d_i \\ 0 & 0 & 0 & 1 \end{bmatrix}, \quad (2) $$

where c and s denote the trigonometric functions of the sine and cosine, respectively.

Kinematic equations are determined after a proper coordinate system is assigned to each link. Given all the parameters, a transformation matrix is made for each of the links. By multiplying all the matrices, a transformation that can be used to express the final frame of reference {N} in the initial one {0} is found as:

$$ {}_N^0T = {}_1^0T\, {}_2^1T \ldots {}_N^{N-1}T \,. \tag{3}$$

With given final transformation it is possible to determine the coordinates, as well as the orientation of the end link in the original Cartesian coordinate system. The solution of the IKP consists in finding the unknown deviation angles $Q_1, Q_2, Q_3, \ldots Q_{N-1}$ of the links. Next, an analytical approach to solving this problem will be considered.

2.3. Analytical method for solving IKP

Firstly, in order to solve the IKP by means of analytical method all parameters for each link of the RS should be specified. When, the matrix ${}_N^0T$ is found according to the expression (2) and the link parameters. This matrix allows to express the position of the final link in the initial coordinate system, according to the formula 3, its general form is presented below:

$$\begin{bmatrix} a_{11} & a_{12} & a_{13} & a_{14} \\ a_{21} & a_{22} & a_{23} & a_{24} \\ a_{31} & a_{32} & a_{33} & a_{34} \\ a_{41} & a_{42} & a_{43} & a_{44} \end{bmatrix} . \tag{4}$$

Matrix components $a_{11} \ldots a_{44}$ are the result of multiplication of matrices ${}_1^0T\, {}_2^1T \ldots {}_N^{N-1}T$. In order to solve the IKP, it is necessary to find the unknown angles $Q_1, Q_2, Q_3, \ldots Q_{N-1}$ from the matrix equation, the general form of which looks as follows:

$$ {}_N^0T = \begin{bmatrix} a_{11} & a_{12} & a_{13} & a_{14} \\ a_{21} & a_{22} & a_{23} & a_{24} \\ a_{31} & a_{32} & a_{33} & a_{34} \\ a_{41} & a_{42} & a_{43} & a_{44} \end{bmatrix} = \\ = {}_1^0T(Q_1)\, {}_2^1T(Q_2) \ldots {}_N^{N-1}T(Q_N) \,. \tag{5}$$

Equation 3 may be reduced by division by ${}_1^0T(Q_1)$ to the form:

$$\left[{}_1^0T(Q_1) \right]^{-1} {}_N^0T = {}_2^1T(Q_2) \ldots {}_N^{N-1}T(Q_N) \,. \tag{6}$$

For matrix ${}_1^0T$ the inverse matrix ${}_0^1T$ is defined. And equation (4) is converted to the form:

$$\begin{bmatrix} c_1 & s_1 & 0 & 0 \\ -s_1 & c_1 & 0 & 0 \\ 0 & 0 & 1 & 0 \\ 0 & 0 & 0 & 1 \end{bmatrix} \begin{bmatrix} a_{11} & a_{12} & a_{13} & a_{14} \\ a_{21} & a_{22} & a_{23} & a_{24} \\ a_{31} & a_{32} & a_{33} & a_{34} \\ a_{41} & a_{42} & a_{43} & a_{44} \end{bmatrix} = \\ = {}_2^1T(Q_2) \ldots {}_N^{N-1}T(Q_N) \,. \tag{7}$$

Multiplying the matrices from the left side of expression (7) and further equating the corresponding elements of the resulting matrix equation we obtain twelve nonlinear equations, only six of which are independent. Consider the matrix ${}_N^0T$ without taking into account the fourth column and the fourth row:

$$\begin{bmatrix} a_{11} & a_{12} & a_{13} \\ a_{21} & a_{22} & a_{23} \\ a_{31} & a_{32} & a_{33} \end{bmatrix} . \tag{8}$$

Matrix (8) represents the relative position of the end coordinate system to the initial one, from which it is possible to identify three independent equations. The elements of the matrix ${}_N^0T : a_{14}, a_{24}, a_{34}$ are the coordinates of the radius vector, which connects the reference-starting points of the coordinate systems associated with the final link and the initial frame of reference, so three other independent equations can be obtained.

2.4 Appearance of multiple solutions of IKP

As it was shown earlier, the number of independent kinematic equations for any RS does not change and is equal to 6. Thus, finding specific solutions of the IKP by an analytical method is possible for a RS with number of DoF not exceeding 6. If the system has a greater number of joints (for example the one presented in [8]), using this method leads to the infinite number of solutions, as an attempt is made to solve a system with n (n> 6) unknown variables using only 6 equations. Moreover, since the kinematic equations are nonlinear, to obtain specific solutions, even for 6-joint RS, it is required that its construction meet certain criteria.

For example, the axes of a part of adjacent joints must be parallel or perpendicular to each other. In general, for a system with 6 joints it is impossible to guarantee analytic solvability. Thus, it should be noted that the analytical method of solving the IKP has a limited field of application. For complex RS, where there are no such structural features [9], alternative methods of IKP solution should be applied [10 - 13].

There is another feature of analytical methods for solving the IKP, called multiple solutions. When solving kinematic equations for a RS with a three-dimensional working space using these methods, in the end, several equivalent alternative solutions can be obtained. The ambiguity of the proposed positioning is explained by the presence of redundant DoF in the RS. These DoF make equal positioning of the end joint of the RS possible via several physical system configurations.

This situation in the concept of the analytical method leads to the fact that when solving certain kinematic equations, the values of the unknown angles can be expressed only as the values of the arguments of some trigonometric function with a known value. This fact leads to ambiguity. In particular, for a RS with 6 DoF for each target position can exist up to 16 alternative IKP solutions.

3 Multiple analytical solution

In presence of multiple solutions for some RS, it seems obvious that, depending on several factors, various solutions may be the most preferable. These factors, in particular, are: the shift of the positions of the joints, the presence of obstacles, the time of transition to an updated position and an energy consumption. Thus, random selection of some IKP solution is inexpedient, and the development of algorithms for finding optimal solutions that take into account various additional factors is a promising and urgent task.

In order to identify the key factors for this task it is necessary to analyze each of them. The causes of their occurrence, the degree of influence on the quality of the IKP solution, as well as the existence of an interrelation with other factors were researched.

When investigating the presence of obstacles, firstly it should be mentioned that the obstacles undoubtedly have an extremely high degree of influence on the quality of any proposed solutions to the IKP, since they are taken into account not at the stage of formation of the solution, but only after it. Thus, obstacles in essence are nothing but external constraints for RS.

Unfortunately, the obstacles from the viewpoint of kinematics are an absolutely unpredictable factor, and therefore its use goes beyond the problem of mechanical motion. Taking into account that such "external" constraints do not have interrelation with other factors of quality of IKP solutions, the obstacles were not chosen as a key factor, and are not further considered in the work.

Another factor is energy consumption, the causes of which are associated with a very high share of energy attributable to mechanical motion, or rather to a shift of the position of the RS joints. At the same time, the amount of consumed energy can be unstable, since energy costs fall not only on functioning of the propulsion devices, but may largely depend on the load value. Thus, the energy consumption function can be generally represented as:

$$E = F(M, R), \tag{9}$$

where F is a certain function, M is the value of the current load, R is the shift of the position. The presence of a direct relationship between the magnitude of the joints shift and the amount of used energy, was also reflected in the formula above.

Further, a study of the time consumption factor was made, the occurrence of which is associated with the structure of motion devices, which require a certain amount of time to perform some movement. The most important parameter of this type of device in this case is the rate of rotation, which mainly characterizes the time spent on carrying out a certain movement. Thus, the function, determining time consumption in general form can be represented as follows:

$$T = F(R) + T_r, \tag{10}$$

where F is a certain function, R is the shift of the position, T_r is the response time of the motion device. It should be noted that the time consumption dependence on the magnitude of the shift is direct, although it can have a nonlinear view. The influence of this factor on the quality of IKP solutions can vary greatly depending on the task, but in any case, it is essential.

The influence of the shift factor of the joint positions, the occurrence of which is determined directly by the presence of mechanical motion, was investigated. At the same time, taking into account the above dependences (9, 10), it can be concluded that this factor has a certain degree of influence, both on the energy consumption and on time costs. A shift increasing leads to increase of other factors. Thus, in this study, the key factors in assessing the quality of the IKP solution were energy and time costs, depending on the values of the shift of the joints.

The next stage is the formation of a total cost index for each joint separately. The expression reflects both the magnitude of the shift and the influence of the energy and time consumption factors:

$$e = R * (k_E + k_T), \tag{11}$$

where e is the total cost index, R is the position shift, k_E is the energy consumption factor, k_T is the time cost factor. Cost factors should be calculated for each specific RS, for this purpose a specific type of energy and time consumption functions or an empirical estimate of their values can be used.

Moreover, depending on the current target priorities, the influence of certain factors will be more or less significant, therefore, additional "importance" indices should be introduced, thus the expression (11) will take the following form:

$$e = R * (\beta_E k_E + \beta_T k_T), \tag{12}$$

where β_E is the energy consumption impact factor, β_T is the time cost impact factor.

Based on the research, an algorithm was developed to find optimal solutions for the IKP using the analytical approach. The algorithm is shown in Figure 1.

This algorithm receives information about all proposed solutions of the IKP, some of which may not satisfy the structure features of the RS, thus, at the first stage, the elimination of unacceptable solutions is implemented. Each proposed solution is considered separately: first, the shift of all joints of the system is calculated relatively to their current positions. Then, each joint is checked, whether the obtained shift value satisfies the allowable interval for the given joint. If this condition does not satisfy at least one of the joints, the current IKP solution is rejected as unacceptable.

Then the second key step of the algorithm is executed, which considers all the solutions that have passed the test for compliance with the constraints of the RS. Similar to the previous stage, each solution is considered separately: first of all, for each of the links the value of the total cost index is calculated in accordance with (12). The total cost index is then calculated for the whole solution, by summing the total cost indices for the individual joints.

After these calculations were carried out for all feasible solutions, the optimal solution of the IKP is defined as the solution with the minimum value of the total cost index.

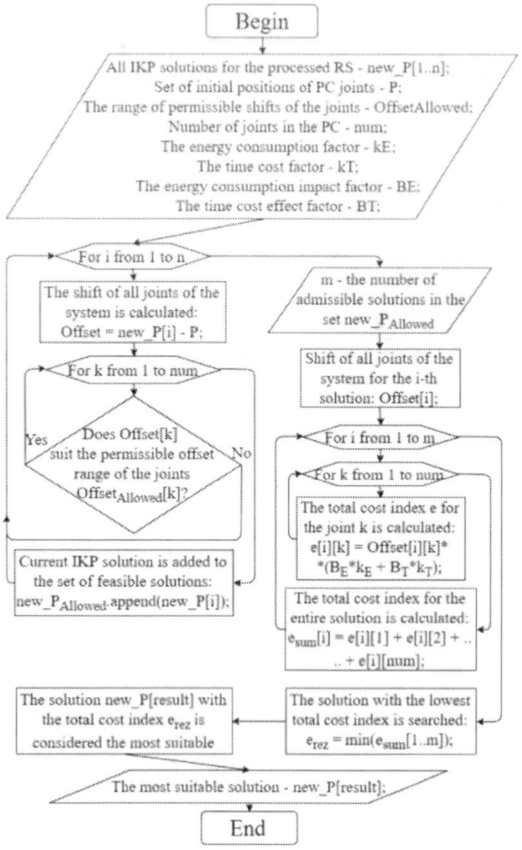

Fig. 1. Algorithm for finding optimal solutions.

4 Conclusion

The paper discusses a new algorithm for improving the quality (energy efficiency) of a RS using the modification of the selection process among multiple solutions using an analytical approach to solve the IKP.

Further research will involve series of practical experiments to evaluate the effectiveness of the developed algorithms on a number of RS, including anthropomorphic robots and multi-joint manipulators [14-16].

The reported study was funded by RFBR according to the research project №17-58-04110_Bel_mol_a and by BRFFR according to the research project №F17PM-035.

References

1. D. L. Peiper, Stanford Artificial Inteligence Laboratory, AIM-72 (1968)
2. R. P. Paul, B. Shimano, *1978 IEEE CDC*, 1398-1406 (1979)
3. R. L. Williams, *ASME 2012 IDETC/CIE*, 1187-1196 (2012)
4. I. Ha et al., *SICE Annual Conf.*, 2178-2181 (2011)
5. S. Kucuk, Z. Bingul, *Applied Mathematical Modelling*, **38**(7), 1983-1999 (2014)
6. N. Kofinas, E. Orfanoudakis, M. G. Lagoudakis, *IEEE 13th ICARS*, 1-6 (2013)
7. D. Gouaillier et al., *IEEE ICRA*, 769-774 (2009)
8. N. Pavluk, A. Denisov, A. Kodyakov, A. Ronzhin, MATEC Web of Conferences, EDP Sciences, **77** (2016)
9. N. Pavluk, A. Ivin, V. Budkov, A. Kodyakov, A. Ronzhin, *Interactive Collaborative Robotics, First International Conference ICR 2016, Budapest, Hungary, August 24–26, 2016, Springer International Publishing. LNAI 9812*, 113-123 (2016)
10. A. Aristidou, J. Lasenby, *Graphical Models*, **73(5)** (2011)
11. A. Colomé, C. Torras, *2012 IEEE/RSJ ICIRS*, 5333-5340 (2012)
12. L.Yu. Vorochaeva, A. S. Yatsun, S. F. Jatsun, SPIIRAS Proceedings, **3(52)** (2017)
13. A.A. Ivanov, O.A. Shmakov, SPIIRAS Proceedings, **49**, (2016)
14. A.S. Kodyakov, N.A. Pavlyuk, V.Yu. Budkov, R.A. Prakapovich, *IOP Conf. Series: Journal of Physics: Conf. Series*, **803**, 012074 (2017)
15. I I. Vatamaniuk, G. Panina, A. Saveliev, A. Ronzhin, *IEEE Int. Conf. on Autonomous Robot Systems and Competitions*, 306-310 (2016)
16. A. Denisov, V. Budkov, D. Mikhalchenko, *Interactive Collaborative Robotics, First International Conference ICR 2016, Budapest, Hungary, August 24–26, 2016, Springer International Publishing. LNAI 9812*, 69-78 (2016)

Generalized approach to bilateral control for EMG driven exoskeleton

Valery Gradetsky, *Ivan* Ermolov, *Maxim* Knyazkov and *Artem* Sukhanov*

Institute for Problems in Mechanics RAS, Laboratory of Robotics and Mechatronics, 119526 Moscow, Prospect Vernadskogo, 101-1, Russia

Abstract. The paper discusses a generalized approach to bilateral control for EMG driven exoskeleton systems. In this paper we consider a semi-automatic mechatronic system that is controlled via human muscle activity (EMG) level. The problem is to understand how the movement of the exoskeleton effects on the control. The considered system can be described in terms of bilateral control. This means the existence of force feedback from the object via the exoskeleton links and drives to operator. The simulation of the considered model was held on the MATLAB Simulink. The mathematical model of the bilateral system with exoskeleton and operator was developed. Transient functions for different dynamic parameters were obtained. It was shown that force feedback is essential for the R&D of such systems.

1 Introduction

The nature of the human exoskeleton function supposes operating in terms of the interference between links and the operator's extremities. The exoskeleton control meets the particular requirements for accuracy and safety of the movement. Motion of exoskeleton's links must ensure safety for operator and implement desired movement in response to specifying muscle activity as a function of a number of parameters.

Safety is essential for the development of an exoskeleton with high payload. Such systems are characterized by the presence of external forces and torques as disturbances that exceed human capabilities. Also the development of such type of exoskeleton system involves load limits [1, 2].

In this paper we consider a semi-automatic mechatronic system that is controlled via human muscle activity (EMG) level. The problem is to understand how the movement of the exoskeleton affects its control. The considered system can be described in terms of bilateral control. That means the existence of force feedback from the object via the exoskeleton links and drives to operator.

To improve the performance of an exoskeleton its drive system should give the operator the opportunity to feel reduced forces experienced by the exoskeleton itself. This is extremely important when the exoskeleton is used to move fragile or dangerous objects.

Unlike Master-Slave systems where two related manipulators are divided in space, the exoskeleton is always connected to an operator via construction elements. Thus it is essential to prevent oscillation in joints due to sudden change of rotation direction.

The PID-control is the most common technique to control the designed parameters in robotics. It is easy implemented and applied for different tasks such as force control, velocity control or angle rotation control. It does not require information of the system dynamic effects [3-5].

PID-control implies measurement between desired parameter and actual parameter like angle between links of an exoskeleton or current in its motor. The proportional part of the PID implementation describes an error with proportional factor K_P. It is used to reduce the difference between the desired parameter q_{des} and the actual one q_{re} with the rate of the factor.

$$P = K_P(q_{des} - q_{re}). \quad (1)$$

The integral part is used to reduce the residual steady-state error and it depends on the error rate.

$$I = K_I \int_0^t (q_{des} - q_{re})dt. \quad (2)$$

Here K_I is integral factor.

Derivative part predicts the behavior of the system and improves its stability and settling time.

$$D = K_D(\dot{q}_{des} - \dot{q}_{re}). \quad (3)$$

Here K_D is differential factor.

The final PID application in the control is obtained as sum of all three parts together:

$$Q = K_P(q_{des} - q_{re}) + K_I \int_0^t (q_{des} - q_{re})dt + K_D(\dot{q}_{des} - \dot{q}_{re}). \quad (4)$$

* Corresponding author: sukhanov-artyom@yandex.ru

2 The exoskeleton design and muscle model analysis

The construction is based on the exoskeleton designed in the laboratory of robotics and mechatronics of the Institute for Problems in Mechanics of RAS (Fig.1). The considered design of the exoskeleton contains a joint system of two links based on the DC-motor. Links of the exoskeleton are connected via clamps to the operator's arm at the forearm and the wrist [6].

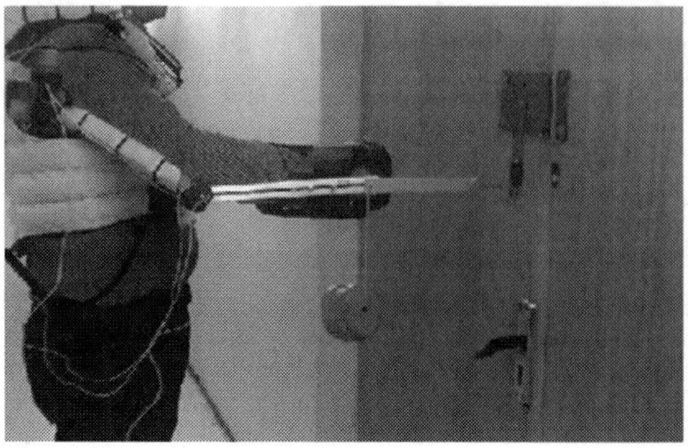

Fig. 1. The designed exoskeleton.

Every DC-motor is controlled by means of EMG-system. This system obtains data from operator's muscles and processes it according to Hill's model. The equation for tetanic contraction for the skeletal muscle is:

$$(v+b)(F+a)=\frac{av_0(F_0+a)}{F_0}. \quad (5)$$

Here F denotes the load in the muscle, v denotes the rate of contraction, a is experimental factor of shortening heat, v_0 denotes the maximum contraction rate ($F=0$) and F_0 is the maximum payload for the given muscle. This equation describes the tension as a function of contraction rate. Hill's muscle viscoelastic model is a representation of the muscle mechanical response on stimulus. It contains three elements (Fig. 2). The first one is contractile element (Bs) that means the acting of the force generated by the myosin and actin at the sarcomere. The parallel element (K_{PE}) is responsible for the skeletal muscle passive behavior when it is stretched. It represents the elastic fascia separating bundles of muscle fibers. The series element (K_{SE}) denotes the tendon of the myofilaments and provides energy storing for the muscle model.

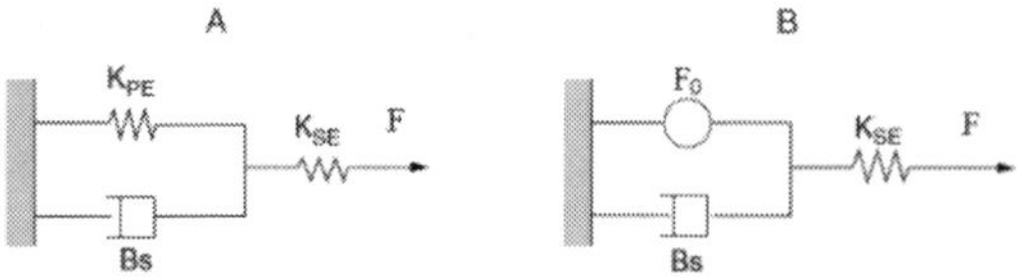

Fig. 2. (A) Hill's three-element muscle model. (B) Simplified Hill's muscle model.

In the simplified model the contraction force on damper can be obtained with the next equation:

$$F_0-F=Bv. \quad (6)$$

Here the parameter B is damper factor that can be obtained in the next equation:

$$B=\frac{a+F_0}{V+b}. \quad (7)$$

The experimental results for parameters a and b show their relation to the maximum contraction force and maximum velocity, respectively.

$$a=\frac{F_0}{4}, b=\frac{v_0}{4}. \quad (8)$$

The series element (K_{SE}) can be obtained by means of quick stretch experiments. In these experimental studies a pull is applied to a resting muscle at its resting length. It also depends on the area where strain is applied and differs from muscle to muscle.

The active skeletal muscle response can be found by obtaining the transfer function for the muscle model. The skeletal muscle force becomes the externally measured tension, F (Fig. 3).

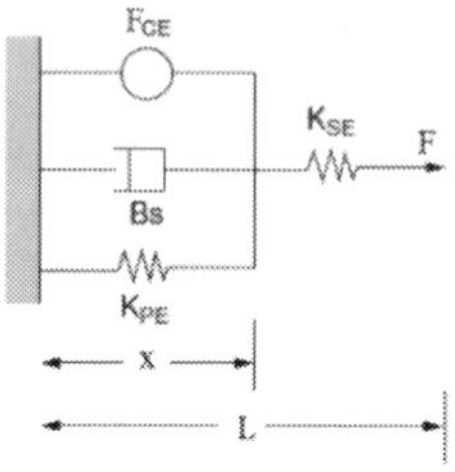

Fig. 3. Active muscle mechanical model.

The equilibrium equations need to be determined as follows.

$$\begin{cases} F_{CE}=K_{SE}L-(Bs+K_{PE}+K_{SE})x \\ x=L-\frac{F}{K_{SE}} \end{cases}. \quad (9)$$

Solving the system 9 gives the next equation:

$$F_{CE}=LK_{SE}-L(Bs+K_{PE}+K_{SE})+\frac{F}{K_{SE}}(Bs+K_{PE}+K_{SE}). \quad (10)$$

Thus we can obtain the external muscle force F:

$$F=\frac{LK_{SE}(B_s+K_{PE})}{Bs+K_{PE}+K_{SE}}+\frac{F_{CE}K_{SE}}{Bs+K_{PE}+K_{SE}}. \quad (11)$$

Let us assume that the maximum tetanic stimulus in the input to be a step function with amplitude F_{CE}. Applying Laplace transform to the step function we can obtain the next transfer function:

$$F(s) = \frac{F_{CE}K_{SE}}{Bs+K_{PE}+K_{SE}}\frac{1}{s} = \frac{1}{s}\frac{F_{CE}K_{SE}}{B}\left(\frac{1}{s+\frac{(K_{PE}+K_{SE})}{B}}\right). \quad (12)$$

In stress relaxation the contraction force is presented by transfer function:

$$F(s) = \frac{xK_{SE}}{K_{PE}+K_{SE}}\left(\frac{K_{SE}}{s+\frac{(K_{PE}+K_{SE})}{B}} + \frac{K_{PE}}{s}\right). \quad (13)$$

The time constant for the muscle activity can be performed as:

$$\tau = \frac{B}{K_{PE}+K_{SE}}. \quad (14)$$

3 Simulation of the model

The simulation of the considered model was held by means of MATLAB Simulink system. The model of the human arm in the elbow joint was implemented by two muscles – biceps brahii (Fig. 4) and triceps brahii (Fig. 5).

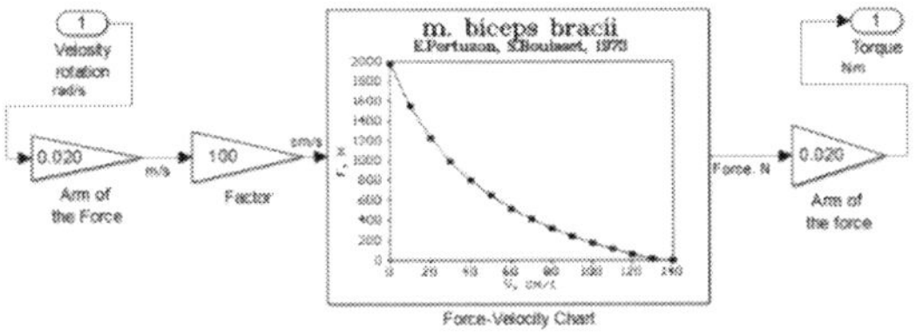

Fig. 4. Biceps brahii model.

Each muscle model represented experimental data obtained by different researchers and approved by similar experiments held in the laboratory of robotics and mechatronics of the Institute for Problems in Mechanics of RAS [9].

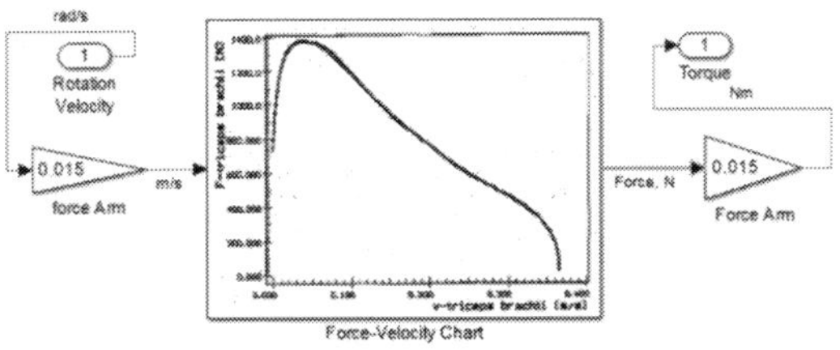

Fig. 5. Triceps brahii model.

Input data for each model was the desired rotation velocity represented by applying first order filter to position error. The idea was to initiate movement on existing position error with visual feedback. The first order filter was needed to simulate operator's reaction rate (Fig. 6).

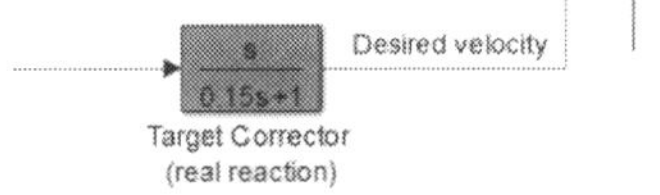

Fig. 6. Filter in reaction simulation.

Both muscles were joined in subsystem with condition checking. A muscle remained active until the desired movement corresponds to it (Fig. 7).

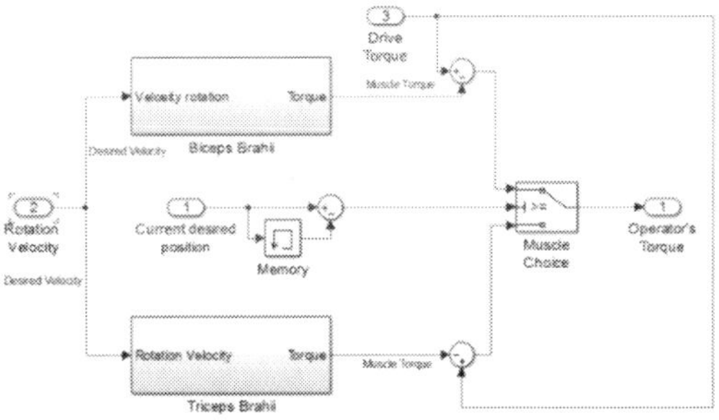

Fig. 7. Subsystem of two muscles.

The operator's torque is also affected by the drive's torque. This means that operator tries to reduce his efforts under the superior torque from the drive of the exoskeleton.

The drive is built with P-control of the shaft position, PI-control of the velocity and PI-control of the current. It has several loops, three internal feedbacks and two external feedbacks (Fig. 8). The internal feedbacks include position, velocity and current data. The external feedbacks include visual data and force information which is presented directly to operator. The visual channel transmits the position error which is transformed into desired velocity in muscle models [7, 8]. The force channel effects on the operator's force, reducing it.

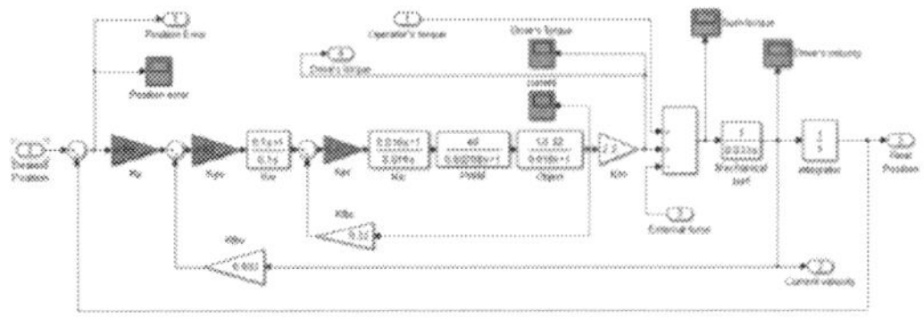

Fig. 8. Multiloop drive.

For the experiment it was decided to form a desired motion via trajectory as a function of time. The controlled parameter was the joined coordinate in the elbow. This angle was changing from 0 degrees to 50 degrees in 5 seconds and went back with similar principle to check work of different muscles (Fig. 9).

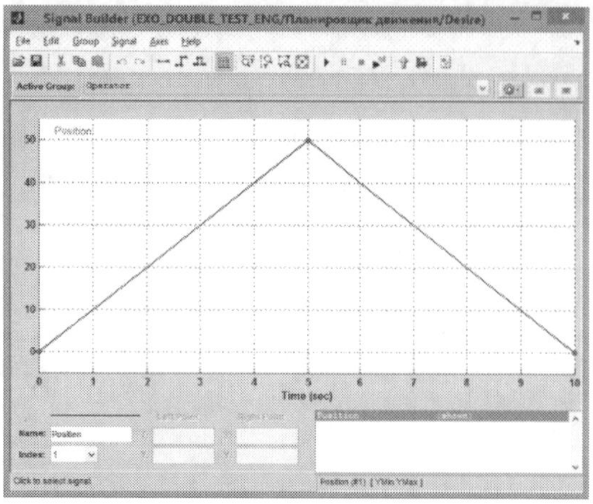

Fig. 9. The desired motion.

The results of the experiments are presented in Figure 10. They show that the real position of the drive repeats the desired parameter. The settling time for the system is 0.5 seconds which is a satisfying result. The overshoot is less than 1 percent which is also a satisfying result for the position parameter.

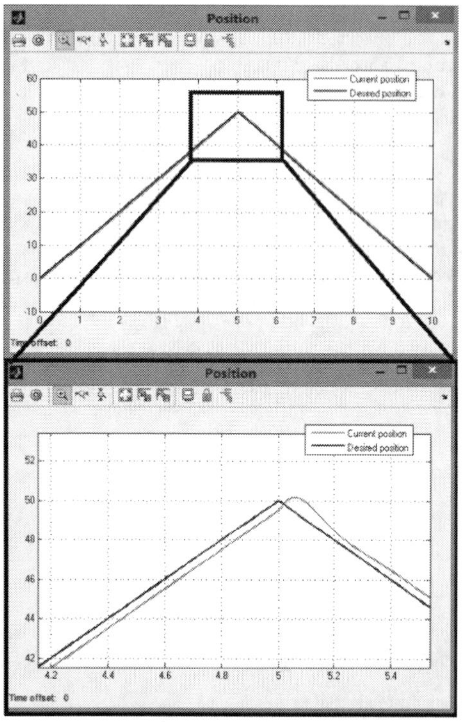

Fig. 10. The real motion in compare with desired motion.

Velocity transient functions are presented in Figure 11. The blue line denotes the desired velocity. That parameter shows how operator would feel the dynamics without exoskeleton device from his experience. Here the reaction filter is also used as a parameter. The orange line represents the resulting velocity for biceps brahii assistance (0-5 sec) and triceps brahii assistance (5-10 sec).

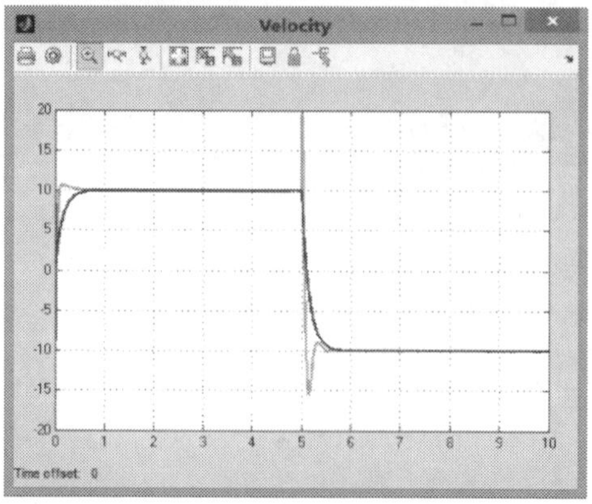

Fig. 11. The desired velocity (blue) and the real velocity of the drive (orange).

Figures 12 and 13 show the dynamics of drive and operator. Here peaks on the fifth seconds mean the change of the direction and sudden increase of position error. After half a second system turns stable.

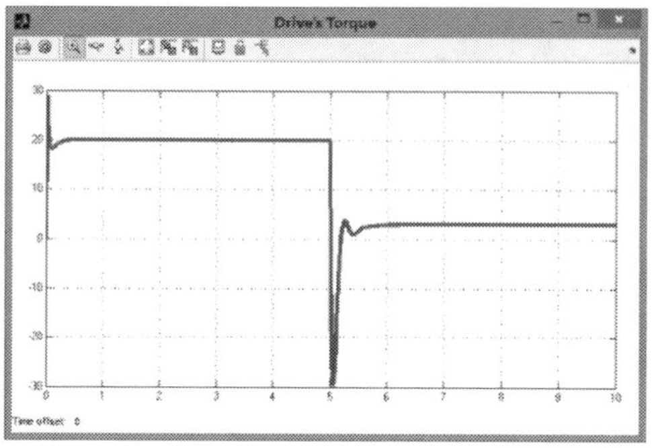

Fig. 12. The drive's torque.

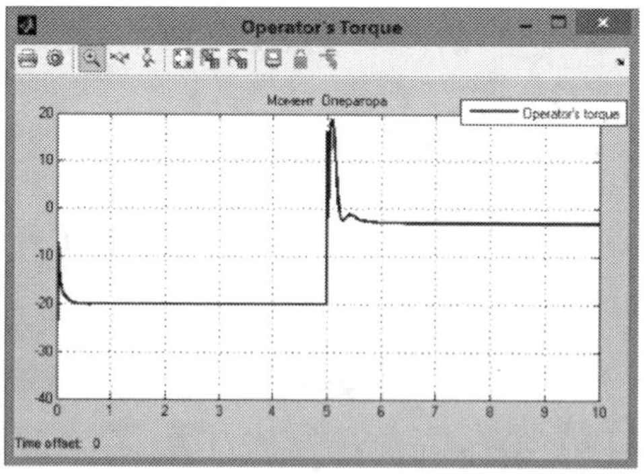

Fig. 13. The operator's torque.

Without the force feedback the system turns unstable and dangerous for the operator (Fig. 14). That is caused by oscillation in a joint. It happens when operator decides to change the direction of the movement. The controller gets the control signal from active muscle.

Physical interaction between operator and exoskeleton allows operator to feel the exoskeleton movement.

If the exoskeleton's link moves wrong direction due to inertia it may cause sufficient position error which leads to increase of a muscle activity. Thus the link with high acceleration and high moment of inertia tries to reach the desired position. From that moment an operator and the exoskeleton are moving in the same direction.

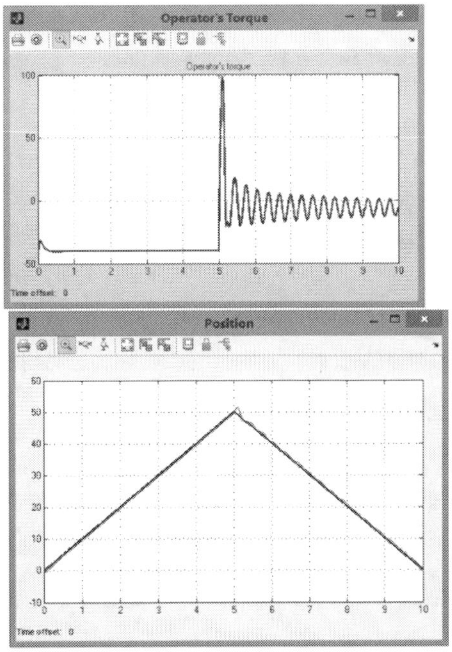

Fig. 14. The operator's torque (top) and the position (bottom).

Without force feedback the considered link of the exoskeleton outstrips the movement of the human extremity and causes pushing of the extremity. This may cause dangerous situation where operator would try to draw his extremity back from the exoskeleton link straining the active muscle. An increase of activity level of the muscle will lead to increase of the exoskeleton impact on the operator (Fig. 14 top) which is dangerous. That is why force feedback is essential for the R&D of such systems.

4 Conclusion

The paper discusses a generalized approach to bilateral control for EMG driven exoskeleton systems. The simulation of the considered model was held on the MATLAB Simulink. The mathematical model of the bilateral system with exoskeleton and operator was developed.

Transient functions for different dynamic parameters were obtained. It was shown that force feedback is essential for the R&D of such systems.

The reported study was funded by Program "Emerging Problems of Robotics" I.31, funded by Presidium of Russian Academy of Sciences.

References

1. K.J. Astrom, T. Hagglund, *Int. Soc. for Measurement and Control, 343p. Craig, J.J., 1986, "Introduction to Robotics: Mechanics and Control"*, (3.ed., Prentice Hall, 575, 1995)
2. C.-C. Yu, *Autotuning of PID Controllers: A Relay Feedback Approach*, (2ed. Springer, 261, 2005)
3. St. Clair, *Controller Tuning and Control Loop Performance*, (Straight-Line Control Co., 100, 1990)
4. P.Y. Li, *Amer. Control Conf.*, 3838–3842 (1998)
5. D. Lee, P.Y. Li, IEEE Trans. on Robotics **21**(5) (2005)
6. I.L. Ermolov, A.N. Sukhanov, M.M. Knyaz'kov, A.A. Kryukova, B.I. Kryuchkov, V.M. Usov, 22nd Saint Petersburg International Conference on Integrated Navigation Systems, ICINS 2015 – Proceedings, 181-185 (2015)
7. V.S. Kuleshov, N.A. Lakota, *Remotely controlled robots and manipulators*. Moscow: Maschinostroenie, 320 (1986)
8. V.S. Kuleshov, N.A. Lakota, *Dynamics control systems manipulators*, (Moscow: Energiya, 1971)
9. V.G. Gradetsky, I.L. Ermolov, M.M. Knyazkov, E.A. Semyonov, A.N. Sukhanov, *Mechanism and Machine Science*, Springer, Heidelberg, 275-281 (2014)

Analysis of balance control methods based on inverted pendulum for legged robots

Alexander Denisov*[1], *Roman* Iakovlev[1], *Ilhshat* Mamaev[2] and *Nikita* Pavliuk[1]

[1]SPIIRAS, Laboratory of Autonomous Robotic Systems, 199178 Saint-Petersburg, 14-th line of V.I., 39, Russia
[2]KIT, Institute for Anthropomatics and Robotics, Intelligent Process Automation and Robotics Lab, Engler-Bunte-Ring 8, D-76131, Karlsruhe, Germany

Abstract. Methods of balance control for a legged robot, the model of which is presented as a two-section inverted pendulum, are considered. The following balance methods for humanoid robots are analysed: the parallel algorithm of the network operator method; the method of natural synergies; the method of fuzzy control, the spherical inverted pendulum mode, a dual length linear inverted pendulum method. The best of these methods will be used in the development of the Russian anthropomorphic robot Antares.

1 Introduction

The development of walking robots is a process consisting of many tasks, among which one of the most important is to ensure the stability of the position and movement of walking robots. This task is especially important in the development of anthropomorphic robots that have only two points of support when moving in space.

The balance problem is actual for legged robot. A lot of different methods of solving this problem exist for robots with different number of legs. There are some of them:

In the paper [1] general Segway model under control based on Riccati equation for solving stabilization problems of four-legged robot RQuad during of walking, running, jumping is used.

In the paper [2] the transitional process of two-legged robot, which begins when the style of move is changed from walking to running and conversely is considered. Proposed approach based on anthropomorphic kinematics provides smooth move speed increase and robust movement when transition from walking to running appears.

In the paper [3] the walking control method for hexapod robot is considered. This method provides robust movements to different sides.

In the paper [4] modeling of movement for four-legged robot and visualization of its walking parameters, which provide to define main balance problems for legged robots, are executed in MatLab, SimMechanics.

In the paper [5] the planning of humanoid robot walking based on solving of contact planning problem is described. The results of testing this method on humanoid robot HRP-2 is a multi-contact motion by relaxing the contacts to occur with any robot and environment.

In the paper [6] an approach for online stabilization of anthropomorphic robot walking on uneven surface is presented. Robust control design for balance based on Zero Moment Point (ZMP), reference map, robust trajectories of center of mass are proposed. The result of method simulation is stability of walking and decreasing of trajectories errors.

In the paper [7] walking trajectory generation method, which includes free leg trajectory and ZMP methods, is considered. The method was tested on humanoid robot HRP-2. The test results show that balance was improved due to changing of ZMP at the expense of changing the foot position.

The purpose of this paper is to analyze methods for ensuring the sustainable position and movement of humanoid robots in space. On the basis of the analysis, the most suitable method will be proposed for the subsequent use in the Russian anthropomorphic robot Antares (Fig. 1).

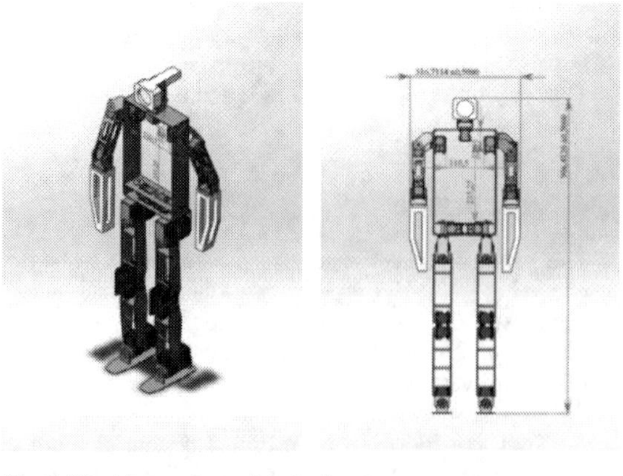

Fig. 1. The kinematic model of robot Antares.

* Corresponding author: denisov@iias.spb.su

MATEC Web of Conferences 113, 02004 (2017) DOI: 10.1051/matecconf/201711302004
12th International Scientific-Technical Conference on Electromechanics and Robotics "Zavalishin's Readings" - 2017

The robot Antares is unique in that it has two-motor knees, which provides larger angles of rotation to the engines, flexibility and plasticity of the leg assembly in comparison with analogues [8]. The two-motor knee greatly complicates the kinematic scheme, so it is necessary to carefully select the solutions ensuring the stability of the position of this robot [9].

2 Analysis of methods to ensure the stability of the walking robot

The result of contemporary balance control methods analysis for legged robots is five more perspective methods, which are considered in this section. This methods formalization, solved problems and necessary computing power are analyzed further.

2.1 The parallel algorithm of the network operator method

In the work [10] for solving the problem of stabilization of a bipedal walking robot on the basis of its inertial properties, a two-link inverse pendulum model was used to represent the robot. The idea of controlling such a pendulum lies in the redistribution of the angular momentum of the translational motion into rotational motion and vice versa. This approach allows reducing the moment on the side of the support and to control the sign and magnitude of the moment of gravity. The authors of the article regard a two-link inverse pendulum (shown in Fig. 2) as the control object, the hinge axes of which are parallel and allow the pendulum to move in the vertical plane.

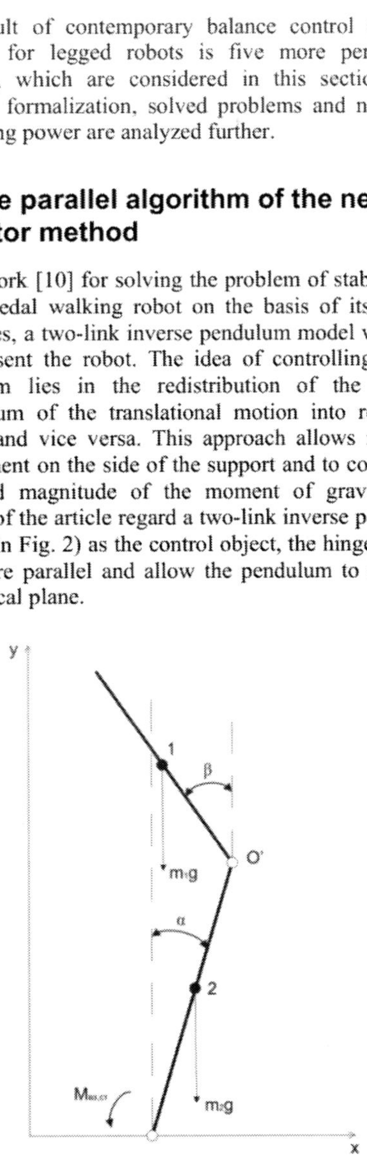

Fig. 2. Model of a two-link reverse pendulum.

In Figure 1, *l*1, *l*2 is the length of the rods; m1, m2 are the masses of rods 1 and 2, respectively. The mass of each rod is distributed evenly along the entire length. The control action is the angle of rotation of the rod 1 with respect to the vertical β.

The authors of the article proposed a formula for the mathematical model of the research object:

$$\ddot{\alpha} = \frac{1}{K_\alpha(\alpha,\beta)}\Big[-K_\beta(\alpha,\beta)\ddot{\beta} + K_{cm\alpha}\alpha + K_{g\alpha}\sin(\alpha) + \\ + K_{g\beta}\sin(\beta) + M_s + m_1\frac{l_1 l_2}{2}\left(\dot{\alpha}^2 - \dot{\beta}^2\right)sin(\alpha-\beta)\Big] \quad (1)$$

where M_s is the moment from the side of the support surface; K is the angular momentum; A, β - angles of rotation; $K_{\text{стα}}$ is a constant coefficient of stabilization with respect to the angle α.

The authors of the work assume that the kinematic control of the reverse pendulum is carried out by setting the desired angle (β_d). β_d differs from the real angle (β_r). The dynamics of the actuator drawing the angle β is described by the following formula:

$$\ddot{\beta}_r = -\frac{2\varepsilon}{T}\dot{\beta}_r - \frac{1}{T^2}\beta_r + \frac{K}{T^2}\beta_d \quad (2)$$

The authors of the paper constructed a pendulum model in the state space and carried out a control search in the form of u=g(x), which satisfies the constraints $u^- \le u \le u^+$ and ensures a minimum of functionals.

$$J_1 = \sqrt{\textstyle\sum_1^4 x_i^2(t_f)} + \textstyle\sum_{i=1}^4 \int_0^t x_i^2(t_f) \to min, \quad (3)$$

$$J_2 = t_f \to min, \quad (4)$$

$$t_f - \text{control time } t_f = \begin{cases} t, if\ \|x(t)\| \le \varepsilon \\ t^+, or\ else \end{cases},$$

where ε is the small positive value, t^+ is the maximum time of the control process.

To construct a network operator, the authors introduce four finite ordered sets of: the variables $X=(x_1,\ldots,x_p)$, the parameters $Q=(q_1,\ldots,q_R)$, the unary operations $O_1=(p_1(z),\ldots,p_w(z))$ and the binary operations $O_2=(x_0(z',z''),x(z',z''),\ldots,x_{v-1}(z',z''))$.

A network operator is an oriented graph that describes mathematical expressions. The elements from sets of parameters Q or variables X correspond to any node. A binary operation from the set O_2 corresponds to any node not being a source. An unary operation from the set O_1 corresponds to any curve of the graph.

In this paper, a network operator matrix was used. It was constructed using the adjacency matrix of the network operator graph to represent the network operator in the computer's memory. In the network operator matrix, off-diagonal non-zero elements indicate the numbers of unary operations, and the diagonal elements indicate the numbers of binary operations [11]. To calculate the result of a mathematical expression from the matrix of the network operator, the vectors of the node numbers of the input variables were introduced. The search for the network operator matrix was made using a genetic algorithm.

A numerical experiment was performed, as a result of which a parameter vector and a network operator matrix were obtained. The matrix corresponds to the following control law:

$$y = -x_1q_1 - x_2q_2 + x_3q_3 + x_4q_4 \quad (5)$$

As a result of using a parallel algorithm for synthesizing the stabilization system of a biped walking robot by the network operator method, a nonlinear control law was obtained with respect to the state coordinates, which satisfies the given quality.

2.2 The spherical inverted pendulum mode

In the paper [12], the spherical inverted pendulum model for the biped robot balance control is considered. It describes the system using the two rotational degrees of freedom at the ankle.

The authors described an energy-based control architecture for push recovery on humanoid robots. The idea of this method consists in transforming the SIP model into a critically damped system with the use of a single tuning parameter.

A stable walking gate was generated and controlled by the SIP model. The authors used the principles of passive dynamic walking. The SIP model was described as a hybrid dynamic system. Two stages of this dynamic system are continuous dynamic model for the single support phase and during the exchange of support instance and impact of the swing leg. To restore any lost energy an internal impulse was used.

The result of robot walking simulation is the vertical motion with the same periodicity as that of the human walk. This approach is different in that the motion of the CoM is not constrained to a constant height of the plane during a single-step motion resulting in a more natural overall motion for the entire humanoid body.

2.3 The method of natural synergies

The paper [13] considers the possibility of controlling the movement of a two-link anthropomorphic robot on the basis of coordination of natural synergies. The authors explore an approach demonstrating the advantages of motor control, which is based on the management of their own movements, which is a natural synergy. Motions occur along the eigenvectors of the dynamic equation of the mechanical system. These are combinations of joint angles and moments, which is an advantage, since it greatly simplifies the development.

To analyze the robot's dynamics, its model is considered in the form of a two-link reverse pendulum, where the upper link represents the body, head and hands, and the lower link - legs [9, 14]. In numerical calculations, the following parameters of the anthropomorphic robot were used: weight - 70 kg, height - 170 cm.

The authors consider the control of the robot motion in terms of natural synergies. With the help of the linear approximation, it is possible to decompose the motion into components directed along two eigenvectors, which allows us to derive two scalar equations describing the motion of a single-element inverse pendulum controlled by its PD controller.

$$\lambda_i \ddot{\xi}_i(t) = \eta_i^{con}(t), i = A, H \quad (6)$$

$$\eta_i^{con}(t) = \eta_i^{con}(t-\tau) + S_i^{eig}\left(\xi_i^{des}(t) - \xi_i(t-\tau)\right) - \\ -V_i^{eig}\dot{\xi}_i(t-\tau) - \lambda_i \ddot{\xi}_i(t-\tau) \quad (7)$$

In the case of a single-faced inverted pendulum – $\lambda = J/mgh$, $\eta = T/mgh$,

$S_i^{eig} = S/mgh$, $V_i^{eig} = V/mgh$, where J is the moment of inertia of the pendulum relative to the axis of rotation; m is its mass; h is the height of the position of the center of mass above the axis of rotation; g is the acceleration of gravity; T is the total moment of forces (except gravitational) relative to the axis of rotation; S and V are parameters of the PD controller.

In the work, stiffness and viscosity matrices were obtained in terms of joint angles. The resulting matrices are symmetric, but not diagonal. Thus, the authors conclude that the management of articular angles depends not only on the current values of a given angle, but also on the angles in all other joints, which determines the implementation of the PD controller, in which correction for each degree of freedom is determined by the values of the coordinates and their derivatives for all degrees of freedom.

The authors of this article carried out an analysis of the stability of control and an experiment on stability of the robot PostuRob II, the parameters of which are close to human. PostuRob II consists of a body and two legs. Connection between the elements of the robot is carried out by means of the hip and ankle joints [15]. During the experiment, signals from mechatronic sensors of joint angles came to the computer in real time.

The independence of the PD controller of each of the natural synergies allowed us to analyze the stability of the entire model through a separate analysis of the stability of the movement of each of these synergies.

The parameters of the PD controller have been calculated and tested at work. The selected parameters provide stability of movements for both the robot and the model. Thus, the movements of the lower segment, as required, compensated for the displacement of the center of gravity associated with the movements of the upper segment.

The study confirmed sustainable management on a real robot by the principle of independent control of each of the natural synergies with feedback loop parameters during a real experiment.

2.4 A dual length linear inverted pendulum method

In the paper [16] the method for obtaining walking parameters for biped robot moving along a slope is considered. The authors investigate a strategy to realize biped robot walking along a slope, when robot's centre of mass (CoM) is moved up or down during movement. This is because the height of the pendulum is kept at the same length on the left and right legs. Thus, one of this method purposes is to bring the CoM up to higher ground. In this method a dual length linear inverted pendulum method (DLLIPM) is used. DLLIPM is when a different height of pendulum is applied on the left and right legs of a robot.

As a result of using the method of walking along a slope based on DLLIPM and Newton Raphson algorithm, symmetrical motions and smooth walking trajectories were achieved and the maximum force fluctuation was reduced. In the future, the authors will expand the concept of DLLIPM for diagonal walking, turning on a slope and stairs.

2.5 The method of fuzzy control

The method of overcoming the "distance" between the theory of adaptive and robust control with the practice of its application, based on the transition to multi-alternative systems, is considered in [17].

The ideas of multi-alternativeness have biological analogies, defined in W. Ashby's Law of Requisite Variety [18], and also correspond to fuzzy control procedures of L. Zadeh [19].

The authors regard the inverted pendulum model as a typical member of structurally unstable objects. The model is fixed on the platform moving in a horizontal plane with some friction under the action of the external force (Fig. 3).

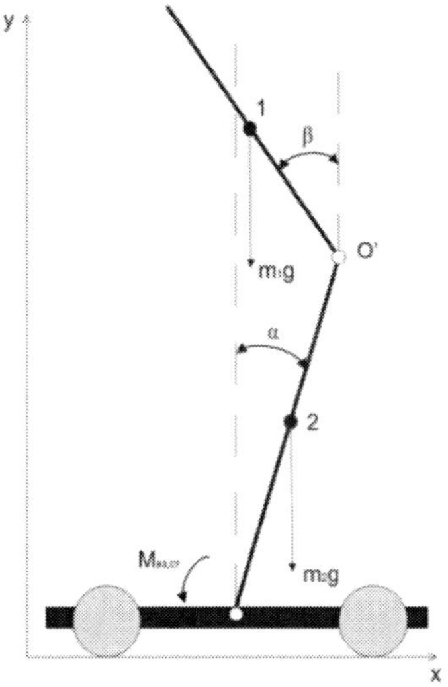

Fig. 3. Model of a moving platform with pendulum.

The movement of an unstable object is given as:

$$\begin{cases} (m_1 + m_2)\ddot{x} - m_2 L(\ddot{\alpha}\cos\alpha - \dot{\alpha}^2\sin\alpha) = F - F_{fr} \\ F_{fr} = \mu[m_1 g + m_2 cos\alpha(g \cdot cos\alpha - \ddot{x}\sin\alpha - \dot{\alpha}^2 L)\mathrm{sign}(\dot{x})], \\ -\ddot{x}\cos\alpha + L\ddot{\alpha} - g\sin\alpha = 0 \end{cases} \quad (8)$$

where m_1, m_2 are the masses of platform and inverted pendulum, kg; L is the length of pendulum, m; x is the coordinate of platform's horizontal location, α is the lack of perpendicularity angle, rad; g is the acceleration of gravity, m/s2; ω is the angular rate, rad/sec; μ is a dry friction coefficient of the platform; F_{fr} is the dry friction force, N; F is the imposed force to the platform, N.

For stabilization of the inverted pendulum the authors used the structure of the system with a regulator, on which the data about object condition is received at the input, and at the output - a signal generated into the control action on the object.

For the synthesis of the fuzzy regulator structure the authors of this article crate the catalogue (list) of linguistic and fuzzy input and output variables of the regulator; form the functions of these variables attachment; create solving rules of the fuzzy logic output determining the object control.

Multicriteriality is the main feature of creating the control algorithm. It is the control of two object coordinates, pendulum angular deviation α and the position of the platform x by one actuation. Thus, the idea of the movement control separation is used: when the angle and speed reach small values z_1^2 and z_2^2, platform condition control is being realized. In this case the angle and speed control becomes a priority, if α and α˙ go beyond the small range. As a result of applying this method, ten fuzzy solving rules were obtained. These rules form evident signs and modules for control values, which lead pendulum coordinates to desired condition, realize forced pendulum angular deviation to desired platform movement, lead platform to desired location x_0, form result control actuation which corresponds to the present object state.

As a result of the solved rules the authors received a fuzzy set:

$$\mu_u^*(y) = \sum_{j=1}^{10} \mu_{uj}^*(y) \quad (9)$$

Balance and robust properties of the synthesized system were tested on the inverted pendulum model, described earlier, with the parameters: m_1 = 100 kg, m_2 = 20...200 kg, L = 5 m, g = 9,81 m/sec2, μ = 0,05, where mass m_2 changed from 20 to 200 kg. Initial pendulum angular position is α(0)=0.2 rad, platform position - x(0)=0.

Stabilization is realized by corresponding platform movements. The time of the platform returning to initial position after ending of angular stabilization is 35 sec in both cases.

The received results confirm efficiency of fuzzy control by using multialternative systems in object control problems in critical modes.

3 Discussion

As a result, five different methods of balance control for legged robots were considered in the research. Also during the survey the main parameters, which should be taken during kinematic modelling was discovered (Table 1). Most of them are generally used for solving balance problem of legged robot.

We reviewed the parallel algorithm of the network operator method, the method of natural synergies, the method of fuzzy control, the spherical inverted pendulum mode and a dual length linear inverted pendulum method (DLLIPM).

Table 1. General parameters.

Physical parameter	Symbol	Unit
Body mass	m	kg
Body length	l	m
Body inertia	J	$Kg*m^2$
Gravity	g	m/s^2
Friction	F_{fr}	N
Angle	β	rad
Turning angle	$β_r$	rad
Angular velocity	ω	rad/sec
Coordinates	x, y	m

The robot model was presented as an inverted pendulum, which corresponds to an unstable position in the space. All the methods were tested on the robot model and proved to be efficient. Conclusions and summary about analyzed methods are presented in Table 2.

Table 2. The analyzed methods of balance control.

№	Method	Solving problem	Algorithm and solving complication (resources consumption)	Time of inverted pendulum balancing
1	the parallel algorithm of the network operator method	Balance control of humanoid robot walking	normal	1.6 sec
2	the spherical inverted pendulum mode	Balance control of humanoid robot walking	normal	2.1 sec
3	the method of natural synergies	Balance and motion control of humanoid robot standing	high	1.2 sec
4	the method of fuzzy control	Balance control of humanoid robot standing	high	1.7 sec
5	DLLIPM	Balance control of humanoid robot walking along slope	normal	1.8 sec

The method presented in [13] was tested on the real robot and showed the best balance control of a humanoid robot. However, this method is very resource-intensive one. It requires a lot of computational capability, so it cannot be used in embedded systems such as the anthropomorphic robot Antares. Given that, the method presented in [10] will be used in the development of Antares. This method showed the second result in balance control tests. The advantages of the method include operation speed; the ability to control the moment value of gravity force and to decrease the support moment.

4 Conclusion

The balance problem for walking legged robots is considered. The perspective methods for kinematic control are selected. The main parameters influencing for stable motion of legged robot are discovered. The future work will be related to the integration of the programming solutions of balance control method to the robotic core of the robot Antares [20-22].

The presented work was supported by the Russian Science Foundation (grant No. 16-19-00044).

References

1. E. Najafi, G.A.D. Lopes, R. Babuska, *19th World Congress IFAC*, 2177-2182 (2014)
2. M. Shahbazi, G.A.D. Lopes, R. Babuska, *19th World Congress IFAC*, 2171-2176 (2014)
3. D. Sarkara, S.K. Dubeya, A. Mahapatrab, S.S. Roy, *2nd International Conference ICIAME 2014,* 93-99 (2014)
4. Z. Shia, Q. Zhang, J. Tu, Z. Wang, X. Zhang, *International Workshop IWIEE*, 3250-3255 *(2012)*
5. A. Escande, A. Kheddara, S. Miossec, Robotics and Autonomous Systems. **61**, 428-442 (2013)
6. N. Wu, C. Chew, A. Poo, R. Li, *6th IFAC Sym. on Mech. Sys*. Hangzhou, China, 577-582 (2013)
7. K. Nishiwaki, S. Kagami, *IFAC Proceedings Volumes*. **39**, 675-700 (2006)
8. A.S. Kodyakov, N.A. Pavlyuk, V.Yu. Budkov, R.A. Prakapovich, IOP Conf. Series: Journal of Physics: Conf. Series. **803**, 012074 (2017)
9. N. Pavluk, A. Ivin, V. Budkov, A. Kodyakov, A. Ronzhin, *First International Conference ICR 2016,* 113-123 (2016)
10. B.B. Kulakov, A.S. Zharkova, D.E. Kazaryan, RUDN Journal of engineering researches. **4**, 26-36 (2010)
11. I. Vatamaniuk, G. Panina, A. Saveliev, A. Ronzhin, *IEEE Int. Conf. on Autonomous Robot Systems and Competitions,* 306-310 (2016)
12. A. Elhasairi, A. Pechev, Front. Robot AI, **2**(21) (2015)
13. A.V. Aleksandrov, T. Mergner, A.A. Frolov, G. Hettich, I.P. Muravev, Russian Journal of Biomechanics. **17**, 1(59), 94-109 (2013)

14. A. Ronzhin, I. Vatamaniuk, N. Pavliuk, *2016 Int. Conf. and exposition on electrical and power engineering*, 926 (2016)
15. L. Assländer, G. Hettich, T. Mergner, Human Movement Science. **41**, 147-164 (2015)
16. F. Ali, A.Z. Hj. Shukor, M.F. Miskon, M.K.M. Nor, S.I.Md Salim, International Journal of Advanced Robotic Sysems. **10**(377) (2013)
17. E.M. Vasiljev, D.M. Prokofjeva, Fuzzy control of structurally unstable objects. VSTU Journal. **8**, №10 (1), 8-12 (2012)
18. W.R. Ashby, *Publishing of foreign literature*. 432 (1959)
19. L.A. Zadeh, *Outline of a new approach to the analysis of complex systems and decision processes.* (M.: Znanie. 64 1974)
20. L.Yu. Vorochaeva, A.S. Yatsun, S.F. Jatsun, SPIIRAS Proceedings. **3**(52) (2017)
21. A. Ronzhin, A. Saveliev, O. Basov, S. Solyonyj, Springer-Verlag Berlin Heidelberg, A. Ronzhin et al. (Eds.): ICR 2016, LNCS/LNAI, 9812 (2016)
22. A. Ivin, D. Mikhalchenko. *21st Fruct Conf.* (2017)

Algorithm for automated calculation of a segmented electromechatronic module of robotic system motion

Mikhail Shepelenko* and *Roman* Meshcheryakov

Tomsk State University of Control Systems and Radioelectronics, Lenina str. 40, 634050 Tomsk, Russia

Abstract. The paper deals with the design of digital models of the elements of the electromechatronic part of motion module of robotic systems, which provides increasing productivity of the electromechatronic components of robotic and mechatronic manipulators. The developed software provides 3D-modeling of the electromechatronic part; calculation of electromechatronic and electrical elements and visualization of the electrical machine part. It reduces the cost of designing electromechatronic components of manipulators of robotic and mechatronic systems.

1 Introduction

Modern studies of the creation of electromechatronic modules of motion (EMM) of robotic systems with a segment inductor (inductors) have shown that it is possible to automate development of the EMM assortment range quickly and with a minimum of labor. In modern conditions, when choosing computer-aided design (CAD) systems, users are increasingly paying attention to the possibilities provided by the system in the field of parameterization, and enterprise managers already imply not only the transfer of the process from manual production of drawings to the computer.

The requirements for modern computer-aided design systems are not limited to the convenience of creating graphic primitives and drawing up drawings in accordance with GOST; on the agenda there appear the capabilities of systems in the field of constructing associative models and assemblies; the degree of simplicity in creating their own parametric libraries is estimated; Use the enterprise's existing know-how. Today, designing on the basis of existing blanks (design based on knowledge bases) makes it possible to significantly reduce the time of the project and the number of errors, which often plays a decisive role in a highly competitive environment.

A range of electromechatronic modules of motion has been developed by Elektromechatronic Systems LLC: segmental, arc and linear EMM, which are the basic elements of a gearless electric drive that rotates along an arc or a circle, as well as linear movements of the operating elements of various devices and mechanisms (Fig. 1).

The research task was to develop modern methods of automated calculation, since the main requirement in the modernization of industry and import substitution is the accelerated design of robotic and mechatronic systems for products used in such spheres as engineering, medicine and radio electronics [1-7].

2 Parametric modeling

Methods of parametric modeling (design) and prototyping in CAD, PDM-systems were considered. Precisely parameterization allows solving constructive schemes and variants in a short time, which in future will help avoid fundamental mistakes in the design process by changing variables in functions, parameters and geometrical relationships. Parametric modeling is significantly different from 3D modeling. In the case of parametric design, the designer creates a mathematical model of elements with parameters that change the configuration of the part, the mutual movement of parts in the assembly, etc [8-10].

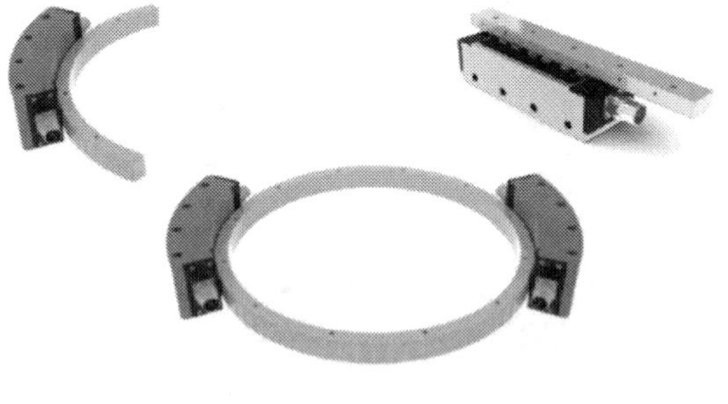

Fig. 1. EMM assortment range.

The idea of parametric modeling appeared at the early stages of the development of CAD systems, but for a while it could not be implemented because of insufficient computer capacities in those years. The first known CAD with the possibility of parameterization came out in 1989. Pioneers were Pro / Engineer (three-dimensional solid-state parametric modeling) of

* Corresponding author: mixan_19@mail.ru

Parametric Technology Corporation and T-FLEX CAD (two-dimensional parametric modeling) of Top Systems.

There are several types of parameterization: tabular, hierarchical, variational (dimensional) and geometric. In our case, all types are involved. But for the most part - variational and geometric parameterization. Because parameterization is based on the construction of sketches (with overlapping of various parametric links on the sketch objects) and the imposition of constraints by the user in the form of a system of equations that determine the relationships between the parameters. This in its turn makes it possible to easily change the shape of the sketch or the size of the parameters of operations; this makes it convenient to modify the digital model. As for the geometric, in it the geometry of each parametric element is recalculated depending on the position of the parent objects, its parameters and variables. It should be noted that geometric parameterization allows for more flexible editing of models. If there is a need to make an unplanned change, then it is not necessary to delete the original construction lines in the geometry of the model. After the design concept is selected, the designer begins to create the original model of the product. Here, along with parameterization, another invention, solid-state modeling, comes to its aid. The basic idea of solid-state modeling is to always guarantee a physically consistent representation of geometric volumetric bodies and to operate with the concepts of adding / removing material. A correct solid body contains an internal volume bounded by the outer surface of the body. This representation allows you to determine the body volume, its mass, moments of inertia, center of gravity, etc. These parameters are often critical in evaluating the design effectiveness of the product and are extremely important for the designer [11-13].

Based on the method of parametric construction of the model of the elements of the EMM, it was decided to develop a subsystem for CAD Autodesk Inventor for the automated calculation of the segmented EMM (SEMM). In it, it will be necessary to work out the algorithm of interaction of the subsystem itself with CAD, and also to envisage and develop each stage of creation of all elements that are part of the EMM.

3 Generation of element parameters of EMM

The developed program "Automated calculation of the product range of the electromechatronic part of electromechatronic modules of motion" is operated in Windows 7/8 x64 with Autodesk Inventor 2014 installed. The software package is developed in the language of Visual Basic .NET (VB.NET). The program is designed to interact with CAD Autodesk Inventor 2014. Input data are models. Reference models are parts and assemblies made in Autodesk Inventor 2014 using the Ilogic tool.

The models consist of 6 parts: the inductor case, the coil, the covers, the magnet, the rotor housing, the rotor magnet and four assemblies: the inductor, the rotor magnet core, the rotor and the SEMM calculation of the product range of the electromechatronic part of electromechatronic motion modules" designed to increase the degree of automation of the design of EMM for manipulators of robotic systems. The subsystem of the automated calculation of the electromechatronic part extends the functionality of the basic CAD Autodesk Inventor - a system of three-dimensional solid and surface parametric design, designed to create digital prototypes of industrial products and provide: the calculation of electrical elements and automatic input to the basic CAD Autodesk Inventor algorithms for calculation of electromechatronic parts with inductor for 3D visualization of the layout of electromechatronic parts (Fig. 2).

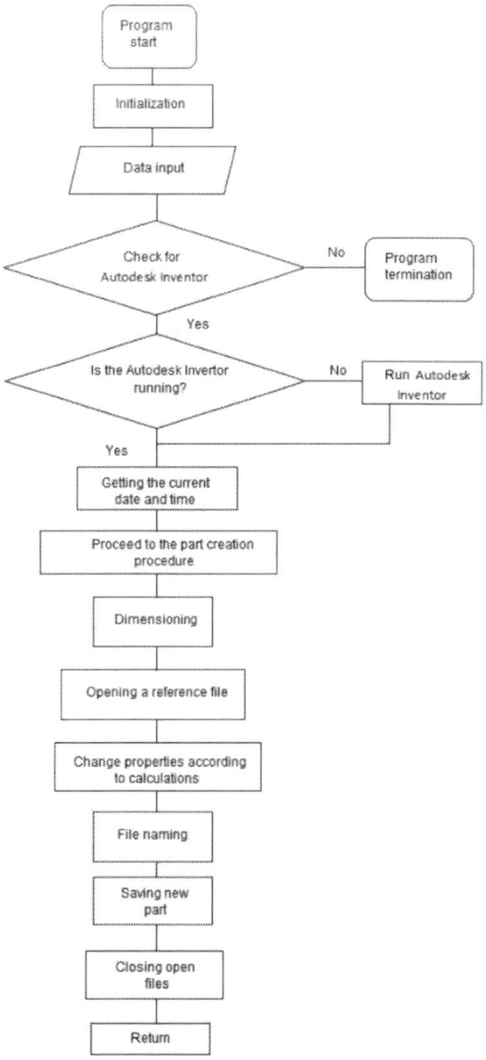

Fig. 2. The block diagram of the program.

The main function of the program is the generation of 3D models of parts and assemblies of parts of electromechatronic motion modules within a given range of products. The program starts with the launch of the

MyInventorPlug-in.exe program file. When the program starts, the main program window is displayed (Fig. 3).

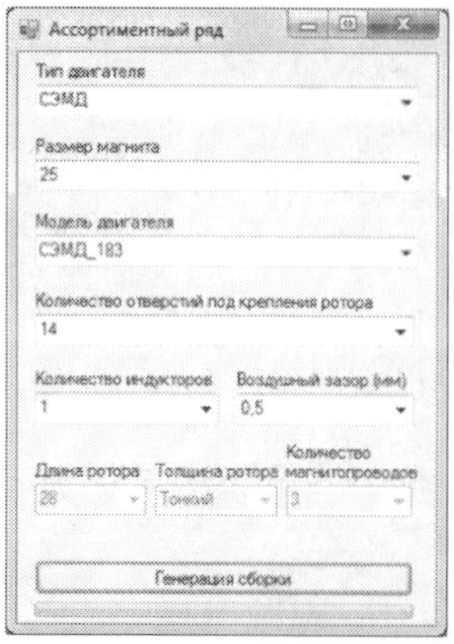

Fig. 3. The main program window.

To generate parts and assemblies of the range of the electromechatronic part of the electromechatronic motion module, the user can first select a number of parameters for generating. To do this, in the "Magnet size" field from the drop-down list, you must select one of the size values for the magnet: 25 or 50. Next, in the "Engine type" field, you can select the appropriate engine type from the drop-down list. The program will automatically set the necessary geometric parameters corresponding to the selected type. If the user did not select anything or chose not all values, the default values are used as the remaining values. To generate details and assemblies with the set values, the user should click the "Generate assembly" button [1].

For example, consider the creation of an inductor housing SEMM. Figures 4 and 5 show the window of the parameters of the functions of the inductor housing of the SEMM and the digital model of the inductor housing.

3.1 Generation of the inductor housing

Part generation occurs in the CreateInductor () procedure (Fig. 6). Based on the user-selected magnet size, the actual size is calculated. Then, a reference sketch of the inductor case is opened (all reference sketches are stored in a folder named InvModel). From the array of standard sizes, the dividing and outer diameters of the inductor housing are programmed programmatically. The internal diameter of the inductor case corresponds to the divisor diameter, increased by 0.5.

A sketch of the inductor case with the required dimensions is created by software making changes to the master sketch: the parameter that is responsible for the height of the inductor case, and the previously dimensioned height is assigned, and the required external and internal diameters are assigned to the sketch. Then, the received sketch of the inductor case is saved with the file name assigned to it in the format: Part name_type of the engine_the current date on this computer in the format yyyy.MM.dd-HH.mm.ss. The file is assigned the .ipt extension.

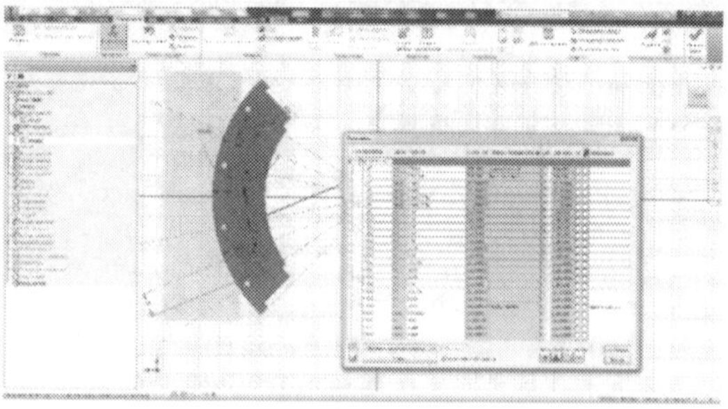

Fig. 4. Creating inductor housing.

Fig. 5. Inductor housing.

```
    ' Создание индуктора
    Public Sub CreateInductor()

        Dim _magnetSize As Double = cmbMagnetSize.SelectedItem / 10

        Dim _fullPath As String = System.IO.Path.GetFullPath(_openModelPath &
"Inductor.ipt")

        Dim _oDoc As Document
        _oDoc = _invApp.Documents.Open(_fullPath)

        Dim _oPartDoc As PartDocument = DirectCast(_invApp.ActiveDocument, PartDocument)

        Dim _oParam As Inventor.Parameters = _oPartDoc.ComponentDefinition.Parameters

        _oParam.UserParameters(5).Value = CObj(_magnetSize)
        _oParam.UserParameters(1).Value =
CObj(_engineTypes(cmbEngineType.SelectedIndex).outerRad)
        _oParam.UserParameters(2).Value =
CObj(_engineTypes(cmbEngineType.SelectedIndex).dividingRad + 0.5)
        _oParam.UserParameters(3).Value =
CObj(_engineTypes(cmbEngineType.SelectedIndex).pairsOfPoles)

        _oPartDoc.Update()

        _scatchName(4) = System.IO.Path.GetFullPath(_saveSketchPath) & "\" &
FileNameDate("Индуктор") & ".ipt"
        'Сохранить и закрыть производную деталь.
        Call _oDoc.SaveAs(_scatchName(4), False)

        _oPartDoc.Close()

    End Sub
```

Fig. 6. Fragment of the program listing.

3.2 Generation of the steel ring of rotor

Part generation occurs in the CreateRotor () procedure. Based on the size of such a large number of rotor rings. From the array of standard sizes, the dividing and internal diameters of the rotor are programmed programmatically. The outer diameter of the rotor corresponds to the dividing diameter, reduced by 0.5. Then the reference sketch of the rotor is opened (all reference sketches are stored in a folder named InvModel). A sketch of the rotor with the required dimensions for making changes to the master sketch: to obtain the required size and the required size, as well as the required outer and inner diameters. Number of holes for fastening. Now it stores the data in a format. The file is assigned the .ipt extension. The file is saved in the folder with the name SEMD_size of the engine \ thumbnails (Fig. 7).

Fig. 7. The steel ring of rotor.

After the generation of all the details has ended, the generation of assemblies of the SEMM that are part of the main assembly begins, such as: the rotor, the inductor magnet core, the inductor and the SEMM itself.

3.3 Generation of the rotor

Generation of the assembly occurs in the procedure CreateAssemblyRotor (). Opens the master assembly of the rotor, containing links to the sketches of the rotor and magnet (all reference assemblies are stored in a folder called InvModel). The reference to the assembly elements containing the rotor and magnet sketches is assigned new values from the array _scatchName, after which the corresponding assembly elements begin to reference the newly created sketches. Then the parameter responsible for the number of magnets on the rotor is assigned the current value corresponding to the number of magnets for this type of motor. The resulting rotor assembly is saved with the corresponding file name assigned to it (Fig. 8). The file is assigned the extension .iam. The file is saved in the folder with the name SEMM_size engine \ Assemblies.

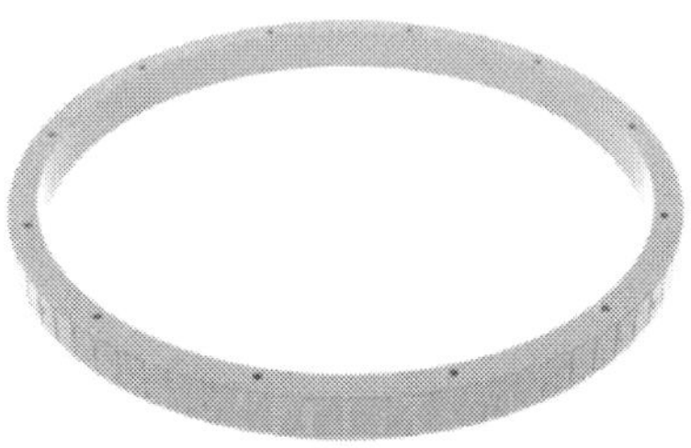

Fig. 8. Rotor SEMM.

3.4 Generation of the magnetic core of the inductor

Generation of the assembly occurs in the procedure CreateAssemblyMagnetInductor (). A reference assembly of the inductor magnetic core is opened, containing references to the coil and core sketches (all reference assemblies are stored in a folder named InvModel). A reference to the assembly elements containing coil and core sketches is assigned new values from the array _scatchName, after which the corresponding assembly elements begin to reference the newly created thumbnails. The obtained assembly of the magnetic circuit is stored and assigned the extension .iam. The file is saved in the folder with the name SEMM_size engine \ Assemblies. The name of the assembly files is stored in the _assemblyName data set for later use.

3.5 Inductor generation

Assembly generation occurs in the CreateAssemblyInductor () procedure. A reference assembly of the inductor is opened, containing references to the sketch of the inductor case, the cover of the magnetic coil and to the assembly of the inductor magnetic core (all reference assemblies are stored in a folder named InvModel). The reference to the assembly elements containing the inductor body and the rotor magnet lead are assigned new values from the _assemblyName array, after which the corresponding assembly elements begin to reference the newly created models (Fig. 9).

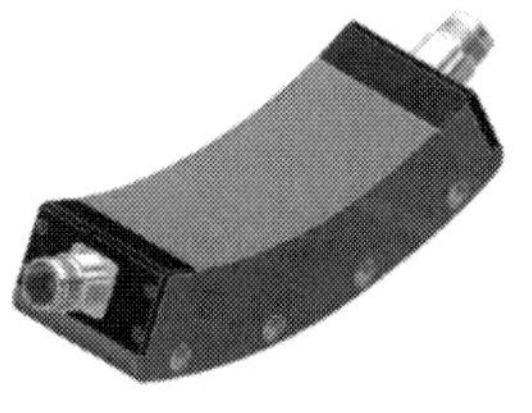

Fig. 9. SEMM inductor.

The obtained inductor assembly is saved with the file name assigned to it in the format: Assembly name_type of the engine_the current date on this computer in the format yyyy.MM.dd-HH.mm.ss. The file is assigned the extension .iam. The file is saved in the folder with the name SEMM_size engine \ Assemblies. The assembly file name is stored in the _assemblyName data set for later use.

3.6 Generation of SEMM

Assembly generation occurs in the procedure CreateAssemblySEMM (). The reference assembly of the SEMM is opened, containing references to the inductor and rotor assemblies (all reference assemblies are stored in a folder named InvModel). A reference to the assembly elements containing the inductor and the rotor are assigned new values from the array _assemblyName, after which the corresponding assembly elements begin to reference the newly created assemblies. After that, the obtained ECM assembly is saved with the appropriate file name assigned to it. The assembly file name is stored in the _assemblyName data set for later use (Fig.10).

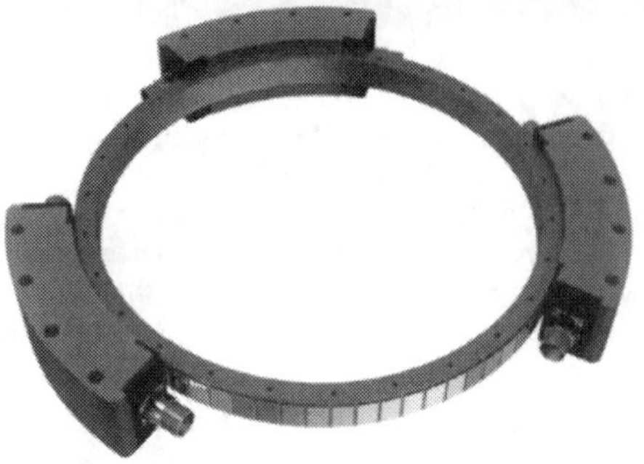

Fig. 10. The main assembly of SEMM.

After executing these procedures, the program waits for further user actions. The user can select new parameters or leave the previous ones and regenerate all the parts and assemblies or can close the program.

4 Conclusion

The software designed for calculating the assortment of EMM elements is developed. The program allows you to specify the input data and calculate the optimal values of the elements of the EMM at the output. User input of information about the initial parameters of the projected electromechatronic parts is provided by the interface module, which allows the user to create assembly structures. The program allows for: 3D-modeling of the electromechatronic part; calculation of electromechatronic and electrical elements and visualization of the electrical machine part; reduction of the cost of designing electromechatronic components of manipulators of robotic and mechatronic systems. The program allows subsequent docking with CAD Autodesk Inventor Professional to automatically obtain design documentation for ESKD.

References

1. L. Chubraeva, A. Ronzhin, A. Shyshlakov, Al. Ronzhin and V. Shyshlakov, SPIIRAS Proc., **33**(2), 207-226 (2014)
2. O.Yu. Osipov, Yu.M. Osipov, S.V. Shcherbinin, *Multicoordinate electromechatronic motion systems: monograph*, Tomsk: Publishing house TUSUR, - p. 320 (2010)
3. *General idea of CAD Autodesk Inventor* [Electronic resource]. - Access mode: http://www.autodesk.ru/adsk/servlet/index?siteID=871736&id=22741018
4. Yu.M. Osipov, P.K. Vasenin, D.A. Medvedev, Reports of TUSUR, **1**(17) (2008)
5. Yu.M. Osipov, B.Ya. Vologdin, Reports of TUSUR, **2**(22) (2010)
6. Yu.M. Osipov, S.V. Shcherbinin, Mechatronics, automation, management. **8** (2012)
7. I. Kalyaev, A. Dordopulo, I. Levin, A. Fedorov, SPIIRAS Proc., **50**(1), 5-31 (2017)
8. *Using the PDM-system in the design of technological processes* [Electronic resource]. - Access mode: http://de.ifmo.ru/bk_netra/page.php?dir=2&tutindex=4&index=54&layer=1
9. A.G. Kiselev, *Project management for the development and implementation of information systems (PMI)*: a textbook, for undergraduates. – (Novosibirsk, 2010)
10. M.G. Shepelenko, A.N. Lunin, *Scientific session of the TUSUR. Materials of the reports of the All-Russian scientific and technical conference of students, graduate students and young scientists*, Tomsk, - Tomsk, V-Spectrum (2008)
11. M.G. Shepelenko, *Scientific session of TUSUR. Materials of the reports of the All-Russian scientific and technical conference of students, graduate students and young scientists*, Tomsk, V-Spectrum, (2010)
12. E. Yares, Design World. WTWH Media, LLC. **8**(4) (2013)
13. N. Golovanov, *Geometric Modeling: The mathematics of shapes*. (CreateSpace Independent Publishing Platform 2014)

Effects of rotation and systematic occlusion on fiducial marker recognition

Artur Sagitov[1], *Ksenia* Shabalina[1], *Hongbing* Li[2] and *Evgeni* Magid[1*]

[1]Intelligent Robotics Department, Higher Institute for Information Technology and Information Systems (ITIS), Kazan Federal University, 35 Kremlyovskaya street, Kazan, 420008, Russian Federation
[2]Department of Instrument Science and Engineering, Shanghai Jiao Tong University, Shanghai, China

Abstract. Fiducial marker systems consist of patterns that are placed in environment for miscellaneous applications and are further automatically detected with cameras. A variety of applications determines the criteria, which characterize qualitative properties of a marker and include such evaluation benchmarks as resilience to occlusion, distance to a marker, false positive and false negative rates, sensitivity to illumination, and others. The paper compares existing ARTag, AprilTag, and CALTag systems utilizing a high fidelity camera, which is a main vision sensor of a full-size Russian humanoid robot AR-601M. In experiments the comparison of the three marker systems reliability and detection rate in occlusions of various types and intensities was verified. Finally, a preferable for AR-601M robot visual applications marker system was selected.

1 Introduction

Coordinate marks are important elements of camera tracking systems that are used in robotics and augmented reality applications to stabilize camera field of view. In robotics, it allows to calibrate a camera and mechanical parts of a robotic system, which are required for industrial applications [1], human-robot interaction and humanoids [2], SLAM [3], rescue robotics [4] and other fields. Many marker design variations exist, e.g., a marker could have a simple form, such as a pattern of dots, or a compound form, such as a barcode of a rectangular shape.

ARToolKit marker system is one of the first popular fiducial marker systems that uses a square shape with a black outline boundary and a unique image inside. During operation, outer black square edges enable localization of a potential fiducial marker and then the potential marker interior allows identification utilizing a set of expected markers. While using square markers, four corners of a located fiducial marker enable unambiguous evaluation of fiducial position and orientation relatively to a camera that is being calibrated.

This paper focuses on comparing ARTag, AprilTag, and CALTag marker systems in a presence of occlusion. We test their reliability and detection rate in the presence of occlusions of various types and intensities using humanoid robot AR-601M mono camera acA640-90gc by Basler AG. In particular, these preliminary experiments are important prior to verification of real occlusions within AR-601M robot field of vision, which often occur due to self-collisions and self-occlusions by its manipulators. The experimental analysis helped us to select a best-fit option among the three marker systems.

2 Fiducial marker systems

Most markers, also referred as tags, have a common design feature – an outlining square shape frame with a pattern image inside, which encodes information. The square shape is popular due to four special points (that correspond to the square corners) detection that allow camera calibration and marker position calculation. To reduce the sensitivity to illumination configuration, planar marker systems use monochrome markers, i.e., a bitonal image [5]. Many markers use some of bits to encode redundant information that allows identifying and correcting errors during detection and decoding process.

This section introduces ARTag, AprilTag and CALTag marker systems. These markers have two stages of detection: unique features detection and identification (or recognition) stage [6]. First, the environment image is searched for the unique feature (quadrilateral shapes). Then, at identification stage, the selected image interior is verified in order to determine if it is marker or some other general object of the environment.

Different tag design, detection and recognition algorithms determine strengths and weaknesses of tags in various situations. There are several criteria for evaluating tag performance: resilience to occlusion (full or partial overlapping of tag by another object), resistance to illumination changes, minimal size of marker (distance to marker), false positive effect, false

* Corresponding author: magid@it.kfu.ru

negative effect and others [7]. Figure 1 demonstrates examples of ARTag, AprilTag and CALTag. A large number of other fiducial marker systems exist, and some of them were briefly introduced in [8-9].

Fig. 1. ARTag, AprilTag and CALTag (from left to right).

2.1 ARTag marker system

ARToolKit fiducial marker system, which had a first release in 1999, is a classical open-source visual tracking library for calculating orientation and position of a camera relatively to markers in real time [10]; this system is popular for projecting virtual objects onto a real world environment. ARTag is an inspired by ARToolKit alternative [11], which uses more complicated data processing in order to increase reliability and resistance to illumination changes. ARTag is supposed to replace ARToolKit due to improved 2D-based markers as it succeeds avoiding a known false positive effect and inter-marker confusion. The ARTag system uses the same concept of squares with an internal image inside, but unlike ARToolKit, the system uses a digital approach for an internal pattern reading (i.e., a binary code).

ARTag system contains 2002 individual tags of square shape: 1001 tag has a white frame and 1001 tag has a black frame with an image (barcode) inside. Marker system detects tags with edge points based approach: an edge point detector finds edge pixels that form segments, which are then grouped into quadrilaterals. A tag internal image forms a 6x6 cell grid, which is composed of black and white cells, each representing 36 bit-values of "1" and "0". The first 10 of 36 bits form marker ID, the remaining 26 bits are redundant and are used to detect and correct errors, and to insure the uniqueness of the four possible marker orientations [7]. ARTag system ensures fast marker identification as it does not require matching its internal tag image with a library of stored templates, as it was previously implemented in ARToolKit.

2.2 AprilTag marker system

AprilTag is applicable for a wide range of tasks, including camera calibration, robotics and augmented reality. AprilTag detection system calculates exact position, orientation and identity of a marker relatively to a camera [12]. Representing a two-dimensional barcode, these tags look similar to QR code. However, AprilTag is designed to encode less data (4 to 36 bits), which allows quick and accurate detection of the tag.

The detection process consists of several stages: searching for linear segments, detecting squares, calculating the position and orientation of the tag, decoding the barcode. Directed linear segments search uses a similar to ARTag approach, and then sequences of segments are processed to form a square. Square detection applies a recursive 4-level depth search and at each level, a search tree adds one side of the square. At the identification stage, the validity of the barcode inside the discovered tag is verified. To encode an internal picture, AprilTag uses a lexicode system characterized by two parameters: a number of codeword (internal pattern) bits n and minimal Hamming distance between any two codes d. Lexicode generates codes for tags, which allow detecting and fixing bits' errors [13]. AprilTag has several tag families that differ in these two parameters, n and d. For example, "Tag36h11" refers to a 36-bit tag (6x6 array) with $n = 36$ and $d = 11$, and "Tag16h5" refers to a 16-bit tag (4x4 array) with $n = 16$ and $d = 5$.

2.3 CALTag marker system

After analysis of classical chessboard-based camera calibration and fiducial markers approach, CALTag marker was proposed as a specially designed for camera calibration solution [14]. This system consists of two components: a marker and a detection algorithm. A calibration marker is used in a proposed calibration grid, which is externally identical to a chessboard. The tags layout within this grid has two variations on tags density. A grid with a highest markers density provides a larger number of calibration points and is thus more reliable and efficient for recognition.

Each marker consists of M x N matrix of black and white squares, which are encapsulated within a boundary that consists of strictly black or white pixels. After initial detection of potential markers, they are filtered and verified by accessing their binary codes. Any missed calibration points of the template are restored as the chessboard layout is known by CALTag system. The binary tag code is validated by calculating the first P bits checksum and comparing it with a checksum that is obtained from four possible marker positions.

3 Experimental design

We define occlusion as a partial overlapping of a marker with other objects. Experimental work was performed in order to compare ARTag, AprilTag and CALTag markers with regard to occlusion resilience. Image capturing during experiments was performed with AR-601M humanoid robot (Fig. 2), that uses Basler acA640-90gc camera as front facing camera (Fig. 3).

For each marker we provided identical conditions of room illumination and camera posture with respect to the tag. To analyze the effect of occlusion on tag recognition, we selected four samples of ARTag tags, four AprilTag tags and two CALTag tags. ARTag and AprilTag marker IDs were selected randomly (Fig. 4) with a unique ID being encoded into an internal pattern of a tag. For CALTag markers we selected 4x4 and 9x6 grid size.

ARTag ID was encoded in 10 bits of 36 bits and tags with ID 2, 3, 6, 34 were selected. All AprilTags were selected from "36h11" tags family, i.e., each of tag ID is encoded in a $n = 36$ bit codeword with $d = 11$ bits; we have arbitrarily selected tags with IDs 4, 6, 8, and 9 (Fig. 4). Four types of experiments were conducted: systematic type occlusion and random type occlusion (arbitrary overlap with an object), rotation with regard to 3 axes and a combination of rotations with systematic occlusion.

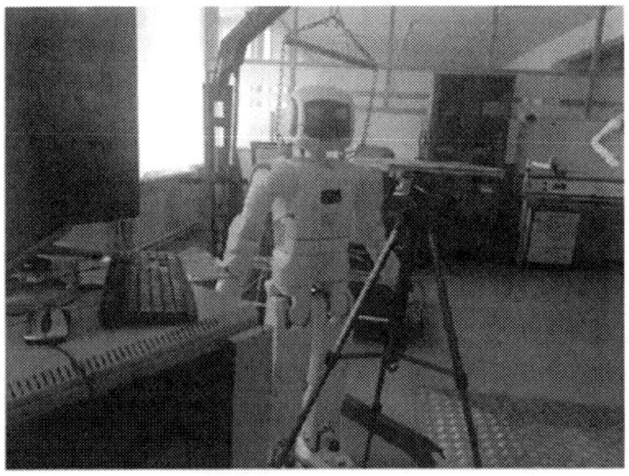

Fig. 2. AR-601M robot.

Fig. 3. Basler acA640-90gc camera.

Fig. 4. ARTag with IDs 2, 3, 6, 34 (top set of images) and AprilTag with IDs 4, 6, 8, 9 (bottom set of images).

3.1 Type 1 – Marker rotation

For the experiments, we set up rotation axes with regard to the marker plane (Fig. 5). X-axis is orthogonal to the marker plane and points into the camera direction; it was responsible for roll rotations of the marker in Type 1C experiments. Y-axis (horizontal) and Z-axis (vertical) are in the marker plane. Y-axis was responsible for pitch rotations of a marker in Type 1B experiments, Z-axis was responsible for yaw rotations of a marker in Type 1A experiments.

In Type 1A experiments, each marker was fixed at two points that formed a vertical rotation axis Z being drawn from a bottom to a top of a marker and passed through the marker center. The rotations were performed clockwise and counter-clockwise for 10, 20, 30, 45, 55 and 65 degrees in both directions around Z-axis (Fig. 6 demonstrates Type 1A experiment with ARTag marker). In Type 1B experiments, for each marker the rotations were performed clockwise and counter-clockwise around Y-axis for 10, 20, 30, 45, 55 and 65 degrees in both directions (Figure 7 demonstrates an experiment with AprilTag marker).

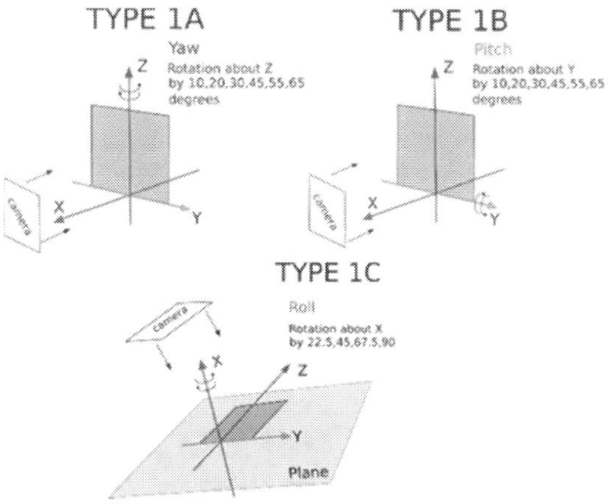

Fig. 5 Types of marker rotation with regard to the camera and the marker plane.

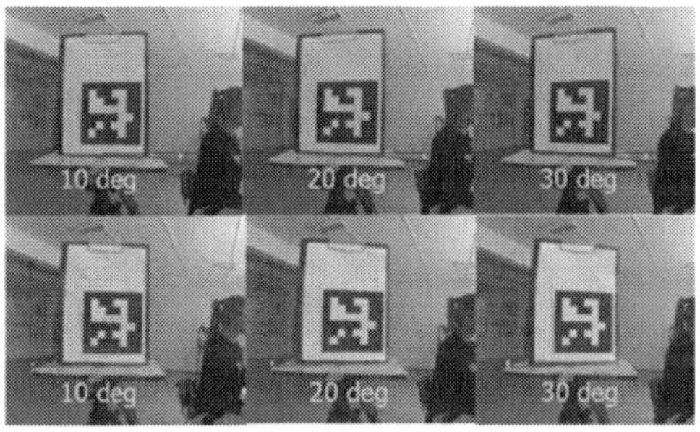

Fig. 6 Yaw rotations (Z-axis) of ARTag marker clockwise and counter-clockwise for 10, 20, 30 degrees in both directions.

Fig. 7 Pitch rotations (Y-axis) of AprilTag clockwise and counter-clockwise for 10, 45 and 65 degrees in both directions.

In Type 1C, each marker was fixed at its central bottom point and the rotation was performed clockwise and counter-clockwise around X-axis for 0, 22.5, 45, 67.5, and 90 degrees in both directions.

3.2 Type 2 – Systematic occlusion

Each tag was covered with a white paper rectangular template starting from the bottom to the top of tag so that the template was occluding K% of the marker's area, where K are 0%, 10%, 20%, 50%, 70% of the marker's area.

3.3 Type 3 – Marker rotation with systematic occlusion

In marker rotation with systematic occlusion experiments Type 1C and Type 2 experiments were combined together. The marker was rotated clockwise and counter-clockwise around X-axis in its plane for 0, 22.5, 45, 67.5, and 90 degrees and simultaneously occluded with the white paper rectangular template for 0%, 10%, 20%, 50%, and 70% of its area. Figure 8 demonstrates an example of rotated CALTag 4x4 marker by 22.5, 67.5, and 90 degrees in both directions and occluded by 20%.

3.4 Type 4 – Arbitrary overlap with an object

Each tag was randomly overlapped with one of three different objects so that an object was entirely located within tag's area and thus the overlap percentage was always kept constant. The first object was a white thick paper strip of 13 cm width and 2.5 cm length with 32.5 cm area. The second object was a metal scissors with 7.99 cm area. The third object was a black plastic strip of 15,7 cm width and 2.6 cm length with 40.82 cm area. This way, for each experiment the constant overlap percentage was always known in advance. In the black strip case, it covers the interior and crosses the boundaries of each tag (Fig. 9). In the white strip case and scissors, they cover only the internal pattern of each tags (Fig. 10). With each of the three object for each tag, 25 experimental trials were conducted. We emphasize a special case for the 4x4 size CALTag: if the strip is placed strictly along the marker side, the occupied area percentage decreases as strip's width exceeds marker size. In this case, the overlap percentage varies between 25.5% and 33.84% in the white strip case and 26.53% and 40.20% in the black strip case.

4 Experimental results

For the experiments, we used publically available official software of AprilTag and CALTag. For ARTag we used ArUco library, which allows detecting and recognizing various kinds of tag families [15]. These (tags were printed on white paper with the following sizes: ARTag: 15.2 x 15.2 cm, total area 231.04 cm^2; AprilTag: 13.5 x 13.5 cm, total area 182.25 cm^2; CALTag 4x4: 9.8 x 9.8 cm, total area 96.04 cm^2; CALTag 9x6: 21.7 x 14.7 cm, total area 318.99 cm^2.
It is important to emphasize that different types of occlusion had slightly different experimental implementation. Systematic occlusions (Types 2 and 3) reflect a very typical real world occlusion situation when an object overlaps both an interior and marker's boundaries, and also becomes partially invisible due to its rotations and inclination in 3D space. Moreover, since tag boundaries serve as a unique tag feature that is necessary for its detection, these types of occlusion affect both tag detection and recognition stages. Type 4 occlusion has an overlapping object completely within internal pattern of the tag, which has effect only on the recognition stage of tag pattern detection.

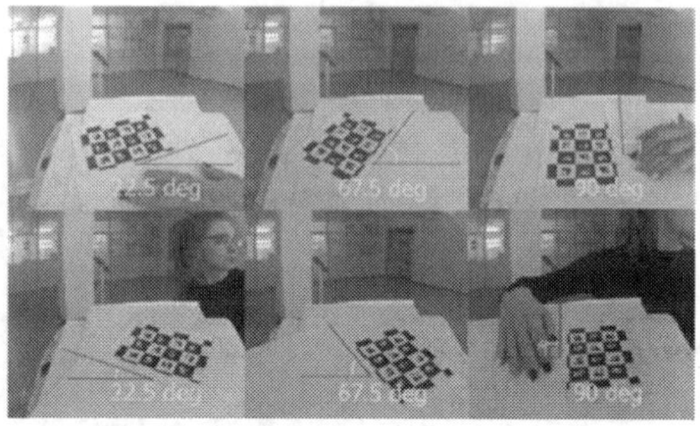

Fig. 8. Roll rotation of CALTag 4x4 on 22.5, 67.5 and 90 degrees in both directions with the 20% occlusion percent.

Fig. 9. Arbitrary overlapping of ARTag with a black strip.

Fig. 10. Arbitrary overlapping of two AprilTag markers with a white strip (top set of images) and metal scissors (bottom set of images).

The experimental results are summarized in Tables 1-7. Table 1-4 present the results of systematic occlusion experiments. All tags were completely resistant to rotations about vertical Z-axis in Type 1A experiments (Fig. 6) and rotations about horizontal Y-axis in Type 1B experiments (Fig. 7). The markers were detected and recognized at any given angle for both rotation axes, and thus because of the paper space limitations we excluded the tables, which showed these results. Tables 1 and 3 demonstrate results of marker rotation around X-axis with a simultaneous systematic occlusion, that correspond to Type 3 experiments (Figure 8 shows experiments of rotations together with 20% of marker area occlusion). ARTag and AprilTag demonstrated identical results, and their success rates are summarized in Table 1. In Tables 1 and 3, "2/2" denotes a successful detection of the marker under rotation about an axis in both clockwise and counter-clockwise directions; "1/2" corresponds to successfully detection only for one of rotation directions, while "0/2" marks unsuccessfully detection for both clockwise and counter-clockwise directions.

Table 1. Systematic approach: occlusion and rotation around X-axis of AprilTag and ARTag.

AprilTag, ARTag	**Rotation about X-axis (degrees)**				
Occlusion (%)	0°	22.5°	45°	67.5°	90°
0 %	2/2	0/2	0/2	0/2	0/2
10 %	2/2	0/2	0/2	0/2	0/2
20 %	2/2	0/2	0/2	0/2	0/2
50 %	2/2	0/2	0/2	0/2	0/2
70 %	2/2	0/2	0/2	0/2	0/2

Table 2. Successful detection rate with regard to the marker occlusion percentage.

Tag	**Occlusion percent (%)**				
	0%	10%	20%	50%	70%
ARTag	100%	0%	0%	0%	0%
AprilTag	100%	0%	0%	0%	0%
CALTag 4x4	100%	100%	100%	100%	100%
CALTag 9x6	100%	100%	100%	90%	40%

Table 3. Systematic approach: occlusion and rotation around X- axis of CALTag 9x6 and CALTag 4x4.

Occlusion (%)	**Rotation around X- axis (degrees)**				
CALTag 9x6	0°	22.5°	45°	67.5°	90°
0 %	2/2	2/2	2/2	2/2	2/2
10 %	2/2	2/2	2/2	2/2	2/2
20 %	2/2	2/2	2/2	2/2	2/2
50 %	2/2	2/2	2/2	2/2	2/2
70 %	2/2	0/2	0/2	1/2	1/2
CALTag 4x4	0°	22.5°	45°	67.5°	90°
0 %	2/2	2/2	2/2	2/2	2/2
10 %	2/2	2/2	2/2	2/2	2/2
20 %	2/2	2/2	2/2	2/2	2/2
50 %	2/2	2/2	2/2	2/2	2/2
70 %	2/2	2/2	2/2	2/2	2/2

Table 4. Successful detection rate with regard to the marker rotation degree around X- axis.

Tag	Rotation degree around X-axis				
	0°	22.5°	45°	67.5°	90°
ARTag	100%	0%	0%	0%	0%
AprilTag	100%	0%	0%	0%	0%
CALTag 4x4	100%	100%	100%	100%	100%
CALTag 9x6	100%	70%	80%	90%	90%

Table 2 summarizes results of successful detection rate with regard to the marker systematic occlusion percentage in experiments of Type 2 for all markers. Table 4 shows success rate (percentage) with regard to the marker rotation degree around X-axis for all markers.

Table 5 demonstrates the results of arbitrary overlap with an object Type 4 experiments for the black strip object. As the tags have different sizes while the strip size is constant, the percentage of occluded tag area differs between the tags. The strip was arbitrarily placed within an internal part of the tag. For each tag, twenty-five experiments were performed so that the position of the strip on the tag was different in each experiment. The black colour of the strip makes it difficult to read binary code of the tag, because all of tags are monochrome, while the strip crosses the tags boundaries. A special case was CALTag 4x4 tag, because due to its small size the overlapped by the strip area varied from 25.5% to 33.84%. ARTag and AprilTag demonstrated particular sensitivity to tag edges overlapping with a black strip. Edge overlapping disables tag unique feature (edge) detection, which in turn results into a failure of tag discovery stage. Yet, CALTag 4x4 and 9x6 markers were successfully detected in all 50 experiments with black strip.

Table 6 shows the result of arbitrary overlap with an object Type 4 experiments for the scissors object. These experiments demonstrated the dependence of tag recognition algorithm on overlap of a marker interior only with a complex object. Again, CALTag 9x6 and 4x4 were successfully detected and recognized in all 50 experiments. At the same time, ARTag and AprilTag turned to be sensitive to the interior overlapping within a complex object. While scissors took only 4.38 (for AprilTag) and 3.45 (for ARTag) percent of the tag area, its complex shape significantly affected the recognition.

Table 5. Results of arbitrary overlap experiments with a black strip object.

Tag	**Occlusion percent**	**Success rate (%)**
ARTag (ID 2, 3, 6, 34)	17.66	0%
AprilTag (ID 4, 6, 8, 9)	22.39	0%
CALTAg 4x4	26.53-40.20	100%
CALTag 9x6	12.79	100%

Table 6. Results of arbitrary overlap experiments with a scissors object.

Tag	**Occlusion percent**	**Success rate (%)**
ARTag (ID 2)	3.45	80%
ARTag (ID 3)	3.45	96%
ARTag (ID 6)	3.45	88%
ARTag (ID 34)	3.45	88%
AprilTag (ID 4)	4.38	92%
AprilTag (ID 6)	4.38	88%
AprilTag (ID 8)	4.38	88%
AprilTag (ID 9)	4.38	96%
CALTAg 4x4	8.32	100%
CALTag 9x6	2.5	100%

Table 7 shows the results of arbitrary overlap in Type 4 experiments for the white strip object. CALTag 9x6 and 4x4 showed impressive results: within 50 Type 4 experiments, the detection algorithm had failed only once.

Table 7. Results of arbitrary overlap experiments with a white strip object.

Tag	Occlusion percent	Success rate (%)
ARTag (ID 2)	14.06	0%
ARTag (ID 3)	14.06	0%
ARTag (ID 6)	14.06	0%
ARTag (ID 34)	14.06	0%
AprilTag (ID 4)	17.83	4%
AprilTag (ID 6)	17.83	0%
AprilTag (ID 8)	17.83	20%
AprilTag (ID 9)	17.83	0%
CALTAg 4x4	25.5-32.5	100%
CALTag 9x6	10.18	96%

5 Conclusions and Future Work

This paper is focused on ARTag, AprilTag and CALTag marker systems evaluation for systematic and random types of occlusion sensitivity as well as verification of their resistance to rotations. Four types of experiments were performed: systematic occlusion, random occlusion, rotation about all three principal axes of the reference frame and a combination of systematic occlusion with a marker rotation around its centre in marker's plane (i.e., orientation in 2D plane).

Systematic occlusion considered partial occlusion with a white paper template for 0, 10, 20, 50, 70 percent of marker area. While CALTag was successfully detected up to 70% of occlusion, ARTag and AprilTag markers demonstrated high sensitivity to edge occlusions, which limits their effective use. In random occlusion experiments, two occluding objects were used: a black plastic strip, a metal scissors and a white paper strip. The first object crossed the boundaries of markers and overlapped interior, while the second and the third objects overlapped only marker's interior and thus affected only the recognition stage of marker pattern detection. For random occlusion, ARTag and AprilTag showed good results for the second object, poor for the third and failed all experiments with the first object. CALTag showed high resilience to all objects in random occlusion type experiments. All of tags show high results of normal and lateral rotation. Yet, only CALTag markers succeeded to resist a longitudinal axis rotation, both without occlusion and with systematic occlusion of up to 70% of marker area. To summarize, we have selected CALTag marker system as a preferred marker system for our applications.

For our future work, we plan to conduct more occlusion resilience experiments, including rotations around all three principal axes, and to increase number of experiments. We plan to verify marker resistance with regard to other criteria such as resistance for illumination conditions changes, inter-marker confusion. Special attention will be paid to the behaviour of CALTag marker, which have demonstrated the best performance in our current empirical research.

This work was partially supported by the Russian Foundation for Basic Research (RFBR, project ID 17-48-160879). Part of the work was performed according to the Russian Government Program of Competitive Growth of Kazan Federal University.

References

1. A. Klimchik, E. Magid, and A. Pashkevich, *IFAC Conf. on Manufacturing Modelling, Management and Control*, Elsevier Ltd., 49(12), 967-972 (2016)
2. R. Khusainov, I. Shimchik, I. Afanasyev, and E. Magid, *Int. Conf. on Informatics in Control, Automation and Robotics*, 2, 141-148 (2015)
3. A. Buyval, I. Afanasyev, E. Magid, *Int. Conf. on Machine Vision*, 141-148 (2016)
4. E. Magid, T. Tsubouchi, *Lecture Notes in Artificial Intelligence*, 6472, 423-435 (2010)
5. M. Hirzer, *Seminar/Project Image Analysis Graz*, 1-25 (2008)
6. M. Fiala, *IEEE Int. Workshop on Haptic Audio Visual Environments and their Applications*, 148-153 (2005)
7. M. Fiala, *IEEE Computer Society Conf. on Computer Vision and Pattern Recognition*, 2, 590-596 (2005)
8. A. Sagitov, K. Shabalina, E. Magid, *Int. Conf. on Mechanical, System and Control Engineering, Saint-Petersburg, Russia,* (2017, to be published)
9. K. Shabalina, A. Sagitov, E. Magid, H. Li, *10-th Int. Conf. on Developments in eSystems Engineering*, France, (2017, to appear)
10. H. Kato, M. Billinghurst. *IEEE and ACM Int. Workshop on Augmented Reality*, 85-94 (1999)
11. M. Fiala, National Research Council Publication, **47419**, 1-47 (2004)
12. E. Olson, *IEEE Int. Conf. on Robotics and Automation*, 3400-3407 (2011)
13. A. Trachtenbert, *Computational methods in coding theory. MS thesis. University of Illinois at Urbana-Champaign*, (1996)
14. B. Atcheson, F. Heide, W. Heidrich, *Vision, Modeling, and Visualization Workshop*, 10, 41-48 (2010)
15. S. Garrido-Jurado, R. Muñoz-Salinas, F.J. Madrid-Cuevas, M.J. Marín-Jiménez, Pattern Recognition, **47**(6), 2280-2292 (2014)

Controlling spherical mobile robot in a two-parametric friction model

Georgy Terekhov[1*] and *Vladimir* Pavlovsky[2]

[1]Moscow State University, Moscow, Russia
[2]Keldysh Institute of Applied Mathematics of RAS, Moscow, Russia

Abstract. The theoretical mechanical model of the spherical robot with the flywheels located in a sphere is considered. The described design is intended for the motion control of a sphere on various trajectories. In the paper, control algorithms of a sphere on basic trajectories are investigated; the movements from rotation on the point and the movements on a straight line piece before the basic movements on curvilinear trajectories are considered. The paper describes dynamical models of a sphere robot, different models of friction, as well as a two-parametrical model as the main model of friction. Technical possibility of realization of a sphere robot with flywheels is proved.

1 Introduction

At present there is a significant variety of robots, differing in design and purpose. Among them, it is interesting to consider robots of spherical shape that have such advantages over other implementations as the tightness of the robot and the absence of junctions and articulations that are most vulnerable to various kinds of adverse effects. This form of the device contributes to the practical use of the robot for research and reconnaissance purposes.

Such devices are actively being developed in different countries of the world - Sweden [1], the USA, Israel, Russia, etc. A large number of constructions use the principle of moving the center of the mass of system, such as various pendulum robots, or robots with moving internal masses. The devices moving due to deformation of the case are also of interest. This paper is devoted to another realization of the gyrostat principle, which implies the presence of a system of drives inside the sphere, which provides the creation of an internal kinetic moment. In particular, we consider an example of a robot, inside of which three flywheels are located on three mutually orthogonal axes. These flywheels produce a control moment.

An investigation of the motion and planning of the trajectories of such a device on an absolutely rough plane was performed in [2]. Then, in [3] the controllability of the system for the nonholonomic system was proved and the robot control algorithms were explicitly indicated. In [4] the results were generalized to the case of a plane with friction, in particular, Coulomb's dry friction, viscous friction are considered.

At the same time, recently, a number of papers appear, suggesting essentially altered models of friction, which supplement and generalize the model of Coulomb's dry friction. The generalization is due to the fact that the point of Coulomb's interaction of the rolling body with the plane is replaced by an interaction along a certain region (contact spots). An example of this approach is the consideration of the contact spot in the form of a circle, this model was first proposed by Contensou [5] and further developed by V.F. Zhuravlev [6]. With this approach, there are plane deformations of the spherical shell of the ball near the contact with an absolutely rigid plane. However, there is also an approach, according to which the contact spot is not a flat region, but a spherical segment. Thus, it is assumed deformation not only areas but also the reference plane. Such a model was proposed and then developed by A.V. Karapetyan in [7, 8]. In this paper we use a two-parameter model of friction [7].

2 Theoretical mechanical model

Consider on a fixed plane $Sxyz$ a sphere of mass M and radius R, in which three flywheel disks with radius r and mass m with centers at the points A_i $(i = 1,2,3)$ are located on mutually orthogonal axes. The distance from the point of intersection of the flywheel axes to each of A_i is equal to a. The center of mass of the whole system is assumed at point O. Next, we move to dimensionless linear, mass and inertial parameters by dividing them $by\, R, M + 3m$ and $ma^2/3$, respectively. We introduce the movable coordinate system $Oe_1e_2e_3$, with the axes directed parallel to the flywheel axes. Theorems on the change in momentum and angular momentum take the form:

$$\dot{v} = f. \tag{2.1}$$

* Corresponding author: arctangens90@gmail.com

$$\frac{d}{dt}(I\omega + C\alpha) = D\mu \ . \qquad (2.2)$$

$$\mu = \mu_0 - b[e_z; f] \, . \qquad (2.3)$$

Here the first equation is written in the absolute coordinate system, and the second - in the movable coordinate system. The matrix I is the inertia tensor of the system in a movable coordinate system. It is assumed that this matrix has the form:

$$I = \begin{pmatrix} \nu & 1 & 1 \\ 1 & \nu & 1 \\ 1 & 1 & \nu \end{pmatrix}.$$

where $\nu > 1$. Matrix C-diagonal matrix, consisting of values of axial moments of inertia of flywheels. The vectors $\boldsymbol{f}$ and $\boldsymbol{\mu}^0$ are vectors of external forces and torques. The vector $\boldsymbol{\mu}$— is the resultant vector of the torques. D-matrix of the transition from a fixed coordinate system to a moving one. $\boldsymbol{\Omega}$-vector of angular velocity in a moving coordinate system, $\boldsymbol{\alpha}$-angular velocity of rotation of flywheels around its axis, $\boldsymbol{v}$-velocity of the center of the robot in the absolute coordinate system. Further, to close the system, it is necessary to specify the conditions of contact interaction between the spherical surface of the robot and the reference plane.

2.1. Two-parameter model of friction

Real experiments show that contact between the spherical surface of the robot and the reference plane occurs not at one point, but in some region called the contact spot. The most precise is an assumption [2, 3] that it is a spherical segment, at each point of which the Coulomb dry friction law is locally satisfied. Let the radius of the sphere specifying this segment be R_f, and the radius of the segment is R_s.

Thus, the contact spot depends on the two parameters ε and δ, defined as R_s/R and R/R_f, respectively. Both of these parameters take values from 0 to 1. Expressions for forces and torques of friction take the following form:

$$f_j = \frac{3kN \int_0^{\beta_0} \sqrt{\cos\beta - \cos\beta_0} \sin\beta \, I_{u_j} d\beta}{4\pi(M + 3m)R(1 - \cos\beta_0)^{\frac{3}{2}}}, \qquad (2.4)$$

$$\mu_j{}^0 = \frac{3kN \int_0^{\beta_0} \sqrt{\cos\beta - \cos\beta_0} \sin\beta \, I_{w_j} d\beta}{4\pi\delta(M + 3m)R(1 - \cos\beta_0)^{\frac{3}{2}}}, \qquad (2.5)$$

$$I_{u_j} = \int_0^{2\pi} \frac{u_j}{||u||} d\alpha \ . \qquad (2.6)$$

$$I_{w_j} = \int_0^{2\pi} \frac{w_j}{||u||} d\alpha \, . \qquad (2.7)$$

Here g is the acceleration of gravity, N is the force of the normal support reaction, u is the slipping velocity of the spherical robot, divided by its radius, the first axis is directed along the slip velocity, the third axis is vertical, and the second axis completes the basis to the right triple. This basis will be referred to as auxiliary in the future, and vector will be indexed with Roman numerals to avoid confusion with coordinate system, rigidly connected with the robot. $\boldsymbol{\Omega}$ is the angular velocity vector in this basis. Finally, $sin\beta_0 = \frac{R_s}{R_f} = \varepsilon\delta$, k is the coefficient of friction. u_i, w_i are the components of the vectors $\mathbf{u}$, $\boldsymbol{\Omega}$ at each point of the contact spot in this basis.

It should be noted that if we go over to spherical coordinates

$$\begin{cases} \delta u = r \cos\theta_1 \\ \omega_I = r \sin\theta_1 \cos\theta_2 \\ \omega_{II} = r \sin\theta_1 \sin\theta_2 \cos\theta_3 \\ \omega_{III} = r \sin\theta_1 \sin\theta_2 \sin\theta_3 \end{cases}, \qquad (2.8)$$

it can be seen that the expressions for the forces and torques do not depend on r (which is a variation of the ideas proposed in [9]). Thus, taking into account the dependence of $\mathbf{f}$ and $\boldsymbol{\mu}$ on the angles θ_k (mode angles), we can regard them as functions defined on the sphere S^3.

3 Motion planning

3.1. Robot's rotation around the vertical

Firstly, consider the case when the center of mass of the device is stationary, and the sphere rotates around the vertical. Thus, the problem of changing the course angle without a shift of the center of mass is solved. For mode angles, this assumption corresponds to the conditions $\theta_1 = \theta_2 = \theta_3 = \frac{\pi}{2}$. In this case, a pure spinning mode realizes, where all the components of forces and torques, except for μ_z, are equal to zero. Friction torque of rotation is $\mu_z = \mu_z^0 = const = \mu$. Thus, it is right:

$$\alpha = \frac{1}{c}(\mu t E - \omega I)e = \frac{1}{c} X e \, . \qquad (3.1)$$

Here, the $\boldsymbol{e}$-vector of the vertical, written in the moving coordinate system.

Assuming that the flywheel is driven by a standard electric motor and $||\boldsymbol{\alpha}|| \gg ||\boldsymbol{\omega}||$, we obtain

$$c\dot{\alpha} = \kappa_1 U - \kappa_2 \alpha \ . \qquad (3.2)$$

Where $\kappa_{1,2}$-constant parameters of the motor, vector $\boldsymbol{U}$-vector of control voltages. Consequently,

$$U = \frac{1}{\kappa_1}\left(\dot{X} + \frac{\kappa_2}{c} X\right) e \qquad (3.3)$$

Let us give an example of control voltage graphs for rotating the ball at an angle of $\frac{\pi}{2}$ in 4 seconds, choosing motor's parameters $\kappa_1 = 6, \kappa_2 = 0.06$. Robot's parameters are: $M = 4$, $m = 0.612$, $R = 0.2$, $a = 0.1, \nu = 54.46$. Vector $\boldsymbol{e} = (1/\sqrt{3}; 1/\sqrt{3}; 1/\sqrt{3})$

Note, that a constant friction torque μ can be considered as a moment of static friction. This value depends only on the roughness parameter of the δ plane. For example, this value for motor Maxon RE-10 is around 0.02.

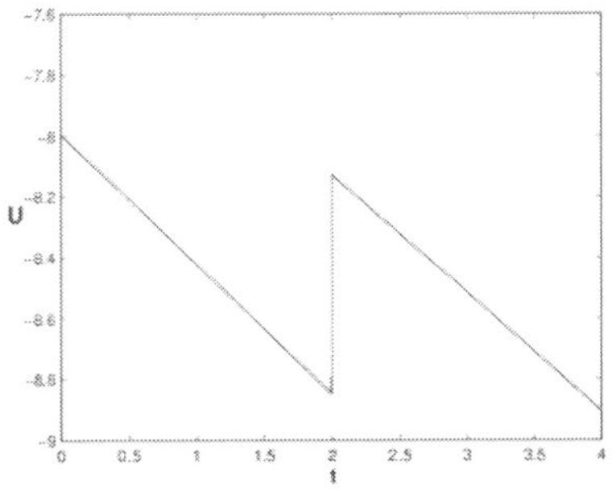

Fig. 1. Control voltage in rotation mode. All voltages are equal to each other.

3.2. Movement along the line

One of the most important tasks is to study the dynamics of the robot, when its center of mass moves along a straight line. Without loss of generality, assume that this line is directed to the axis Sx. Within the framework of the proposed model, slip rolling takes place. This means that the force and the torque are functions of the angle θ_1, which we will denote as θ in this part. The system of equations takes the form ($\boldsymbol{e}$ is vector $\boldsymbol{e_y}$ in the moving coordinate system)

$$\begin{cases} \dot{v} = f(\theta) \\ \dot{\omega} I e + c\dot{\alpha} + \omega[e; \omega I e + c\alpha] = \mu(\theta) e \\ v = r\left(\frac{1}{\delta}\cos\theta + \sin\theta\right) \end{cases} \quad (3.4)$$

For forces and torques, as functions θ, the following statements are right (their proof can be found in [10]):

1. $f(\theta) \in C(0;\pi)$ and increases on this interval;
2. $\exists \theta_f \in \left(\frac{\pi}{2};\theta_*\right)\ f(\theta_f) = 0, \quad \theta_* = \pi - arctg(\frac{1}{\delta})$;
3. $\mu^0(\theta) \in C(0;\pi)$ And increases on interval $\theta \in (\frac{\pi}{2};\pi)$ and decreases if $\theta \in (0;\frac{\pi}{2})$;
4. $\mu(\theta) \in C(0;\pi)$ and decreases on $\theta \in (0;\theta_*)$.

The value $\mu(\theta_f)$ thus can be identified as the moment of static friction. So, the maximum torque on the motor must be greater than this value to allow the motion of the system. This value depends on the contact parameters. An example of areas of possibility of the motion for two real motors (movement possible inside the curve) is below (Fig. 2).

We note that in the framework of this model, the choice of the function v should be done in the class C^1, because the discontinuous function with respect to the accelerations leads to discontinuities in velocities. Therefore, to construct control algorithms, we set:

$$v = A\cos\frac{\eta t}{2}\sin\eta t, \quad t \in \left[0;\frac{\pi}{\eta}\right]. \quad (3.5)$$

Below (Fig.3), we give examples of control voltages for the moving of the segment 2π during 7 s if the vector $e = (1/\sqrt{3}; 1/\sqrt{3}; 1/\sqrt{3})$.

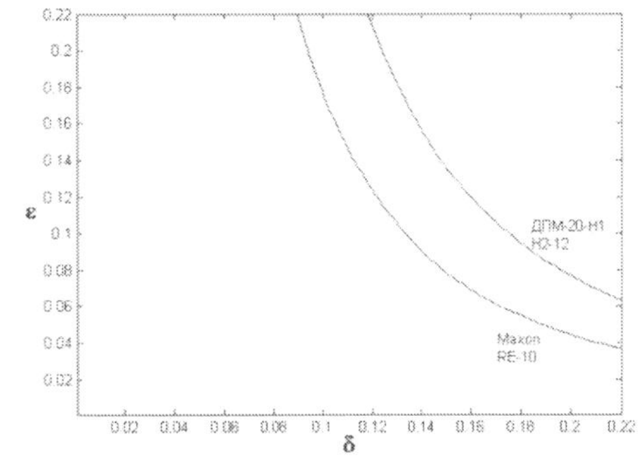

Fig. 2. Area of motion availability.

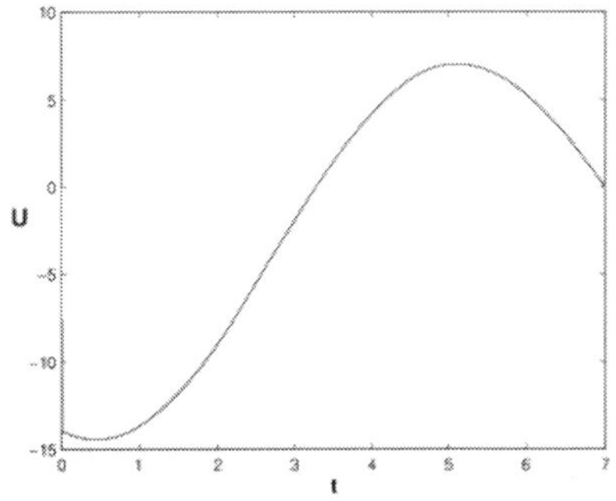

Fig. 3. Control voltage in motion along straight line mode. All voltages are equal to each other.

3.3. Curvilinear motion

In spite of the fact that movement on any rectifiable curve can be approximated by movement along broken lines, from the practical point of view such movement is often inefficient. When accuracy of the approximation of the desired trajectory increases, the more the number of segments will have to pass through the robot, and it is necessary to take into account the stop at the end of each segment. Thus, the curvilinear trajectories should be added to the list of alphabetic motions.

First of all, it should be taken into account that for the motion along a curvilinear trajectory, the condition $\Omega_l = 0$ can be satisfied only on a set of measure zero, since the opposite provides a straight-line trajectory. We associate with the curve along which the center of the sphere moves the natural Frenet basis. Denoting by φ the angle between the vectors e_l and e_τ, we obtain the following system of equations in the projection on the axis of the natural basis:

$$\begin{cases} \ddot{s} = f_l\cos\varphi - f_{ll}\sin\varphi \\ k(s)\dot{s}^2 = f_l\sin\varphi + f_{ll}\cos\varphi \end{cases} \quad (3.6)$$

Here s is a natural parameter, k (s) is the curvature of the curve. The analysis of forces and torques [10] shows that in the first approximation, as before, we can state f and μ determining only by one essential component, for the force f_I, for the torque - μ_{II}. Below, the dependence f and μ functions of θ_1, θ_2 is shown (Fig. 4-8). As a result, we obtain the following system, to find φ, θ_1, θ_2, after obtaining the required values of $k(s)$, $s\ (t)$,

$$\begin{cases} \ddot{s} = f_I(\theta_1, \theta_2)\cos\varphi \\ k(s)\dot{s}^2 = f_I(\theta_1, \theta\_2)\cos\varphi \\ tg\,\varphi = \dfrac{-\delta\sin\theta_1\cos\theta_2}{\delta\sin\theta_2\sin\theta_1 + \cos\theta_1} \end{cases} \quad (3.7)$$

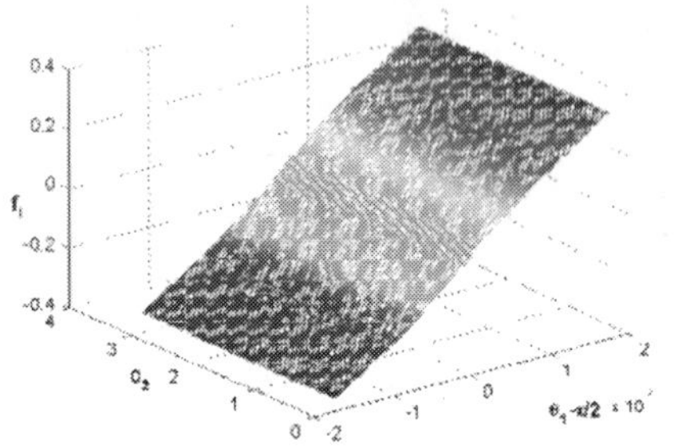

Fig. 4. Surface $f_I(\theta_1, \theta_2)$.

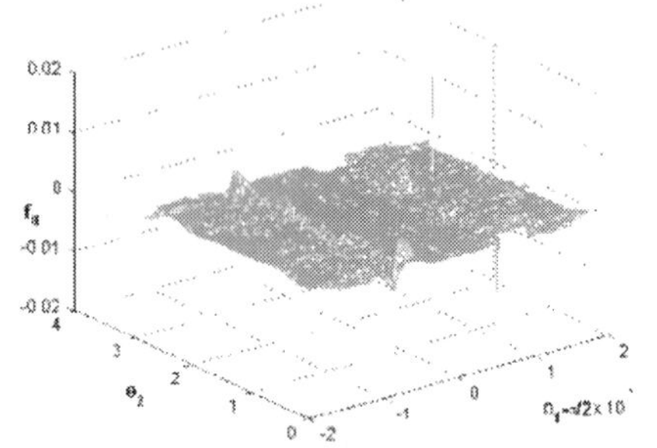

Fig. 5. Surface $f_{II}(\theta_1, \theta_2)$.

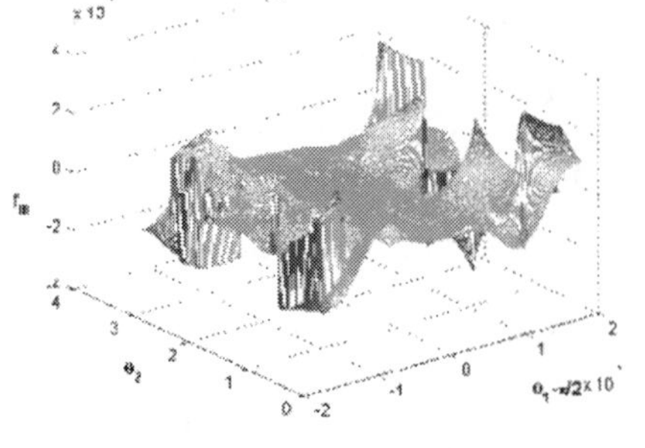

Fig. 6. Surface $f_{III}(\theta_1, \theta_2)$.

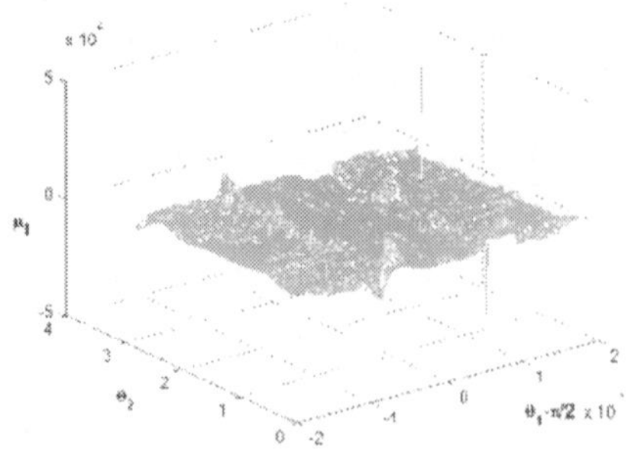

Fig. 7. Surface $\mu_I(\theta_1, \theta_2)$.

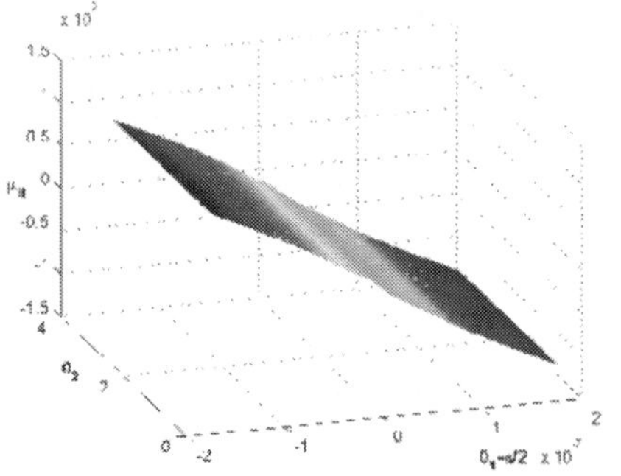

Fig. 8. Surface $\boldsymbol{\mu_{II}(\theta_1, \theta_2)}$.

The s (t) dependence is chosen by the formula:

$$\dot{s}(t) = A\cos\frac{\eta t}{2}\sin\eta t, \quad t \in [0; \frac{\pi}{\eta}].$$

Examples of control voltages when moving along a quarter of a circle of radius ρ = 4 in 7 seconds (Fig. 9).

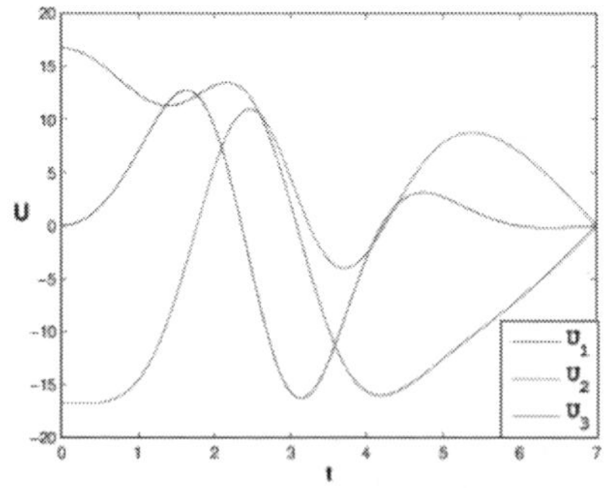

Fig. 9. Control voltages on motion along circle arc.

4 Conclusion

The developed mechanical model of the spherical robot solves the task of motion control of a sphere on various trajectories. The movements from rotation on the point and the movements on a straight line piece before the basic movements on curvilinear trajectories are implemented. Experimental results proved technical possibility of realization of a sphere robot with flywheels.

References

1. Electronic resource http://www.rotundus.se/
2. V.E. Pavlovsky, G.P. Terekhov Special equipment and communication, **3**, 19-25 (2012)
3. A.V. Borisov, A.A. Kilin, I.S. Mamaev, Nonlinear dynamics, **8**(2), 289-307 (2012)
4. A.V. Borisov, A.A. Kilin, I.S. Mamaev, *Mobile robots: a robot-wheel and a robot-ball,* 131-168, (M.-Izhevsk: Institute for Computer Research, 2013)
5. P. Contensou, *Problems of gyroscopy*, 60-77 (M.: Mir, 1967).
6. V.F. Zhuravlev, PMM, **62(5)**, 762-767 (1998)
7. A.V. Karapetyan, PMM, **73**(4), 515-519 (2009)
8. A.V. Karapetyan, PMM, **74**(4), 531-535 (2010).
9. T.V. Salnikova, D.V. Treshchev, S.R. Gallyamov. Nonlinear dynamics, **8**(1), 83-101 (2012)
10. G.P. Terekhov, V.E Pavlovsky. Preprints of KIAM **16**(31) (2017)

Justification of the technical requirements of a fully functional modular robot

Nikita Shlyakhov[1,2], *Vladimir* Dashevskiy[1], *Irina* Vatamaniuk[1], *Milos* Zelezny[3] and *Andrey* Ronzhin[1,2*]

[1]SPIIRAS, Laboratory of Autonomous Robotic Systems, 199178 Saint-Petersburg, 14-th line of V.I., 39, Russia
[2]SUAI, Department of Electromechanics and Robotics, 190000 Saint-Petersburg, Bolshaya Morskaya str., 67, Russia
[3]University of West Bohemia, Pilsen, Czech Republic

Abstract. Modular robots are characterized by limited built-in resources necessary for communication, connection and movement of modules, when performing reconfiguration tasks at rigidly interconnected elements. In developing the technological fundamentals of designing modular robots with pairwise connection mechanisms, we analysed modern hardware and model algorithms typical of a fully functional robot, which provide independent locomotion, communication, navigation, decentralized power and control. A survey of actuators, batteries, sensors, communication means, suitable for modular robotics is presented.

1 Introduction

The complexity of developing robot group interaction algorithms in modular swarm robotics is related to the contact connection of homogeneous robotic elements to single structures in a three-dimensional space, limited by the capabilities of individual robots, due to small overall dimensions, low energy reserves, design features of propellers, etc. When reconfiguring the swarm of modular robots, algorithms for controlling robotic pairing of robots are necessary, both during the formation of the three-dimensional structure and after completion to maintain the shape of the swarm of the required configuration.

The most important specificity of modular robotics (MR) is the solution of kinematics problems of rigidly interconnected elements under conditions of limited built-in resources necessary for communication, connection and movement of modules, when performing reconfiguration tasks, assembling, capturing, moving a set of modular robots. In work [1], several basic tasks solved by MR are distinguished: reorganization on a surface or into a given form; movement; manipulations with external objects; support and balancing of unstable objects.

In self-reconfiguration, problems arise not only in the limited resources of individual modules, but also in collisions in the overall structure, for example: the separation of a modular robot into several parts; the overlap of individual modules in one position when planning a rebuild; formation of hollow closed forms, where a number of the modules remain inside and cannot be used further; and vice versa, when some of the modules remain outside and are not available to form an internal structure.

In work [2], a number of architectures of modular robots are considered depending on their structural parameters. The simplest variant is a group interaction of traditional robots with wheel/caterpillar propulsors, which are interconnected to increase their parameters when solving a particular problem. For robots whose whole body participates in movement and connection, there are three types of architecture: serpentine-like, lattice, hybrid. In this case, individual modules can take the form of a cube, a cylinder, a parallelepiped, a parallelepiped with rounded corners, trapezoidal, triangular, hexagonal, spherical and others.

For a modular robot consisting of homogeneous blocks with identical functional and physical characteristics, it is somewhat easier (from a computational point of view) to perform self-reconfiguration and solve other tasks. However, the use of universal blocks increases the cost of a modular robot, therefore, in a number of works it is proposed to use heterogeneous blocks, including passive elements. In many modular robots two types of main and connecting blocks are used [3]. In [4], 18 types of blocks are used, differing in form, connection mechanisms, methods of motion, capture, the presence of sensors and signal processing facilities.

When developing the mechanism for connecting blocks, one should take into account its necessary strength, reversibility and repeatability, power consumption, permissible geometric dimensions, as well as environmental conditions of operation. The mechanical properties of modular couplings impose certain limitations on the space of possible reconfigurations of the modular robot. In the majority of works mechanical, magnetic and chemical methods of connection are used, methods of connection based on artificial muscles are also actively investigated [5].

* Corresponding author: ronzhin@iias.spb.su

A separate important issue in modular robotics is the communication problem, which in most cases is solved with the use of infra-red transceivers or wired communication means in the event of a permanent physical connection of all robot modules. Now the most widely used wireless communication tools are based on Bluetooth, ZigBee and other standards of the IEEE 802.15 protocol, as well as IEEE 802.11 WiFi. However, with wireless communication, as the number of modules increases, the problem of signal interference and channel capacity arises, given the small size and limited resources of the modules.

2 Related works

In [6], the motion of a closed chain-like modular robot in the rolling style is simulated. The complexity of the problem consists in calculating the position of each block, controlling the effects of the connecting mechanisms between two neighbouring blocks, which are necessary for the formation of motion, including on an uneven surface, and for maintaining the stable position of the entire modular robot. The experiments were carried out with a robot UBot during the self-rolling of a closed circuit in the form of a triangle and an oval along diverse types of surfaces, including the movement of the payload inside the chain.

The work [7] describes a novel approach to planning the reconfiguration of a modular robot based on the graph theory. The computational complexity of generating a graph structure is the logarithm of the number of modular robots involved. Calculations were made for modular robots M-TRAN and SuperBot with the number of involved elements up to 12 pieces.

The work [8] is dedicated to the problem of communication between modular robots which a priori do not have information about their situation and previous communication sessions. In the proposed approach of building a communication network using an infrared channel, neighbouring nearby modules are first identified, and then other robots are identified based on the XBee channel. The experiments were carried out with a maximum removal of 10 ModRED modular robots in an enclosed space of up to 30 meters and in an open space of up to 90 meters.

The review [9] presents the main tasks of rover robotics, behavioural algorithms (clustering, gathering in a swarm, collective food search), algorithms for clustering and sorting swarm elements, applied methods of navigation and communication, ways to distribute tasks within a robot swarm, as well as promising applications, solved on the basis of methods of swarm robotics.

In the work [10], the classification of modular robots is given and trellised and chain-like robots with 1D-3D topology are distinguished. Linear robots can be serpentine-like and serpenoid-like. The latter can move not only due to changes in shape but also built-in propellers (caterpillars, wheels, etc., installed on the outer surface of each block). For serpentine-like robots, three variants of pairwise connections and changes in the configuration of neighboring blocks are described: 1) in the vertical plane, which ensures the movement of the robot-snake only back and forth; 2) in the horizontal plane, which ensures movement in any direction along the plane; 3) in the vertical and horizontal planes, which ensures the movement of the robot-snake in three-dimensional space. The latter option has a more complex mechanism, consisting of two elements that ensure the rotation of the neighboring block, respectively, in the vertical and horizontal planes. Also, the work considers 12 principles of serpentine movement, implemented on the basis of the proposed mathematical models and developed mechanisms of the modular robot-snake.

Thus, despite the new tasks of robotics in the field of medicine and transport, the tasks of modular robotics are still far from the final solution and are actively explored. The proposed technological bases for designing modular robots with pairwise coupling mechanisms describe modern hardware and model-algorithmic tools necessary for the development of promising models of modular robots [11, 12]. These robots are equipped with systems typical of a fully functional robot, which provides their locomotion, communication, navigation, decentralized power and control [13-17].

3 Actuators

All types of the considered actuators have their advantages and disadvantages. Thus, different types of actuators are usually specialized and used for solving specific problems. In the case of large industrial works, where large capacities are required and massive loads are manipulated, pneumatic and hydraulic actuators are usually used, mainly due to greater economy compared to electric ones (which, however, does not negate the possibility of using the latter). As a consequence, large actuators, such as pneumatic and hydraulic actuators, which require external power supply or any other means of transmission, are not suitable when developing autonomous and miniature robotic systems.

For mobile robotic systems, servo drives are used most often because of their serial production (and therefore relatively low cost), reliability and ease of management. The main advantage of stepper drives is their accuracy. The supply of a control pulse to the drive winding causes the rotor to rotate by a strictly defined angle.

Stepper drive is used to automate individual units, where relatively high accuracy is required at a small angular velocity. However, there is a known problem associated with the possibility of "slipping" the rotor, which hinders the use of actuators based on stepper motors.

Actuators based on materials with shape memory and electroactive polymers have many advantages in the field of miniaturization and the repetition of the dynamics of real muscles and therefore have great potential. However, the high cost and the lack of an established industrial production makes the use of such actuators extremely difficult.

In the creation of a robotic module for performing swarm tasks several key aspects should be highlighted.

Firstly, the task of creating the above-mentioned robot module assumes the creation of a module for the formation of three-dimensional structures with the contact connection of other modules, rather than distributed tasks, with considerable removal of the modules from each other. Consequently, the main purpose of the actuator in such a robotic module will be to activate some mechanism for connecting the modules and, possibly, a mechanism that would allow several modules to move in space.

Secondly, the task of modular robotics involves a compromise between the miniaturization of modules, their functionality and their cost. Proceeding from the analysis of the already implemented methods of combining homogeneous robots, the most preferable variant is a hybrid variant of connecting the modules with the help of mechanical and electromagnetic means. This is the most tested and reliable version of the connection with the highest speed of operation, allowing relatively fast connection and disconnection of modules, albeit with great mechanical complexity being a disadvantage.

Based on the analysis of different types of actuators, as well as various projects on modular robotics for the last 30 years, it can be concluded that the servo drive is best suited as an actuator for a mechanical connector. This type of actuator is smaller than a stepper motor and is capable of providing high accuracy in the control of the gripping mechanism, but the obvious disadvantage of this option is its relatively high cost [18]. Concerning the method of displacement in space, it is difficult to draw unambiguous conclusions, since the method of moving modules in space and, accordingly, the type of actuator for implementing the displacement depend on the design of the modules. An analysis of the existing solutions shows that at the moment the most efficient way of moving the modules is rotating a part of the module around the axis (Project Roombots) or using a two-motor movement mechanism of the MTRAN-III modules.

4 Batteries

The choice of type of batteries depends on their energy density, dimensions, cost and other parameters. Lead-acid batteries have a rather long service life, are capable of giving off a large current, without a memory effect and have a very high operating temperature range. Lead-acid batteries are widely used to power automotive systems, emergency power systems, and also for use as uninterruptible power supplies. However, this type of battery has such significant drawbacks in the form of large dimensions and mass, and a rather low specific energy density, which in the case of modular robotics is unacceptable.

Nickel-cadmium batteries were widely distributed in the world as power sources for portable electronics, but then significantly lost ground to nickel-metal hydride batteries. The latter can give relatively large currents and are best suited for powering devices with short and high current consumption. This type of battery is the only one that is recommended to be stored in a discharged state. It has such advantages as a very long service life, relatively short charging time and low cost; however, the disadvantages of these batteries (high self-discharge, relatively low energy density and "memory effect") make them unsuitable in the development of modern electrical devices.

Developed as a replacement for nickel-cadmium batteries, nickel-metal hydride batteries are successfully used in various fields. Having equal dimensions with nickel-cadmium batteries, nickel-metal hydride batteries have capacity three times higher than the former, but half the service life and high self-discharge (1.5-2 times). Also, nickel-metal hydride batteries are practically not influenced by the memory effect and require a full discharge before recharging only with long-term storage (more than 30 days). All these advantages in combination with stable output voltage make nickel-metal hydride batteries a good replacement for nickel-cadmium as a power source for consumer electronics devices. In this case, high-capacity batteries (1500 - 3000 mAh) are more suitable for devices that require high energy consumption for a short period of time (radio-controlled models, players, cameras, etc.). Batteries with a lower capacity (300 - 1000 mAh) are more suitable for cases when the energy consumption does not start immediately but after a considerable amount of time after charging or in the case of long-term or periodic energy consumption (walkie-talkies, hand lamps, GPS-modules, toys). Disadvantages of self-discharge and service life are mostly eliminated in batteries with low self-discharge (LSD), but they have a lower capacity and a higher price. Such batteries are recommended for use in devices with high current consumption, as well as in devices with low consumption (watches, remote controls).

Lithium-cobalt batteries, being a traditional type of lithium-ion battery, surpass nickel-metal hydride batteries in a variety of parameters. They have low mass, high energy density, longer service life, low self-discharge and they lack the "memory effect". These qualities allow lithium-cobalt batteries to find the widest application in electrotechnical devices. However, these batteries are subject to aging and loss of capacity even when not in use, which should be considered when designing the device.

Lithium-manganese batteries are safer than lithium-cobalt batteries and are capable of withstanding rapid charging by high currents, and also give current of up to 5C. Because of this, they are most often used in devices with large current consumption: in high-power lanterns and radio-controlled models. The capacity of lithium-manganese batteries is slightly lower than that of lithium-cobalt batteries, and they are more sensitive to temperature changes; however, under operating conditions of lithium-manganese batteries (high current recoil), lithium-cobalt ones will most likely not be able to work and will break down because of overload.

Lithium-ferrophosphate batteries are very similar in characteristics to lithium-manganese batteries. The main differences lie in the lower cost and an ability to work under heavy loads (up to 10C). The structure of these batteries is such that, at critical loads, oxygen is not formed in the cell, which determines the increased safety of these elements. Unobtrusive, durable, resistant to low temperatures, lithium-ferrophosphate batteries are an excellent solution for powering devices operating under extreme conditions.

Lithium-polymer batteries, in fact, are an improved version of lithium-ion batteries. As an electrolyte, they use a polymer, which allows developers to create batteries of various shapes and sizes (including flexible ones). And although they are subject to aging and explosive during overloads, such advantages as low mass, high energy density, lack of "memory effect" and low self-discharge make this type of battery one of the best solutions for powering compact electrical devices. In the case of modular robotics, the size of the battery, its shape and mass play an extremely important role.

Based on the task of finding a balance between the optimal mass-scale parameters and the energy density, it is recommended to use lithium-polymer batteries for the development of homogeneous robotic modules. Of course, this choice imposes restrictions on the environmental conditions suitable for the operation of the device. Also, we should note new accumulators that have hydrogen, formic acid and other compounds.

5 Communication

When choosing the means of communication, one should consider the fact that in the case of rover robotics, the issue of the method of communication becomes even more significant, since the solution of the swarm tasks, both distributed and with direct contact interaction of the modules, requires a stable and fairly fast communication channel, both between the agents themselves and between the agents and the coordinator of the operation [19].

The communication problem is not trivial due to a number of physical limitations, the presence of a variety of interference, heterogeneity of the environment in which the modules are to work, etc. It should also be taken into account that there are different approaches to the algorithms of communication between agents. For example, the so-called tree topology does not imply a communication channel between all agents and is therefore fairly simple to use, while the network topology requires that each participant can take on the role of a communicator.

This organization is much more difficult to configure, but it provides much more fault tolerance, because failure of one of the network agents does not disrupt the functioning of the network as a whole. Because of this, the choice of the swarm control algorithm also influences the choice of the communication method. From a technical point of view, the chosen method and means of communication must correspond to the principles of compactness and economy, which are the main ones in modular robotics.

The considered methods of communication satisfy these parameters to varying degrees. The infrared channel (IR) is more robust to interference and external noise, safe, has an extremely low level of transmission errors, low power consumption and compactness, which is why it was used in the first mobile phones as a means of transferring media data for a relatively long time. Also, advanced code schemes for transferring data over the infrared channel provide a very high transmission rate. However, such disadvantages as the high cost of equipment (transmitters and receivers), as well as the need to maintain direct visibility between devices, make this method of communication not optimal for use in modular robotics.

The protocol that came to replace the infrared Bluetooth channel has significant advantages over IR technology. So, Bluetooth allows you to communicate at a relatively greater distance and create a connection with multiple devices at the same time, while also having a faster data transfer rate. In combination with low power consumption, these advantages generally make Bluetooth a good option for communication between modules.

However, it should be taken into account that Bluetooth operates under the "master-slave" principle and the interaction between master device and slave devices is possible, but between the slave device and several master devices is only realizable in more advanced devices with support for Bluetooth 4.1, the specification of which is somewhat ambiguous to meet the requirements of distributed networks. This imposes significant limitations on both the size of the swarm and the speed of the network.

The WiFi protocol has already become one of the main methods of wireless communication in contrast, it provides a greater number of simultaneously connected devices, as well as a longer connection range and a higher data transfer rate. For distributed networks, such characteristics are the best, but WiFi modules also have high power consumption, which is a critical disadvantage for the development of a compact robotic module. As a result, the use of WiFi technology is promising in modular robotics together with the development of batteries.

Finally, the radio modules working on the ZigBee protocol, specially developed for the creation of mesh networks, have a sufficiently long range for small dimensions (the range increases only with special antennas, which inevitably increases the overall dimensions), low power consumption, are capable of supporting a large number of modules, which is ideal suitable for a swarm network.

The disadvantages of this method of communication are a relatively low data transfer rate and low noise immunity, which, of course, is a significant problem in the work of modular robots in an extreme environment. However, the compactness, combined with the advantages already mentioned, makes this method of communication the most optimal for the development of modules, under the conditions of interacting at relatively short distances and without the need to work in an environment with a large amount of interference. It should also be taken into account that with the increase in the number of modular robots of up to a thousand or more, there is a serious problem of signal interference with the simultaneous exchange of data by all robots [20].

6 Sensors

The choice of sensor systems and sensors in homogeneous modular robots is, perhaps, one of the most difficult. In the creation of a modular robot there should be modules of a certain shape (cubic, spherical, parallelepipedal, etc.). In this regard, the sensors cannot be placed on the module body, since such arrangement of the sensors will complicate the dynamics of the modules and their interaction among themselves, and will also make the sensor itself vulnerable. As a result, the sensors used must be mounted in the housing, which, of course, will significantly increase the structural complexity of the module, most of which will be occupied by the propulsion unit and the battery. The use of sensors can also significantly increase the dimensions of the module, its mass and cost, since cheap sensors of standard sizes have rather large dimensions, and the reduction in the size of the sensors most often entails their substantial rise in price.

For these reasons, many projects to create homogeneous modular robots eliminated the use of sensors in modules, concentrating on their dynamics and the optimal method of connection. In this case, information about the environment of the module came from external sources (cameras surrounding the area of the robots, using the Kinect touch controller, etc.), or the modules did not have information about the environment and the location of other modules and were controlled directly by the operator. The ultimate goal is to create the most autonomous modules. However, on the basis of the comparative analysis of implemented modular robotic systems, it can be said that the higher the density of filling the module with functional elements (such as sensors), the more complex the module experiences when interacting with other modules.

The density of filling the module will also significantly affect the efficiency of the tasks. When performing tasks such as lifting the series-connected robots, quite large moments can act on the joints of the modules. As the practice of developers of already existing projects in modular robotics shows, the development of a complex self-sufficient self-reconfigurable modular robot results in a significant increase in the module size. For tasks involving holding modules, crawling and holding various objects, a large density will be a problem. Therefore, for problems of moving modules and other objects, an optimal choice between a complex design with a large functional and a small specific density requires a separate simulation. Due to the identified problems, further work on the project requires more precise definition of the working conditions of the robotic modules, a range of tasks that they can perform in view of the presented constraints, and, accordingly, a budget that can be incorporated into the creation of one module with specific parameters and functionality. Nevertheless, there are sensors, generally acceptable in size and price, for prototyping and clarifying emerging design difficulties, required modifications and missing information about the environment, which will require other internal or external sensors, or even a complete change in the design of the robotic module.

To implement the modular robot system, a MEMS gyro is recommended, which allows receiving information about the positioning of the module. Due to the cheaper production technology of MEMS systems, vibration MEMS gyroscopes have also become very cheap and widely available and often used in devices such as smartphones. Mass-dimensional parameters of such gyros are quite suitable for modular robotics. The question remains unclear as to what specific information about the external environment may be needed by the modules to perform specific tasks. However, in centering the modules relative to each other for connection, as well as determining the distance to the modules and obstacles around at the minimum range, an infrared distance sensor is miniature enough for insertion into the module frame. A sensor for determining the angle of rotation for the actuator is also required. In the case of a servo drive (recommended earlier for the development of mechanical connection of modules with hooks), a miniature turn sensor, based on the Hall effect (as in the dynamixel servo drives) and giving sufficient information about the angle of rotation of the motor rotor, is currently used. At the moment, the use of contact, sound and light sensors, as well as laser distance meters and GPS, is seen as inexpedient, since information from these devices at the moment cannot be classified as absolutely necessary for the operation of the modular robot.

7 Conclusion

The actuators, batteries, sensors and communication means, suitable for modular robotics, are considered. Further development of the robotic module systems and their design after the implementation of the declared functional will be possible after the testing of prototypes, and especially modeling and development of algorithmic support. To this end, at the next stage of the project, it is planned to analyze the limitations and justify the requirements for the model-algorithmic support for a single modular robot, which affect the methods of controlling pairwise connections of the ground-based robot. After this, it is planned to develop algorithms for controlling the connections of a swarm of terrestrial modular robots in the configuration of three-dimensional shapes.

This work is partially supported by the Russian Foundation for Basic Research (grant № 16-29-04101-ofi_m and grant № 16-08-00696).

References

1. H. Ahmadzadeh, E. Masehian, Artificial Intelligence, **223**, 27–64 (2015)

2. P. Moubarak, P. Ben-Tzvi, Robotics and Autonomous Systems, **60**, 1648–1663 (2012)
3. J. Baca, S.G.M. Hossain, P. Dasgupta, C.A. Nelson, A. Dutta, Robotics and Autonomous Systems, **62**, 1002–1015 (2014)
4. M.S. Moses, H. Ma, K.C. Wolfe, G.S. Chirikjian, Robotics and Autonomous Systems, **62**, 945–965 (2014)
5. J. Neubert, A. Rost, H. Lipson, IEEE Transactions on Robotics, **30**(6), 1344-1357 (2014)
6. X. Wang, H. Jin, Y. Zhu, B. Chen, D. Bie, Y. Zhang, J. Zhao, Robotics and Computer-Integrated Manufacturing, **39**, 56-67 (2016)
7. Kh. Taheri, H. Moradi, M. Asadpour, P. Parhami, Robotics and Autonomous Systems, **79**, 72-86 (2016)
8. J. Baca, B. Woosley, P. Dasgupta, C.A. Nelson, Robotics and Autonomous Systems, **91**, 284-298 (2017)
9. L. Bayındır, Neurocomputing, **172**(8), 292-321 (2016)
10. J. Gonzalez-Gomez, H. Zhang, E. Boemo, *Bioinspiration and Robotics: Walking and Climbing Robots,* 403-428, (Maki K. Habib (Ed.), InTech, 2007)
11. N.E. Shlyakhov, I.V. Vatamaniuk, A.L. Ronzhin, *IOP Conf. Series: Journal of Physics: Conf.,* **803**, 012146 (2017)
12. N. Pavluk, A. Denisov, A. Kodyakov, A. Ronzhin, *ICPEME 2016*, MATEC Web of Conferences, **77**, 04006 (2016)
13. A. Ivanov, O. Shmakov, SPIIRAS Proceedings, **49**(6), 190-207 (2016)
14. A.I. Motienko, A.L. Ronzhin, N.A. Pavlyuk, Science Bulletin of the NSTU, **3(60)**, 147–165 (2015)
15. R.M. Yusupov, A.L. Ronzhin, From Smart Devices to Smart Space, Herald of the Russian Academy of Sciences, **80**(1), 45-51 (2010)
16. A. Denisov, V. Budkov, D. Mikhalchenko, *ICR*, 70-79 (2016)
17. I. Vatamaniuk, G. Panina, A. Saveliev, A. Ronzhin, *IEEE ICARSC,* 306-310 (2016)
18. N. Pavluk, A. Ivin, V. Budkov, A. Kodyakov, A. Ronzhin, *ICR,* 113-123 (2016)
19. I. Vatamaniuk, D. Levonevskiy, A. Saveliev, A. Denisov, *SPECOM*, 588-595 (2016)
20. A. Saveliev, A. Ronzhin, Pattern Recognition and Image Analysis, **25**(3), 517–525 (2015)

Analysis of main tasks of precision farming solved with the use of robotic means

Vinh Nguyen[1], *Quyen* Vu[1], *Oksana Solenaya*[1] and *Andrey* Ronzhin[1,2*]

[1]SUAI, Department of Electromechanics and Robotics, 190000, Saint-Petersburg, Bolshaya Morskaya Str., 67, Russia
[2]SPIIRAS, Laboratory of Autonomous Robotic Systems, 199178, Saint-Petersburg, 14-th line of V.I., 39, Russia

Abstract. The main classes of problems of exact farming, solved with the involvement of robotics, are discussed. The use of precision navigation systems for robotic agricultural machines allows us to determine the actual soil parameters, calculate the cartograms for fertilizer application, taking into account the spatial variability of the agricultural contour. A promising method is the use of unmanned aerial vehicles for surveying and selecting intra field agricultural contours from a small height. To reduce the variability of soil characteristics in the fields, various robotic means of processing and preparation are used, including for leveling the soil. Given the regularity of planting crops on the fields, automatic systems of parallel driving of agricultural machinery are gaining popularity. Based on the analysis of modern robotic agro systems, promising areas for further research are formulated.

1 Introduction

The use of new technologies and technical means, such as positioning systems, automatic samplers, sensors, on-board computers, measuring systems and devices, mobile information systems and others, allows one to conduct spot survey of soil characteristics; collect, accumulate and analyze the data; draw up a map of the yield forecast of the cultivated plants; introduce mineral and organic fertilizers, chemical ameliorants in a certain field, taking into account the revealed variability of its characteristics in order to obtain maximum yield, reduce pollution and grow organic products; implement and monitor the intended recommendations; further monitor the characteristics of the field.

Agricultural robots (agrobots) have been discussed for a long time and were initially considered as automatic machines consisting of an executive device in the form of one or several manipulators and a device for controlling their movement [1]. Ten years ago, robots in agriculture were already being developed and existed, carrying out various operations such as: harvesting vegetables and fruits; weeding and thinning of vegetables; processing of greenhouses; commodity preparation of fruit and vegetable and berry products, including sorting, packaging and loading into containers; maintenance of agricultural complexes and livestock, including milking, sanitation, haircut; loading and unloading operations with agricultural products; driving of agricultural machinery, etc.

The active introduction of robotic means and new production technologies in agriculture also contributes to the solution of personnel problems. The scarcity of qualified personnel and young cadres in agriculture in Russia and other countries has increased even more in recent years. Specialists who choose to work in agriculture are becoming fewer and significant parts of them have pre-retirement age [2]. The prevailing view that agriculture is unattractive for young qualified personnel due to low pay is gradually changing. The use of modern technologies, including the use of robotics, can increase the prestige of the professions of the agrarian sector.

Now the global market for agricultural robots is experiencing significant growth [3-5]. Robots are used in many stages and in various types of agriculture: fieldwork, livestock, production and collection of food [6-8]. The market of agro robots in 2013 reached $ 817 million and according to expert forecasts it will reach $ 16,3 billion in 2020 [9]. The reasons for this colossal growth are the need for robotization of agricultural production, caused by the following factors: the need to increase labor productivity in agriculture; renewal of equipment and machinery used in agriculture; reduction of personnel risks; the need to facilitate or eliminate workers from physically challenging tasks and others.

For example, milking is one of the most difficult and laborious processes in dairy production, so the problem of automation of milking on farms is being actively explored. In the world market of milking robots, "DeLaval" takes the first place, controlling about 50% of the market, and is the undisputed leader of the Russian market: DeLaval 56%, WestphaliaSerge 6%, Boumatik 5%, Lelly 5%, Impulse AG 4%. Because of the lack of Russian analogues, the problems of import substitution and increasing the competitiveness of domestic products are relevant in agriculture.

* Corresponding author: ronzhin@iias.spb.su

2 Analysis of existing agricultural robots

Examples describing the specific implementation of robots in solving problems of agrarian tasks and precision farming are considered. In work [10], robotics are considered for production and processing of industrial horticultural products, focused on increasing yields, excluding the "human factor" in the production of goods, minimizing the harmful effects of chemicals on humans and the environment. In the All-Russian Institute of Mechanization (AIM) a robotic sprayer has been developed (with a load capacity of 500 kg, dimensions H x L x W not exceeding 1600 x 2300 x 1800 mm, road clearance 600 mm, the smallest turning radius not more than 299 mm, and speeds of translational motion at work 2 – 6 km/h and for transportation of 8 km/h) with an intelligent control system unit. The chassis of the sprayer uses an electric motor, has sensors and video cameras that provide positioning relative to plants using the technology of spectral image filtration.

At the Ural Agrarian University, the work is underway to create domestic robotics for agriculture, including the development of a set of scientific and methodological documentation regulating the creation, departmental acceptance and operation of agricultural robots; research methods of automation of technological processes in the agro-industrial complex (AIC) with the use of robotics; development of a database on the latest achievements in the field of robotics; carrying out mathematical and semi-detailed modeling of robotic agricultural complexes; creation of executive devices, modules and working bodies; creation of auxiliary systems and sensing systems; creation of prototypes of robots and design documentation for them for serial production [9].

In joint research of the All-Russian Research Institute of Agrochemistry and IPMeh of RAS, the task of robotization of agrochemical services is solved which consists in the creation of automated interactive programs for calculation of fertilizer doses in accordance with electronic agrochemical cartograms of soil fertility, planned crop yields, the level of the previous harvest, moisture availability, relief features, agrophysical soil properties and other factors. For the differentiated application of fertilizers, robotic aggregates consisting of a tractor, a mounted or trailed fertilizer spreader (applicator), electronic equipment including an onboard computer, a GPS/GLONASS navigator, an automatic driving system (autopilot), and additional fertilizing of vegetative crops sensors, are used. According to the results of experimental studies, the yield of cereals was increased from the territorial differentiation of fertilizer doses with the variability of agrochemical indicators of fertility, in various soils with humus content and in different soil-climatic conditions [11].

Also, there are prototypes of foreign robots for processing fruit and berry crops. The strawberry harvesting robot, developed at the California Institute of Technology, performs harvesting of berries by scanning the surface of rows from several sides and analyzing the images. Robotic devices Agrobot SW6010 and AGSHydro of the Spanish company AgroBot are used to collect only mature berries on hydroponic plantations with the help of 60 manipulators and a vision system. The robots of "VisionRoboticsCorporation" and the French company "Wall-ye" are used for trimming vines and equipped with a vision system from several built-in cameras. A full-wheel mobile robot NaioTechnologiesOz of the company Naio Technologies (France) is used for loosening the soil and destroying weeds in the inter-rows of cultivated plants.

In work [12] the possibility of using geostatistical methods for constructing cartograms of spatial distribution of parameter values characterizing soil fertility, as well as yield and quality of wheat grain is analyzed. The parameters under study are divided into three groups: soil characteristics; quantitative indicators of yield; indicators of grain quality. Samples were selected using a rectangular regular grid with a pitch of 25 × 50 m. The studies performed at the Menkovsky branch of the Agrophysical Research Institute showed that for all the analyzed soil properties there is a fairly strong spatial dependence, and to a greater extent it is related to the position of the sampling points. When choosing a sampling network, it is recommended to take a grid spacing less than half the semi variogram rank and, in a complex study of the field, use a grid with a step of not more than 30 m. The three-component model was based on the geostatistical studies, following which the studied field characteristics were considered as spatial variables, including: large-scale deterministic variation component, spatially correlated medium-scale component, spatially uncorrelated microcomponents. The method of modeling large-scale spatial changes is the construction of a line or surface regression from data obtained by observing at individual points. Characteristics of the soil have a pronounced spatial structure, reflected in the form of agrocontours. A cartogram of soil properties constructed with the help of regression kriging makes it possible to identify areas with an increased or decreased value of agrochemically important properties, and then, depending on the location and size of these zones, to make a decision on differentiated soil cultivation, in particular, to adjust the doses of mineral fertilizers applied. The variability of the characteristics of grain quality and yield in most cases was greater than the variability of soil properties. For yield indicators, there is also a spatial dependence, but weaker than for soil properties. Mapping yields is also possible with regression kriging. Dependences between soil properties and yield indicators can be established by cokriging or factorial kriging to improve accuracy in further research. Characteristics of the quality of wheat grain did not reveal any significant spatial connectivity, so the mapping of spatial distributions of these indicators is limited to the construction of trend surfaces. In this case, one should use any other method of constructing cartograms (for example, by the method of reciprocal distances), or conduct a repeated geostatistical analysis, but using the grid in steps of several meters, which, however, can lead to additional unreasonable costs.

In work [13] the results of the research of the Precision Agriculture Center of the RGAU-MAAA on the application of navigation equipment for sowing grain crops, differentiated feeding of crops are presented; evaluation of the fertilization of each section of the field with the compilation of electronic yield maps. To accomplish the tasks, the precision seeder D-9-30 was used in the studies with the GPS system and the marker, and the direct seeder DMC-3 using only the autopilot. In the course of the research, the variation in the width of the butt spacing between adjacent passages of seeders during the sowing of grain crops by a marker and autopilot was revealed. The values of the deflection of the butt spacing of the drill D-9-30 on the dumping background when using the marker in the range 3,2 - 4,8 sm, and when using the autopilot, respectively, was within 1,4-1,8 sm. When sowing cereals on zero and minimal background with the DMC-3 seed drill, the width of the butt-end spacing did not exceed 1 sm, averaging 0,1-0,4 sm. The quality of sowing was improved. When sowing with a marker, the qualitative characteristics of sowing were lower, the seed consumption increased by an average of 10-15%. To monitor and differentiate the application of fertilizers, depending on the condition of the crop plants, an optical sensor N-Sensor ALS (Germany) was installed on the John Deere tractor to scan the crops. The map of plant development, compiled on the basis of the NDVI-map, allows determining how the biomass is distributed within the field, and to calculate the coordinates of the application of fertilizers. The differentiated application of fertilizers ensured a saving of 15-20% of mineral fertilizers per hectare. Based on the evaluation of the characteristics of the soil in a particular section of the field, crop yield maps were generated, which allow identifying the problem areas of the field that require the application of fertilizers. The results of the research showed that the yield of cereals depended more on the timely processing and fertilization of the soil than on the technology of its cultivation.

In work [14] a comparative evaluation of parallel driving systems (PDS) for controlling the movement of tractors based on the GPS navigation system was carried out. On the fields of the test range of NTC KubNIITM, a number of PDS were tested, including foreign production: "Geomir-pilot"; RAVEN Cruizer II; EZ guide 250; Topcon X14; S-Lite Outback Guidance; Trimble EZ pilot CFX 750 CenterPoint RTX when carrying out the technological operation of applying solid mineral fertilizers. All of PDS devices were installed on the MTZ-82 tractor in a Bogballe M2 base spreader. The results of the analysis of the performance of the PDS showed the advantage of the Trimble EZ pilot CFX 750 system, which is the most expensive of the systems under consideration. The results of the measurements of the working width of the application of fertilizers indicate that the oscillation of the distance between traces of the tractor's wheels of adjacent passes (expressed in terms of the coefficient of variation) while driving with the PDS (± 0,075 m) is lower compared to the straightness of driving in the traditional way (± 0,175 m). On the whole, it was suggested that the scope of work (4,500 ha) for the proposed options for retrofitting a unit of PDS for spreading mineral fertilizers during the cultivation of winter wheat be limited to one MTZ-82 tractor with a Bogballe M2 base mineral fertilizer spreader and one of the analyzed PDS, while in the base variant, two MTZ-82 tractors with Bogballe M2 base spreaders are needed. Thus, the results of implementation of any of the six options for retrofitting the PDS unit in comparison with the base version will result in a 50% reduction in the demand for tractors and spreaders of mineral fertilizers; decrease in capital investment by 1,18 - 1,36 million rubles. Or by 42,4% - 28,9%; decrease in the cost of mechanized works by 7,8 - 9,0 rubles per hectare (by 15,7% - 18,1%).

In [15] it is proposed to determine the state of the microrelief of rice checks with the aid of an auto-leveler. The main tasks for the implementation of accurate farming on rice pens: to collect and store the initial data on geodetic survey and soil characteristics of rice checks; create a computer database of the collected data array; manage the process of further collection, accounting and processing of incoming information; to compile current forecasts of rice yields and irrigation water consumption on a check, a map for individual farms, districts and the whole for the coming years; determine the dynamics of deformation of the surface of checks as a function of time and other factors and establish the frequency of planning work; to determine when shooting and store design marks of the layout of rice checks that ensure the balance of excavation; justify the choice of digging and planning machines, planning technologies and technical and economic indicators based on the obtained geodetic and soil characteristics of rice checks. The most relevant geodetic and soil characteristics of precision farming on rice pens are: microrelief of rice checks and its calculated indicators; group and density of soil, determining the difficulty of its development by the working bodies of digging machines; the soil moisture, on which the operating conditions of the machines depend; thickness of the fertile soil layer, establishing the type of planning and technology of planning works.

In the research conducted by OJSC Engineering Center "Luch" for the vertical survey of the surface of the rice check, the AN-2 and AN-3 autoliners installed on the vehicle were used. Vertical survey on the check is carried out from the starting point by rectilinear movement of the vehicle along the marked routes by the shuttle way. To increase the productivity of the autogenerator, it is also planned to use ground-based equipment such as "Agronavigator" (EZ-Guide 250 from Trimble), which works with the satellite navigation system GPS and GLONASS. From the soil-soil characteristics of the rice check, it is recommended first of all to choose the density, soil moisture and the thickness of the fertile soil layer. The density is measured with a densitometer (DORNII drummer), the soil moisture is measured by drying samples or by means of moisture meters by immersion in the ground of special probes with stainless steel sensors, and the thickness of the fertile soil layer is determined by drilling wells or digging holes into the depth of the fertile layer in different places of the check. To take samples of soil on the field, the automatic sampler Amity A 2450 (Fritzmeier) is used, which in the laboratory allows determining various soil characteristics, including density, moisture and thickness of the fertile layer. The sampler equipment weighing 100 kg is mounted on a tractor. The sampling time of one sample is 10-12 seconds, the depth is 61 cm, the number of allowable samples per pass is 25 pieces.

Further increase in productivity and introduction of agro robots is associated with the development of local navigation technologies, precise positioning of manipulators, vision technology, onboard calculators and multisensor sensors, decision support systems based on multi-criteria evaluation, strategies for group management of heterogeneous robots in solving agricultural problems [16-20].

3 Main tasks of precision farming

The greatest progress in the robotization of agricultural production is currently visible in the field of precise farming (PF), the distinguishing feature of which is the local differentiation of agrotechnical influences, taking into account the spatial variability of soil and other factors of crop formation within a single agricultural field (agrocontour) [12].

Work in the field of precision farming is divided into 3 main stages: 1) primary exploration, 2) analysis of monitoring results, and 3) development of a field treatment strategy for a specific agricultural operation or in general for the following year. Figure 1 presents a classification of the main tasks of precision farming, solved with the use of robotic agricultural tools, in the three above-mentioned stages.

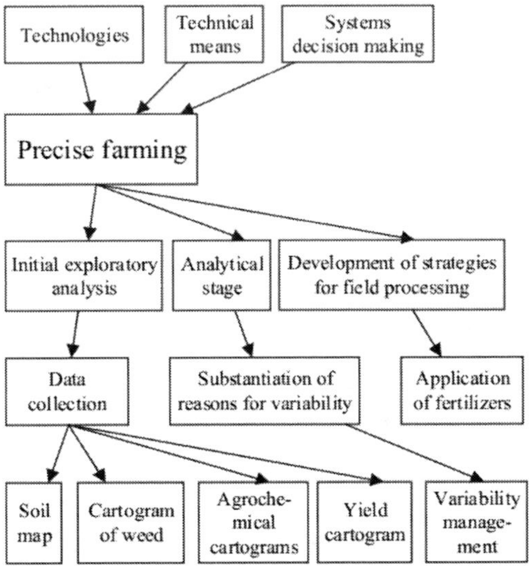

Fig. 1. Classification of the main tasks of precision farming.

The primary exploratory analysis includes the collection of data on conditions on a specific field, as well as various ways to visualize them-drawing up a soil map, plotting cartograms for yields, debris, agrochemical cartograms, and so on. In the studies devoted to precision agriculture, the following classes of problems were solved at the first stage [12; 21]: justification of the expediency of differentiating technological impacts; choice of the optimal step and sampling scheme; analysis of the obtained variability of soil properties under different methods of land use, agrotechnology or on different types of fields; the construction of cartograms of agrochemical and agrophysical properties of fields, as well as their yield and contamination [22].

At the analytical stage of PF, the reasons for the variability of soil properties are identified and priority methods for managing them are established. Then, at the third stage, a strategy is developed for processing the field for a specific agricultural operation or in general for the next year. At the same time, the question is solved which fertilizers should be applied, in what doses and how they should be corrected within the considered inhomogeneous field.

Remote sensing based on aerial photography using unmanned aerial vehicles (UAV) in recent years has been actively used due to technical progress, reducing costs and dimensions of sensors, the development of a global positioning system, intelligent programming systems and flight control. Improved parameters of spatial and temporal resolution of aerial photography using UAVs make it possible to extract more data on the state of the leaf cover of the crops. Figure 2 presents a classification of the main monitoring tasks, solved with the help of UAVs on various agricultural lands.

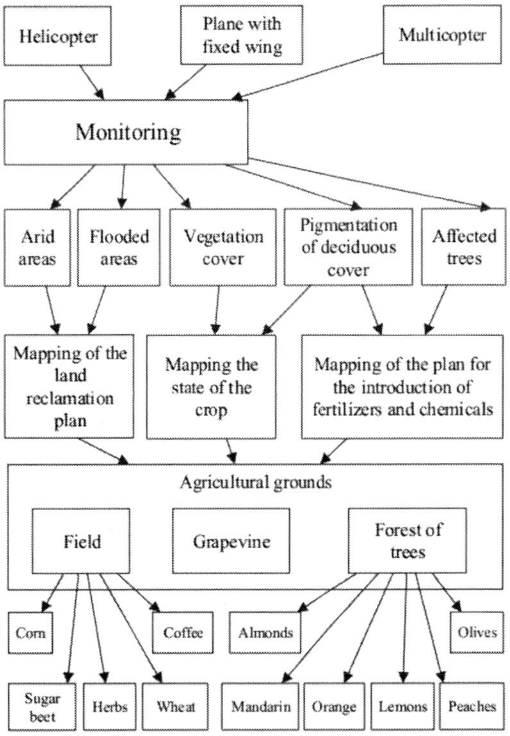

Fig. 2. Classification of tasks of agrarian robotics solved by UAVs.

The work [11] is devoted to precision agriculture, the problem of analyzing the diversity of in-ground agricultural contours is raised, which is difficult to solve using traditional survey methods and field mapping, calculation of fertilizer doses and their introduction, determination of seed sowing rates, other operations for each allocated agro-contour. Introduction of a part of soils on the field of averaged doses of fertilizers significantly affects the yield and ecological compatibility of the cultivated products.

The development of robotics for precision farming is associated with the solution of a number of problems: the creation of systems for the accurate navigation of agricultural machines; determination of the actual level of soil fertility; calculation of fertilizer doses for the planned yield of agricultural crops.

The task of determining the actual level of fertility in agricultural fields is complicated by the need to isolate the under-floor areas that differ in their fertility. Among the remote methods of analysis are: 1) the use of space vehicles, which is quite expensive, inoperative and the quality of the survey is highly dependent on weather conditions; 2) the use of unmanned aerial vehicles (UAV) for surveying and revealing under-floor agrarian contours from a small height, which is a more promising due to increased accuracy, promptness of the diagnostic survey and relatively low cost. To the terrestrial methods of automatic determination of the yield of agricultural crops, first of all, special on-board sensors installed on harvesters and conducting analysis during harvesting are included.

A promising way of allocating agro-soil contours is topographic survey of field relief. Another way to determine the level of fertility is to scan the electrical conductivity of the soil. These methods, combined with automated sampling of soil and the use of GIS-technologies for the processing of field cartographic material, significantly reduce the costs of agrochemical field survey compared to traditional methods [11].

4 Conclusion

The relevance of the application of precision farming in the agrarian sector is caused by the reasons of the transition of traditional agriculture to new production technologies, the solution of personnel problems, increasing productivity and eliminating the human factor from physically challenging tasks. Agricultural robots are used in almost all stages and different types of agriculture. The world market of agricultural robots is developing with significant growth, however, in Russia own production of agricultural robots is still limited.

The use of robotics in precision farming will make it possible to solve a number of problems: the creation of systems for the accurate navigation of agricultural machines; determination of the actual level of soil fertility; calculation of fertilizer doses for the planned yield of agricultural crops. A distinctive feature of accurate farming is the local differentiation of agrotechnical influences, taking into account the spatial variability of soil and other factors of crop formation within the agricultural contour. To the task of determining the actual level of soil fertility, a promising way is to use UAV for surveying and revealing the under-floor agrocontours from a small height. It is connected with the solution of the problem of allocation of agro-soil contours, a topographic method of topography of fields is a promising method.

The analysis of modern works in the field of precision farming has shown that the main achievement of the use of remote and terrestrial methods for monitoring the soil state is the allocation of agro-contours in their fields that differ in their characteristics, and the subsequent processing and application of fertilizers in the agro-contours, depending on their parameters.

To reduce the variability of soil characteristics in the fields, various robotic means of processing and preparation are used, including for leveling the soil. Given the regularity of planting crops on the fields, automatic systems of parallel driving of agricultural machinery are gaining popularity, providing a smaller deviation from straight-ahead driving in comparison with the traditional method. Parallel driving is carried out from the starting point by rectilinear movement of the vehicle along the marked routes by the shuttle way.

This work is partially supported by state research № 0073-2015-0003 and the Russian Foundation for Basic Research (grant № 16-08-00696).

References

1. E.I. Yurevich, *Fundamentals of robotics (*2-nd ed., Revised and additional. – SPb: BHV-Petersburg, 2005)
2. V.I. Nabokov, E.A. Skvortsov, M.K. Saakyan, E.G. Skvortsova, Economic sciences, **4**(31), 149-153 (2015)
3. A. Bechar, C. Vigneault, Biosystems Engineering, **149**, 94-111 (2016)
4. N. Noguchi, O.C.J. Barawid, *Proceedings of the 18th World Congress The International Federation of Automatic Control,* 633-637 (2011)
5. S. Wolfert, L. Ge, C. Verdouwa, M.J. Bogaardt, Agricultural Systems, **153**, 69-80 (2017)
6. M. Perez-Ruız, D.C. Slaughter, F.A. Fathallah, C.J. Gliever, B.J. Miller, Biosystems Engineering, **126**, 45-55 (2014)
7. F. Rovira-Mas, I. Chatterjee, V. Saiz-Rubio, Computers and Electronics in Agriculture, **112**, 172-183 (2015)
8. L. Holloway, C. Bear, K. Wilkinson, Journal of Rural Studies, **33**, 131-140 (2014)
9. E.A. Skvortsov, Vestnik of the State Agrarian University of the northern Trans-Urals, **3**, 113-117 (2015)
10. A.Yu. Izmailov, Z.A. Gojaev, R.A. Afanasiev, Fertility, **5**(92), 9–13 (2016)

11. R.A. Afanasiev, I.L. Ermolov, Mechatronics, automation, management, **12**, 828-833 (2016)
12. V.A. Sidorova, E.E. Zhukovsky, P.V. Lekomtsev, V.V. Yakushev, Agrochemistry and fertility of soils, **8**, 879-888 (2012)
13. A.I. Belenkov, A.Yu. Tyumakov, M.U. Sabo, Bulletin of the Altai State Agrarian University, **4**(126), 5-10 (2015)
14. D.A. Petukhov, M.E. Chaplygin, S.A. Sviridova, V. Voronkov, Agrosnabforum, **1**, 18-21 (2016)
15. A.N. Efremov. *Collection of proceedings of the conference Integrated melioration – a means of increasing the productivity of agricultural lands*, 278-287 (2014).
16. A. Ronzhin, A. Saveliev, O. Basov, S. Solyonyj. Springer-Verlag Berlin Heidelberg, A. Ronzhin et al. (Eds.): *ICR 2016, LNCS/LNAI, 9812,* 32-39 (2016)
17. N. Pavluk, A. Ivin, V. Budkov, A. Kodyakov, A. Ronzhin. Springer-Verlag Berlin Heidelberg, A. Ronzhin et al. (Eds.): *ICR 2016, LNCS/LNAI, 9812,* 113-123 (2016)
18. L. Chubraeva, A. Ronzhin, A. Shyshlakov, Al. Ronzhin, V. Shyshlakov. SPIIRAS Proceedings, **33**(2), 207-226 (2014)
19. A. Motienko, A. Tarasov, I. Dorozhko, O. Basov. SPIIRAS Proceedings, **46**(3), 169-189 (2016)
20. A. Ronzhin, I. Vatamaniuk, N. Pavliuk. *9th International Conference and Exposition on Electrical and Power Engineering (EPE)*, 675-680 (2016).
21. V.P. Yakushev, E.E. Zhukovsky, V.V. Yakushev Bulletin of the RAACS, **3**, 16-20 (2009)
22. P.V. Krasilnikov, V.A. Sidorova. *Geostatistics and geography of soils*. 67–80 (Moscow: Nauka 2007)

Service and multimedia data transmission in IoT networks using hybrid communication devices

Anton Saveliev[1,*], *Dmitry* Malov[2], *Michael* Tamashakin[1] and Victor Budkov[1]

[1]SPIIRAS, Laboratory of Speech and Multimodal Interfaces, 199178 Saint-Petersburg, 14-th line of V.I., 39, Russia
[2]SUAI, Department of Electromechanics and Robotics, 190000 Saint-Petersburg, Bolshaya Morskaya str., 67, Russia

Abstract. Employment of various protocols and technologies in IoT networks leads to the lack of module unification and increase in incompatible technical solutions. Modern IoT networks are not designed for streaming audio/video data, so their application field is limited. Also, modern IoT networks should have connection areas for devices transferring data to the Internet, and consider hardware and software specific characteristics of these devices. We offer one-size-fits-all solution for organization of IoT network, using hybrid modules. These devices provide flexibility, scalability, energy efficiency and multi-use of network for the transfer of various types of data. This approach takes into account software and hardware features of the devices used for data transmission in IoT networks, which helps to automate connecting the modules chosen by user.

1 Introduction

IoT networks are one of the most perspective solutions for increasing the level of automation in different areas of human activity. Currently, IoT technologies are being successfully implemented not only in service business, but also in real sector of economy, e.g. mining and oil & gas industry. One of the most popular application of IoT technologies is “Smart Home” systems, which allow simplifying human interaction with systems and devices of their house [1,2]. Also, such technologies are widely used in medicine [3,4] for extracting, storing and transmitting information about indicators of human vital functions.

Analysis of big data accumulated by the IoT devices allows formulating not only analytical reports, but also recommendations for decision makers [5,6]. Moreover, analytical data collected by the IoT systems can be useful during operation and testing of various devices. For example, in [7], a method for monitoring structural state of air and space craft using wireless sensor networks is presented. The paper discusses algorithm of data transmission using wireless IoT networks organization technology ZigBee. Also, the model of wireless sensor network consisting of gateway unit, routing device and motes (autonomous computational devices) with sensors is described.

Apart from information transmission, IoT networks are able, in particular, to simplify problem solution for indoor localization and navigation of autonomous mobile objects (users, robotics platforms, etc.). For example, paper [8] discusses the combined method for mobile object localization and improving its indoor positioning using wireless technologies. The current location of a person is determined using embedded sensors of his\her smartphone. The authors consider technologies sustained by the majority of modern mobile devices: accelerometer, gyroscope, magnetometer, proximity sensor, luminance sensor, as well as Wi-Fi, Bluetooth, GSM, NFC. The best solution, in authors' opinion, is using the Bluetooth beacon that operates according to the iBeacon technology. Besides, the conducted research shows the long-term benefits of the combined method.

One of the specific characters of wireless IoT networks is common wireless environment, based on single technology of data transmission, which makes multiple access of “smart devices” to the environment more difficult. In [9], the procedure of access regulation for “smart devices” was presented, which allows taking into account the specific characteristics of heterogeneous traffic service. The presented computation example proves the implementation efficiency of access regulation for decreasing delays in data frame service of highest priority subnets.

2 Review of popular audio and video codecs

Devices of IoT system should provide necessary communication channel capacity for transfer of diverse types of data. Codecs are used for more efficient and fast audio/video data transfer; it's possible to reduce amount of transmitted data, using codecs. Let us consider the most popular and widely used codecs. We can determine the minimum parameters of bitrate for IoT system development (Table 1, 2). Table 1 shows that the most demanding of network resources codec is MJPEG, so we

* Corresponding author: saveliev@iias.spb.su

set its bitrate as a minimum required network bandwidth for video data transfer.

Table 1. Bitrate of popular video codecs

Codec	H.264	MJPEG	VP8
Minimum bitrate, Kbps	736	17800	11064

Table 2. Bitrate of popular audio codecs

Codec	MP3	FLAC	ALAC	Opus	iSAC	iLBC
Minimum bitrate, Kbps	320	1200	256	510	56	16

The bitrates of the most popular audio codecs are presented in Table 2. The most resource-consuming codec is FLAC, so we set its bitrate as a minimum required network bandwidth for audio data transmission.

3 Key features of modules for data transfer in IoT networks

To select the appropriate communication devices, consider the key features of popular wireless technologies for IoT system implementation (Table 3).

Table 3. Main technologies for IoT-networks creation

Features	ZigBee	Z-Wave	Insteon	Bluetooth 4.0	Wi-Fi
Frequency, GHz	2,4	0,868-0,908	~0,915	2,4	2,4
Bandwidth, Kbps	Up to 250	Up to 100	1,2-180	Up to 1000	Up to 300000
Signal distance indoor, m	10-20	10-30	Up to 25	Up to 60	Up to 200
Max number of devices	65536	232	Not limited	8 devices for 1 piconet, max 10 piconets	30
Modes	Client, access point (CC2530)	Client, access point	Network made up of independent devices	Client, access point	Client, access point, client and access point
Number of connected devices	8	232	Not limited	7	30

Modern IoT systems can be implemented in many ways, but it is necessary to determine the set of data transmission technologies for unification of the development process. According to the characteristics considered above and the analysis of publications, the most perspective technologies for the IoT systems implementation are Wi-Fi and Bluetooth. First, consider their most popular and affordable embedded hardware solutions. A set of Bluetooth modules is presented in Figure 1: a, b, c. These modules can interact with a sensor or a microcontroller by means of the UART-interface technology.

Fig. 1. Bluetooth and Wi-Fi modules for IoT systems.

The HM-10 and HM-11 modules are low-powered, which makes it possible to use them in systems with low energy consumption. The effective radius of action of these modules is 60 m outdoors. The modules are small in size and weight and have preinstalled microcontroller software. The HM-12 module has the effective radius of action of up to 100 m outdoors. Moreover, it has lower energy consumption, compared with other models. All modules discussed above are compatible with Bluetooth 3.0 and 2.1 technologies, so it is possible to use them with the modules of previous versions: HC-05, HC-06, and HC-07. The main characteristics of the HM-10, HM-11 and HM-12 modules are presented in Table 4.

Table 4. Main characteristics of HM-10, HM-11, HM-12 modules

Name	HM-10	HM-11	HM-12
Bluetooth version	4.0 BLE	4.0 BLE	4.0 EDR and BLE
Operating voltage	3.3 V	3.3 V	3.3 V
Bitrate, bps	9600	115200	115200
Sensibility	≤ -84 dBm on 0.1%	≤-84 dBm on 0.1%	≤ -84 dBm on 0.1%

The Table 4 shows that the Bluetooth 4.0 technology is able to provide transmission of text data packages with low energy consumption, but its bitrate and bandwidth are not enough for audio/video data transfer. Next, consider the Wi-Fi technology that is able to provide enough bandwidth for multimedia streaming.

Apart from the Bluetooth modules, a range of widely used ESP modules can be implemented in IoT systems for data transmission. The ESP8266 module (Figure 1, d) operates with the UART-interface that simplifies the process of connection to a microcontroller or other periphery devices.

The module supports two operational modes: client and access point. Its maximum radius of connection is 400 m. The module energy consumption is 215 mA in transmission mode and up to 62 mA in receiving mode. The operational standard of the module is 802.11 b/g/n. Its bitrate is up to 3 Mbps that is enough for sound and image transmission. However, it's impossible to organize full-featured streaming video data transmission due to the limited bitrate.

It is reasonable to solve this problem by using the ESP32 module (Figure 1, e). The ESP32 is a hybrid (Wi-Fi and Bluetooth) chip, developed for using in systems, where the values of energy consumption are needed to be minimal (wearable electronics, «internet of things» etc.). The device has 6×6 mm size. The advantages of this module are: combining Wi-Fi and Bluetooth technologies in one chip, two processor cores and a wide range of peripheries. The ESP32 supports protocol stack of Wi-Fi standards 802.11n and BT4.2, providing the functionality by SPI/SDIO and I²C/UART interfaces. The Espressif ESP 32 chip can operate as a CPU (support for Open CPU) or as a "slave device", controlled by a microcontroller. The maximum power needed for data transmission over Wi-Fi is 19.5 dBm on the bitrate up to 150 Mbps. The module supports 3 operational modes: "client", "access point" and "client and access point simultaneously". It is possible to provide multimedia data transmission over Wi-Fi using an ESP32 module.

The Bluetooth technology allows organizing piconets in low energy consumption mode and to transfer text data. Thus, it is possible to distribute the load of various types of data transmission between few network standards and to provide the parallel information transmission. Moreover, the Bluetooth technology allows creating systems for space localization of various objects. One may use the Bluetooth modules from the HC range for nodes that are not supposed for multimedia streaming.

4 Mathematical models and algorithms of IoT network functioning

Consider bitrate-distance dependency and determine maximum distances between the modules for stable data transmission. To determine the bitrate-distance dependency for a Wi-Fi network we use data presented in the documentation of standards IEEE 802.11a and 802.11g.

On the basis of primary analysis of the documentation, we can conclude that a linear regression is adaptable for the experimental data approximation. Let us establish the equation of the linear regression of the form $y = ax + b$, where b is an intercept term, and a is a slope coefficient.

The given data are an array of distance values $\mathbf{distance}_{a,g}$ and a bitrate $\mathbf{bitrate}_{a,g}$. The given data for IEEE 802.11a standard are:

$\mathbf{distance}_a$ = [24,45,60,69,76,84,91,100], m;

$\mathbf{bitrate}_a$ = [54,48,36,24,18,12,9,6], Mbps.

The given data for IEEE 802.11g standard are:

$\mathbf{distance}_a$ = [30,53,76,84,100,107,110,114,122,128,134,137], m;

$\mathbf{bitrate}_a$ = [54,48,36,24,18,12,11,9,6,5.5,2,1], Mbps.

To determine the coefficients a and b the method of least squares is used. The found linear regression equations for IEEE 802.11a standard data (1) and IEEE 802.11g standard data (2) are:

$$bitrate_a = -0.71 distance_a + 74.82, \tag{1}$$

$$bitrate_g = -0.54 distance_g + 72.3. \tag{2}$$

The obtained dependencies are presented in Figure 2.

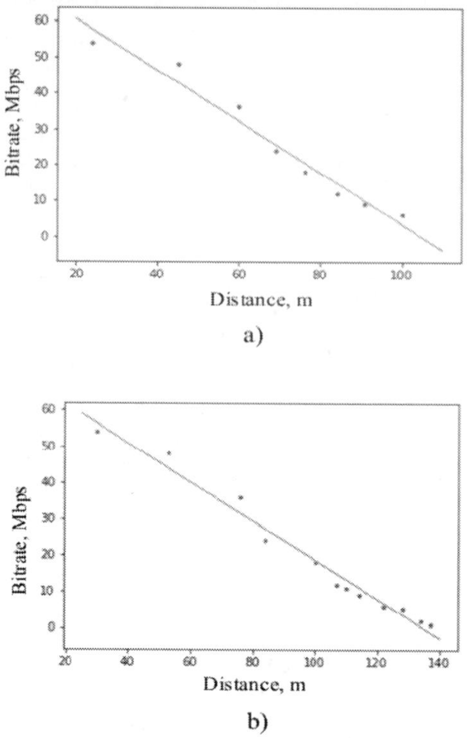

Fig. 2. Bitrate-distance dependency for standards: a) IEEE 802.11a, b) IEEE 802.11b.

As seen from Fig. 2, the considered standards are able to support minimum bitrate 10 Mbps at a distance of up to 80-90 m indoors without considering any obstacles.

To determine the bitrate-distance dependency for a Bluetooth-network we use formula (3) [17], based on the Friis transmission equation, that gives the power received by one antenna under idealized conditions given that another antenna transmits a known amount of power at some distance [18]:

$$RSSI = P_o - 10n \lg\left(\frac{d}{d_o}\right), \tag{3}$$

where d is the distance between device and transmitter, m; d_0 is the distance between device and point, where signal power P_0 was measured (1 m); n is the coefficient of signal energy losses (for air $n = 2$); $RSSI$ is the received radio signal power level, dBm. To determine the regression equation, we use data from the documentation of the Bluetooth IEEE 802.15.1 standard:

distance = [0, 10, 20, 40, 60, 80, 100], m;

RSSI = [-35, -38, -36, -37, -45, -48, -43], dBm.

Let us transform the **RSSI** values from dBm to W using equation (4):

$$P_W = 10^{((P_{dBm}-30)/10)}. \tag{4}$$

Next, compute the channel capacity using Shannon–Hartley theorem (5):

$$C = B\log_2\left(1+\frac{RSSI}{N}\right), \quad (5)$$

where C is the channel capacity, B is the channel bandwidth, N is the average power of the noise and interference over the bandwidth. The bitrate-distance dependency for Bluetooth modules is presented in Figure 3.

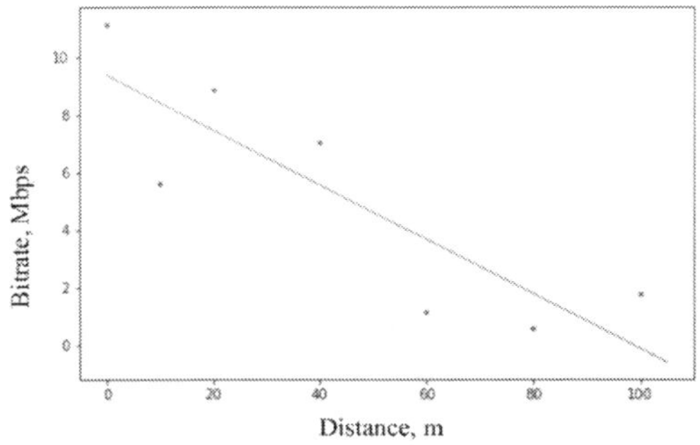

Fig. 3. Bitrate-distance dependency for standard IEEE 802.15.1.

Bluetooth-modules are able to maintain the 3,5 Mbps bitrate at a distance of 65-70 m (indoors without considering any obstacles). This value is enough for text data transmission. Thus, it is necessary to create an automatic system, which is able to take into account the difference between bitrates considered above. Also, the system should provide the end user with tools for simplified configuration of such hybrid network and selecting a gateway unit and devices. The algorithms developed to achieve this goal are presented in Figures 4 and 5.

An array of unprocessed devices is fed to the input of the algorithm presented on Figure 4. Next steps are consistently realized for each device in the set. The type of device is extracting. Based on this information, the system determines which devices can be connected with this module. Then, the system is looking for the closest device for the further connection establishment. If effective radiuses of the devices are disjoint, the necessary number of repeaters will be added into the network to provide connection between the devices. For appropriate algorithm operation, it is necessary to check whether any possible connections of the device are left. If there are no available connections left, the device is put in array of inaccessible devices and will be no longer used for searching for new connections. The last step of the algorithm is checking whether there are any unprocessed devices in the system. After the algorithm has done, the IoT system has fully connected devices.

To set a connection to the Internet, the end device data is fed to the input of the algorithm presented in Figure 5. These data are used to create the optimal areas of gateway unit installation or selecting the closest device with necessary type of connection: the user should point a place in the room or a device of the IoT network. Then, the system checks user's input data. If a device of the IoT network is chosen, the algorithm proceeds to the last step. In case of choosing a gateway unit, the system checks whether all the devices are within the gateway's effective radius. If it's not, some repeaters will be added in the system. The connections of these devices are removed and these devices get status "unprocessed". Then, the data about these devices are passed to the input of the algorithm for setting connections between the IoT devices (Fig. 4). The last step is the connection between the end device and the Internet.

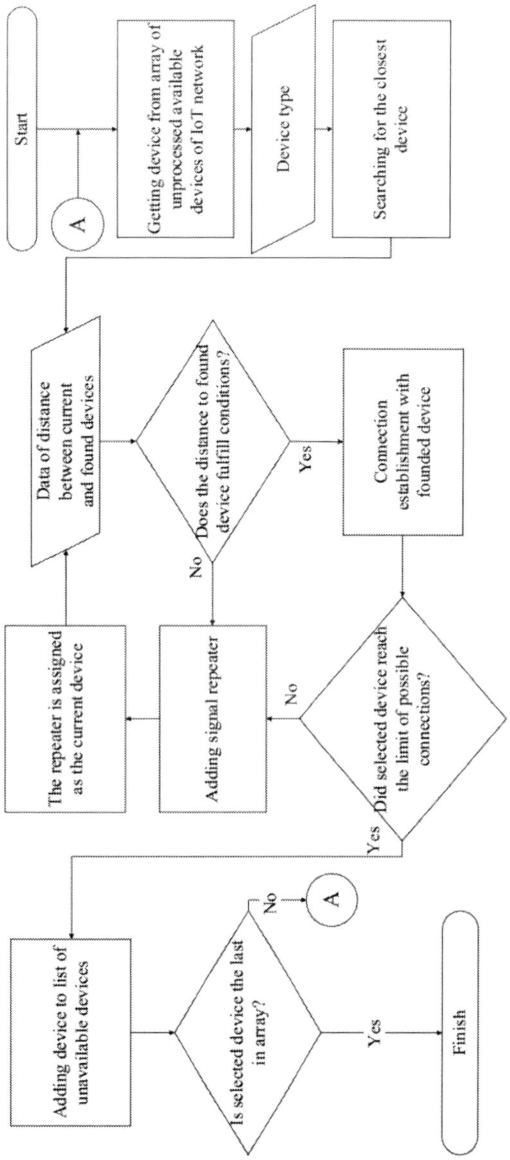

Fig. 4. Algorithm of IoT devices connection.

On the basis of the algorithms discussed above, consider a possible interface of the user-system interaction (Fig. 6).

At the first step (a), all devices are available to the user (except for gateway unit and repeater) for installation in the room. After the user has located all devices, the system shows devices' effective radiuses (b): the Wi-Fi module has green signal coverage, the Bluetooth module has blue signal coverage.

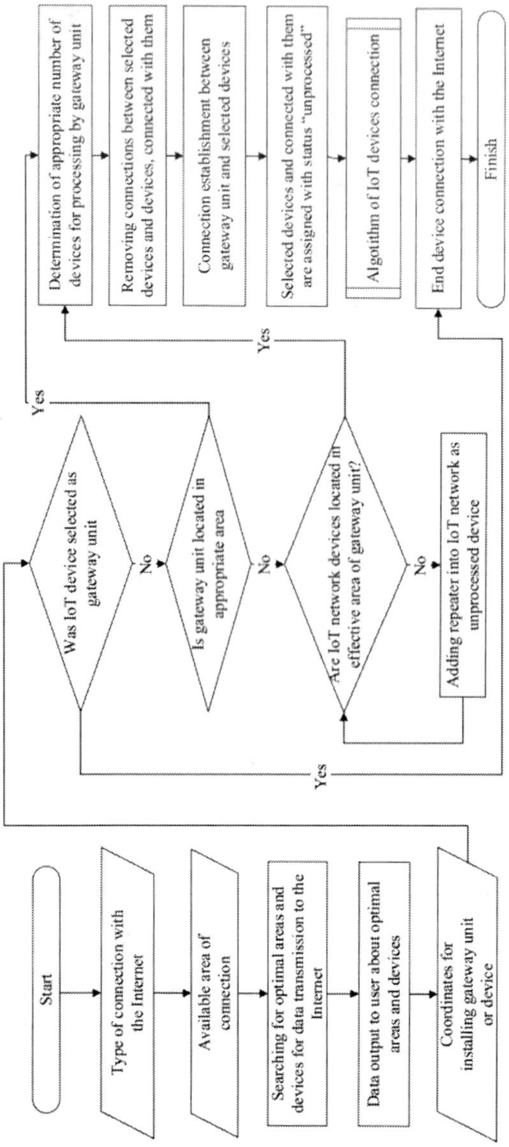

Fig. 5. Algorithm of connection of local IoT network with the Internet.

If user chooses a stand-alone device as a gateway unit, the system offers to install it in the appropriate area ((c): the gateway unit is marked as «GU»). Then the system checks possibility of network organization based on gateway's effective radius. In case it is not enough to cover all devices (d), the system offers to install repeaters where it is necessary ((e): the repeater is marked as «R») and waits for user confirmation of repeaters installation or changing gateway's location. If user decides to use an already processed device as a gateway unit, the system shows an updated effective radius of the gateway device ((f): the gateway device is marked as «GU»). After all the computations that are needed for organizing the IoT network are done, the system is waiting for user confirmation.

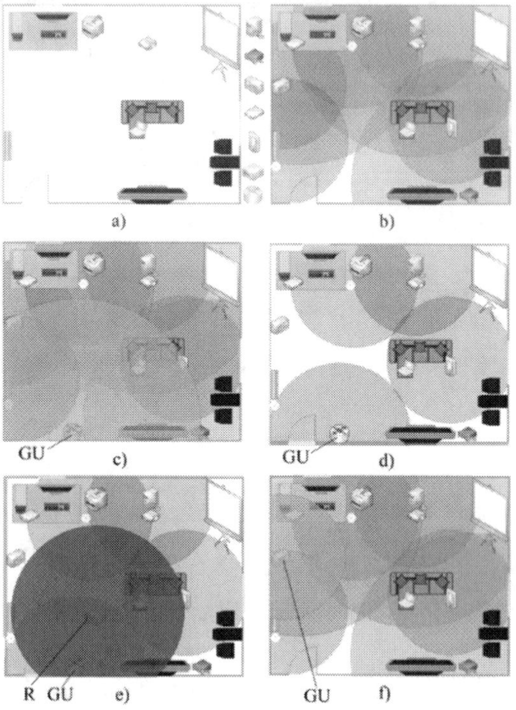

Fig. 6. System interface of hybrid system configuration.

5 Conclusion

In the paper the essential components for IoT network's organization were analysed. The parameters of network devices, necessary to provide multimedia data transmission, are found. Based on these parameters, the existing modules able to provide the necessary bitrate of audio/video streaming data transmission were selected. Moreover, the bitrate-distance dependency was analysed for modules working on Wi-Fi and Bluetooth technologies. These results were taken as a basis for the algorithms being developed for automatic connection of IoT network's modules. These algorithms allow estimating the distance between devices in the IoT system for providing necessary bitrate of data transmission; adding signal repeaters for the outermost devices' connectivity and organizing connections, which provide data transmission from the IoT network nodes to the Internet. The combination of data transmission technologies provides energy efficiency and cost reduction due to low-load nodes with low bitrate. Also, it supports a high bitrate in the areas where multimedia data transmission is needed. Thus, the hybrid communication modules essentially widen the application field of IoT networks and can be integrated in robotics systems [19, 20] for cyber-physical space interaction [21-23] and interaction with users by multimedia information distribution [24, 25].

This work is partially supported by the Russian Foundation for Basic Research (grant No. 15-07-06774-a) and the Council for Grants of the President of Russia (Projects No. MK-7925.2016.9).

References

1. H. Ghayvat, S. Mukhopadhyay, X. Gui, N. Suryadevara, Sensors, **15.5**, 10350-10379 (2015)
2. P. Thinagaran, S. K. Datta, C. Bonnet, *IEEE 4th Global Conference on Consumer Electronics (GCCE)* (2015)
3. Y. Geng, et al., IEEE transactions on industrial informatics, **10.4**, 2180-2191 (2014)
4. S.V. Zanjal, G. R. Talmale, Procedia Computer Science, **78**, 471-476 (2016)
5. F. J. Riggins, S. F. Wamba, *48th Hawaii International Conference on System Sciences (HICSS)*, 1531-1540 (2015)
6. B. Payam, A. Sheth, C. Henson, IEEE Intelligent Systems, **28.6** (2013)
7. M. V. Sergievskiy, S. N. Syroezhkin, Cloud of science, **3.1** (2016)
8. V.A. Rustinov, A.R. Sorokin, Radioelectronics and informatics, **4** (**67**) (2014)
9. N.A. Verzun, M. O. Kolbanev, Omsk scientific news, **4** (**148**) (2016)
10. K. Won-Ho, H. Lee, N. Sanjeewa, Life Science Journal, **11.7** (2014)
11. F. D. Yu, et al., Applied Mechanics and Materials. Vol. **543** (2014)
12. U. Miroslav, J. Bienik, M. Vaculik, *ELEKTRO IEEE*, (2016)
13. V. Jean-Marc, et al., *arXiv preprint arXiv:1602.04845* (2016)
14. J. Oliver, et al., *Elektronische Sprachsignalverarbeitung. Tagungsband der 27. Konferenz*, 81 (2015)
15. J. Paul, et al. *"RTP Payload Format for the iSAC Codec"* (2013)
16. L. Domingo, A. M. Gomez, J. L. Pérez-Córdoba, *22nd European Signal Processing Conference (EUSIPCO)* (2014)
17. P. James, T. Yang, C. Lee, eds., *Future Information Technology: 6th International Conference on Future Information Technology*, 184 (2011)
18. D. Rutledge, *"Encyclopedia of practical electronic"* (2002)
19. A. Ronzhin, I. Vatamaniuk, N. Pavluk, *EPE* (2016)
20. I. Vatamaniuk, G. Panina, A. Saveliev, A. Ronzhin, *ICARSC* (2016)
21. A. Ronzhin, A. Saveliev, V. Budkov, *Internet of Things, Smart Spaces, and Next Generation Networking*, 307-315 (2012)
22. A. Ronzhin, A. Saveliev, O. Basov, S. Solyonyj, *ICR* (2016)
23. I. Vatamaniuk, D. Levonevskiy, A. Saveliev, A. Denisov, *ICR* (2016)
24. A. Saveliev, SPIIRAS Proc., **31**, 61-79 (2013)
25. A. Saveliev, A. Ronzhin, Pattern Recognition and Image Analysis, **25.3**, 517-525 (2015)

MATEC Web of Conferences **113**, 02011 (2017) DOI: 10.1051/matecconf/201711302011
12th International Scientific-Technical Conference on Electromechanics and Robotics "Zavalishin's Readings" - 2017

Structure and basic functions of cognitive neural network machine

Vasiliy Osipov*

SPIIRAS, 199178, Saint-Petersburg, 14-th line of V.I., 39, Russia

Abstract. The structure of a cognitive machine based on a recurrent neural network with controlled elements is considered. The basic functions of this machine are defined and analyzed. The expanded capabilities of the cognitive machine to solve creative tasks are discussed. The results of modelling are given.

1 Introduction

In the coming years, small-scale cognitive neural network machines (CNMs) can form the basis for control systems of autonomous robots operating in difficult-to-forecast and difficult-to-reach conditions. These machines are understood as those that are capable of perceiving and connecting events (signals). Such CNMs should form the corresponding space-time models and develop creative solutions. These machines must quickly solve a wide range of difficult-formalized creative tasks.

There are several approaches to the creation of CNMs. These include digital, analog-digital and analog approaches. Examples of the digital approach are CNMs based on TrueNorth chips developed under the DARPA SyNAPSE program [1, 2]. This chip emulates 1 million neurons and 256 million synaptic connections between them. Among the analog-digital CNMs, it is possible to single out the systems developed within the framework of the BrainScaleS project: Neuromorphic processors. To create them, wafer with 384 HICANN chips is used. Each wafer allows emulating 200,000 neurons and 49 million synaptic connections [3]. As an analog cognitive neural network, there is a resistive processor of IBM [4], and others [5].

Now there is an opportunity to implement large neural networks and cognitive machines on chips with memristors. However, the level of development of artificial neural networks and ways of cognitive processing of information in them largely does not satisfy the needs of practice [6 - 13]. The possibilities of classical models of neural networks and the cognitive machines created on their basis do not go far beyond the recognition and classification of patterns [14].

The most promising models include artificial recurrent neural networks (RNNs) of real time. Unfortunately, the depth of information processing in traditional RNNs of real time is not high.

In recent years RNNs with controlled elements have been developed [15-19], on the basis of which perspective cognitive neural network machines can be created. These networks can be endowed with transparent logical structures and provide not only fast, but also deep processing of heterogeneous information.

Despite the results obtained, many issues related to the structural properties and capabilities of CNMs remain undeveloped. It is necessary to clarify the structural aspects of such machines and to determine their basic functions, allowing to provide expanded opportunities for solving creative tasks.

The article specifies the structure of a cognitive neural network machine based on a recurrent neural network with controlled elements. The system of extended basic functions of such a machine, which allows solving many creative tasks, is formed and discussed.

2 Structure of CNM

The basis of such a machine can be a two-layer pulse recurrent neural network with controlled elements, as well as blocks of forward and reverse transformation of signals (Fig. 1).

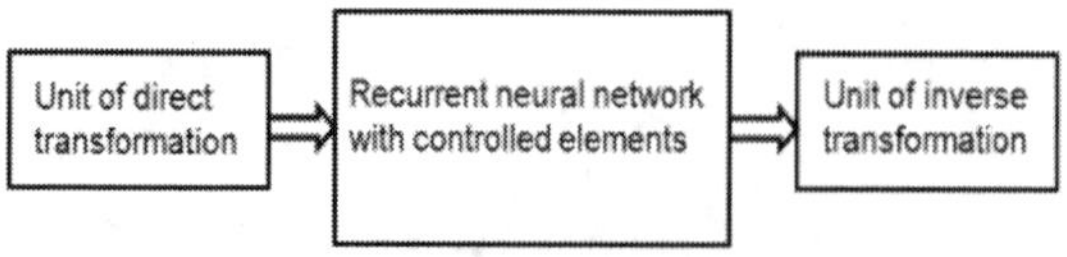

Fig. 1. Generalized structure of CNM.

The direct transformation of signals can involve first decomposing them into space-frequency components. Then each component of the signal is converted into a sequence of single pulses (SP) with a frequency and a phase, as functions of the amplitude and phase of the component. The frequencies of the components are

* Corresponding author: osipov_vasiliy@mail.ru

assigned to the numbers of the generated pulse sequences. As a result, the input of the neural network is given by successive single pulse populations (SPPs), carrying all the information about the input signals. At the output of the neural network, the resulting signals are formed from output SPPs. As part of the cognitive machine, it is proposed to use a recurrent neural network with a refined structure, reflected in Figure 2.

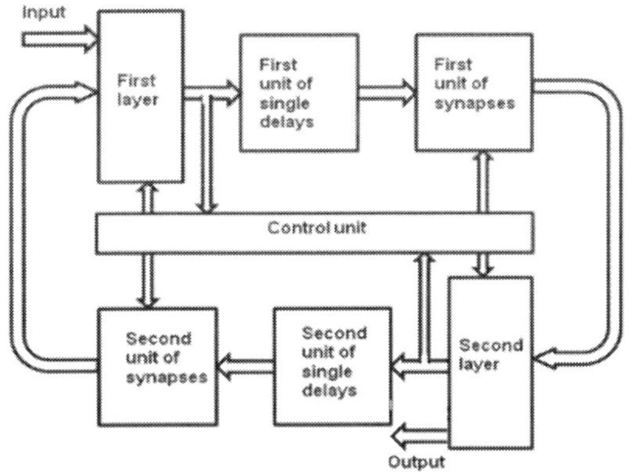

Fig. 2. Recurrent neural network with controlled elements.

The peculiarity of the RNN is that many functions for providing cognitive processing of information are assigned to the control unit. In this case, control actions can affect not only the characteristics of the synapses, but the capabilities of neurons. In this network, each neuron of one layer is in general connected with all the neurons of the interacting layer. Neurons of the same layer of bonds do not. This limitation can be justified by the need to minimize possible conflicts between neurons. The pulse delay time in the formed two-layer network circuits is less than the refractory time of neurons after excitation.

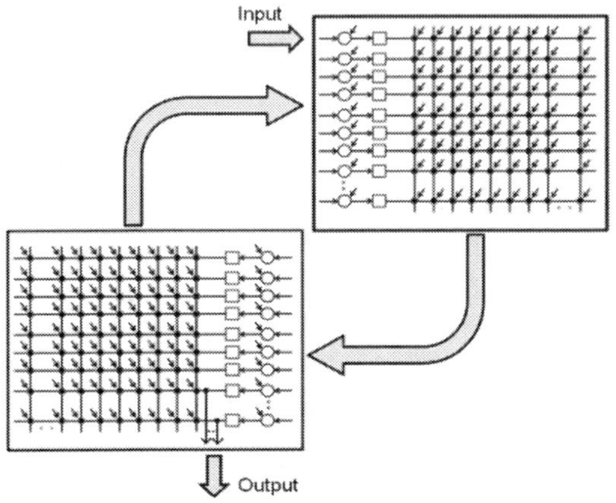

Fig. 3. Detailed structure of the recurrent neural network: neurons are indicated by circles, single delays - by squares, memristors - by fat dots, and control actions - by input small arrows.

At the level of individual neurons, single delays and synapses, such a neural network can be represented in the form of two controlled identical circuits, Fig. 3. The basis of these schemes are the controlled memristor crosbars, columns of neurons and single delays. In such a network, neuron layers are considered as two-dimensional structures. Moreover, due to the spatial shifts of SPPs, when transferring them from layer to layer, the network can be endowed with various linear, spiral, and other logical structures. The unambiguity between the input and output in the RNN is ensured by the priority of short bonds with the lowest attenuation. In addition, in such a network it is possible to rotate the processed SPPs and change the parameters of the divergence of the single pulses themselves. Changing these and other parameters depending on the current states of the layers can significantly affect on signal processing.

Fig. 4 shows one example of the space-time structure of RNN with three signaling systems, the basis of the CNM. Strengthening and weakening of interactions between signal systems is feasible by controlling the spatial parameters of single pulses diverging in the network. The same way it is feasible to form and interrupt the processing cycles of signals in the network, switching from one solved task to another.

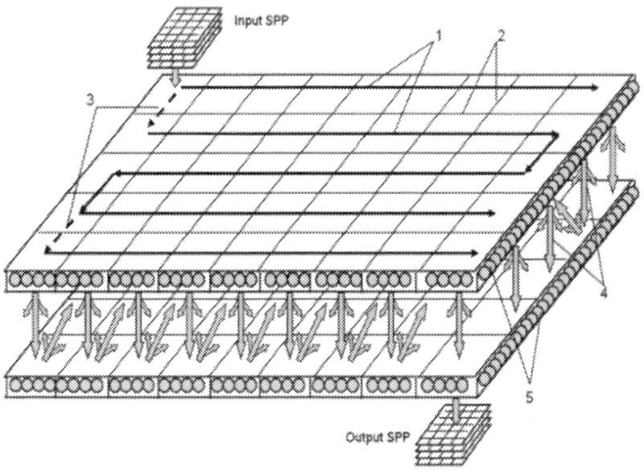

Fig. 4. Space-time structure of the RNN with three signaling systems: 1, 3, 4 — directions of forwarding SPPs along the layers and between them; 2 — lines, breaking down the layers into logical fields due to SPP shifts along the layers; 5 – neurons.

The CNMs created on the basis of such RNN can have wide functional capabilities. These machines are endowed with additional basic functions associated with controlling the spatial and temporal parameters of the signals processed in the network. Let us define the expanded system of basic functions of the cognitive machine below.

3 Basic functions

When forming a system of such functions, we will proceed from possible actions on signals in CNM. If we exclude from them the actions for direct and inverse transformation of signals, it all comes down to the functions realized by a recurrent neural network with controlled elements. Analyzing the capabilities of the

RNN, we distinguish the basic functions of signal processing and reduce them to Table 1.

The realization of F1 can be carried out with significant changes in the space-time relations between single pulses both within a SPP and their sequences. The greatest changes in such relationships are characteristic of spiral structures, and the smallest ones are linear RNNs. The current level of memorizing the results of signal binding depends on many factors. It is affected by the spatial and temporal distance between signals, their surroundings, the weights of the involved synapses, the total network load, and other factors.

Table 1. Basic functions of signal processing

Function Symbol	Function Content
F1	Promotion of SPPs over the network and the binding of these pulses.
F2	Storage of binding results.
F3	Acyclic and cyclic associative extraction of signals from the RNN memory with current signals.
F4	Suppression of false single pulses in signals propagated through the network, taking into account the previously stored information.
F5	Amplification and attenuation of signals.
F6	Spatial compression and expansion of SPPs when transmitting them from layer to layer.
F7	Rotations of such SPPs in the space of layers.
F8	Shifts of SPPs along the layers of the network.
F9	The change in the divergence parameters (shapes and directionality of the cross sections) of the single pulses themselves when they are transmitted from layer to layer.
F10	Compression and stretching of signals in time.
F11	Signal shifts in time.
F12	Space convolutions of SPPs.
F14	Time convolutions of complex signal constructs to simpler ones.
F14	Control of these functions, depending on the current status of the network elements.

According to F3, acyclic and cyclic associative extractions of signals from different memory areas can be simultaneously performed in the network. The type and power of recalling information from memory largely depend on its current direction, the number of pulses in RNN. Suppression of false single pulses according to F4 is feasible at low levels of signal amplification in the network. Rotations (F6) of the SPPs in the RNN along with other functions make it possible to expand the space of possible states of signal processing. An important place in cognitive processing is occupied by the function Φ9, which provides for an operative change in the spatial orientation of the associative interaction of signals in the RNN.

Based on the F14, complex sequences of actions on signals at different space-time levels of their representation can be formed. Some of the behavioral programs can be hard-coded into the memory of a cognitive machine when it is developed. At the same time, it is provided opportunities for self-learning based on these programs in the process of accumulating experience.

The main creative tasks that can potentially be solved when implementing the functions F1 - F14 in CNMs based on large RNN with controlled elements include the following:

1. The tasks of filtering and the reconstruction of distorted signals.
2. Problems of classification and recognition of dynamic one-, two- and three-dimensional signals.
3. Tasks of adaptive forecasting of various events in the interest of autonomous robots.
4. The tasks of looking to the past and analyzing the history of events.
5. The tasks of transforming some signals into others, including associative translation of speech from one language to another.
6. Creative synthesis and realization of the behavior programs of the cognitive machine itself and autonomous robot.

The list of such tasks can be significantly expanded when correlated with specific areas of application of CNMs.

4 Modelling results

By modeling, we studied empowering CNMs with advanced functions. Two-layer recurrent neural networks with linear and spiral space-time structures were implemented programmatically. The size of the layers did not exceed seven thousand neurons, and the number of synapses is ranged from one 1 to nine million. The possibilities of the RNN with various space-time structures and functions were compared. In particular, the two-layer RNN with 2016 neurons in each layer was studied. At the expense of spatial shifts of SPPs, each layer of the network was divided into six lines of 8 logical fields, 6 × 7 neurons each. Subsequent SPPs input to RNN, moved along the layers a spiral way during the signal processing. When transferring SPPs from layer to layer the parameters of the spatial divergence of single pulses changed on the current states of layers. Through this, the RNN generated controlled signal processing cycles.

Examples of smoothed spatial distribution of relative charges carried by currents in a diverging beam without and with taking into account the individual difference in potentials at the synapses are provided in Figure 5. Analysis of Figure 5 shows that the current states of neurons in the receiving layer may substantially alter the charge distribution in the divergent signal beams. Depending on the current states of RNN layers the same input signal retrieves different associated signals from the network memory.

Successful formation of cycles of associative call of information from the RNN memory for this associative and spatial addressing is realized when there are regions of the unexcited neurons after the inducing signals. Otherwise, input signals retrieve information from the memory which must precede them. To deeper process input signals, pauses are necessary between them, or deep processing should be temporarily separated from shallow one.

Figure 6 shows how, depending on the current states of layers, which change the spatial parameters of divergent signal beams, transmitted from layer to layer, the mode of functioning of the network is changed. In Fig. 6a, the RNN cyclically retrieves stored signals. The dips on the curve in Figure 6a reflect the fact that SPPs leave the RNN.

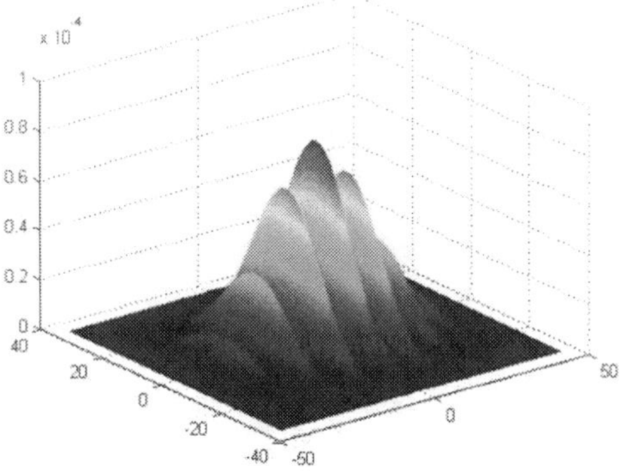

a

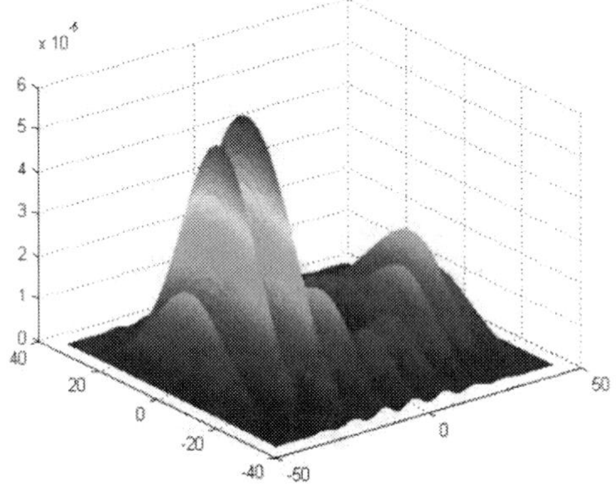

b

Fig. 5. Examples of a smoothed spatial distribution off relative charges carried by currents in a diverging beam: a - distribution without taking into account the individual difference of the potentials on synapses; b - the distribution with taking this into account.

When new SPPs changing the RNN state come to the RNN input, the cyclic information retrieval is interrupted (Fig. 6 b). Note that RNN capabilities on associative information retrieval depend not only on the number of training samples, but on the threshold of neuron excitation as well. The experiments showed that under certain conditions even with a single training sequence of SSPs the neural network is capable to associative retrieval of related signals.

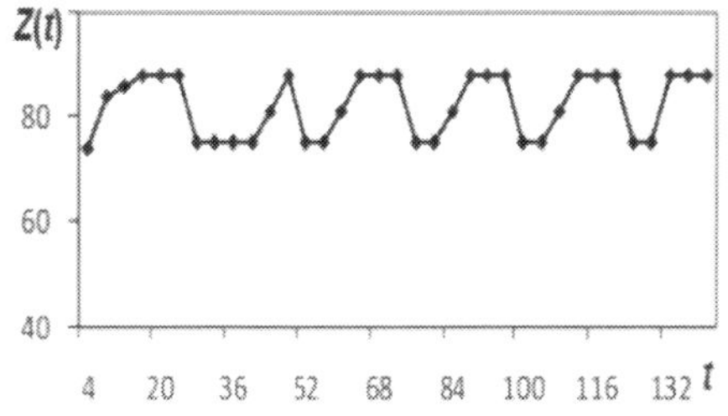

a

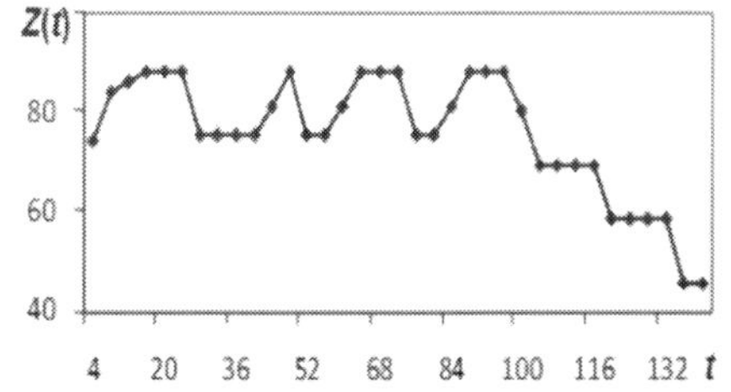

b

Fig. 6. Results of signals processing by a recurrent neural network: a - the dependency of the number Z(t) of excited neurons in the RNN from time at cyclic information retreival from memory; b - the same dependency, when the cyclic information retrieval from memory is interrupted by SPPs additionally introduced into the network.

It was confirmed that the more possibilities for changing the space-time and energy relationships between the signals in the RNN, the wider the range of creative tasks that CNM can potentially solve. For example, controlling only the directionality of the cross-sectional shapes of divergent pulses, depending on the current states of the layers, it was possible to organize and interrupt the processing cycles of the RNN signals in a wide range. By controlling only internal time, the RNN can adaptively predict possible events.

5 Conclusion

As a result of the study it can be argued that implementation of the CNMs with the proposed functionality can significantly increase their space of states and the capabilities of cognitive information processing. Due to additional space-time processing functions and depending on the current state of layers, more selective binding and storing of signals becomes feasible. More elaborated models of processed events are formed and stored in the network due to this mechanism. Opportunities for associative information retrieval from memory are improved as well. In contrast to the known solutions, the extended space-time processing makes available mechanisms to manage interactions of RNN subsystems, the memory load, the processing level, switching from one task to another.

Associative and spatial addressing to memory is proposed to be realized through rapid changes of spatial parameters of the signals being processed. Note that technically this addressing in RNN can also be accomplished by changing the attenuation functions of the diverging signals. Spatial shifts of the processed signals along the RNN layers can be justified by slow changes of the spatial parameters of the elements of the network itself because of minimizing potential conflicts among neurons.

The considered new solutions are focused at analog implementation of RNNs with memristors. Analog implementation allows overcoming the complexity problem of creating large RNNs with the number of artificial neurons and synapses, comparable to characteristics of biological systems.

The potential possibilities of the studied cognitive machine allow solving not only the problems of classification and recognition of various signals. This machine can, on the basis of experience, carry out an operative search for information of interest, adaptively analyze current and predict future events, synthesize and implement complex programs of various actions. Undoubtedly, when creating such machines in hardware, work will be required to provide the necessary experience. Later, this experience can be easily transferred from one machine to another without significant costs.

In the coming years, when creating small-size cognitive machines based on large RNNs with controlled elements, it is possible to go far beyond the existing limitations of cognitive processing of information.

The proposed approach for space-time processing of signals may be used in the development of promising small cognitive analog neural machines.

References

1. J.V. Arthur, P.A. Merolla, F.A. Akopyan, R. Alvarez, A. Cassidy, S. Chandra, S. Esser, N. Imam, W. Risk, D. Rubin, R. Manohar, D. Modha, *Joint Conference on Neural Networks, IJCNN* (2012)
2. S. Modha. *Introducing a Brain-inspired Computer TrueNorth's neurons to revolutionize system architecture*, IBM (2014) www.research.ibm.com
3. BrainScaleS – *Neuromorphic processors.* Heidelberg University, Germany (2014) www.artificialbrains.com
4. T. Gokmen, Yu. Vlasov, *Acceleration of Deep Neural Network Training with Resistive Cross-Point Devices*, https://arxiv.org/abs/1603.07341
5. Neuromorphic engineering, From Wikipedia, (2016) https://en.wikipedia.org
6. T. Madl, K. Chen, D. Montaldi, R. Trappl, Neural Netw., **65**, 18-43 (2015)
7. C. Xiu, C. Liu, Y. Cheng, Neurocomputing, **158**, 204-209 (2015)
8. C. Clopath, L. Busing, E. Vasililaki, W. Gerstner, Nature neuroscience, **13**(3), 344-352 (2010)
9. S. Jeong, M. Lee, Neural Netw., **25**, 130-140 (2012)
10. G. A. Montazer, D. Giveki, Neurocomputing, **168**, 221-233 (2015)
11. D.B. Strukov, G.S. Snider, D.R. Stewart, R.S. Williams, Nature, **453**, 80-83 (2008)
12. R. Berdan, E. Vasilaki, A. Khiat, G. Indiveri, A. Serb, Scientific reports, **6**, 1-9 (2016)
13. I. Vourkas, G. Ch. Sirakoulis, IEEE Circuits and systems magazine, third quarter, 15-30 (2016)
14. P. O. A. Haikonen, Cognitive Computation, **1**, 42-49 (2009)
15. V. Osipov, M. Osipova, Method and device of intellectual processing of information in neural network, RU Patent **2413304** (2011)
16. V. Osipov, Method for intelligent information processing in neural network, RU Patent **2427914** (2011), RU Patent **2502133** (2013)
17. V. Osipov, Informacionnye tehnologii, **21**(8), 631-637 (2015) (In Russian)
18. V. Osipov, *Advances in Neural Networks, LNCS 9719,* 177-184 (2016)
19. V. Osipov, SPIIRAS Proceedings, **30**(7), 141-155 (2013)

Mathematical model and algorithm of operation scheduling for monitoring situation in local waters

Boris Sokolov[1,*], *Vladimir* Palitsyn[2] and *Dmitriy* Nazarov[1]

[1]St. Petersburg Institute for Informatics and Automation of the Russian Academy of Sciences, 14th line 39, St. Petersburg, 199178, Russia
[2]St. Petersburg State University of Aerospace Instrumentation, Bolshaya Morskaya Str. 67A, Saint-Petersburg, 190000, Russia

Abstract. A multiple-model approach to description and investigation of control processes in regional maritime security system is presented. The processes considered in this paper were qualified as control processes of computing operations providing monitoring of the situation adding in the local water area and connected to relocation of different ships classes (further the active mobile objects (AMO)). Previously developed concept of active moving object (AMO) is used. The models describe operation of AMO automated monitoring and control system (AMCS) elements as well as their interaction with objects-in-service that are sources or recipients of information being processed. The unified description of various control processes allows synthesizing simultaneously both technical and functional structures of AMO AMCS. The algorithm for solving the scheduling problem is described in terms of the classical theory of optimal automatic control.

1 Introduction

At present, ensuring the safety of navigation of ships on inland waters of the Russian Federation is an urgent task, which is primarily due to the following factors: the number and intensity of ship movement in coastal marine areas, lakes and rivers, especially small-sized vessels used for active recreation, tourism, transportation of passengers, sports and as special means.

To date, various elements of the regional maritime security system (RMSS) have been developed, or are under development. The existing concept of the RMSS includes the following elements: - automatic monitoring systems (AMS); control systems of ship traffic; automatic identification systems; communication systems in case of disaster; a complex of terrestrial and satellite radio navigation systems; uniform database of courts and monitoring information. At the same time, as analysis shows, now, unfortunately, the automated collection and storage of monitoring information is of a local nature, there is an information inconsistency of monitoring data obtained by heterogeneous automated monitoring and control system (AMCS).

To compensate a contradiction between the existing need and to conduct operational analysis and prediction of the surface situation and the current state of the organizational and technical means ensuring the safety of navigation, methods, algorithms, techniques, software can be integrated for solving problems of analysis, planning and prediction of the surface situation on the basis of modern intellectual technologies. One of the most important tasks, which needs to be solved to achieve a goal, is the task of joint planning and scheduling of the measuring and computing operations providing monitoring of the situation in the local water area and connected to relocation of different ship classes (further the active mobile objects (AMO)) [1-2].

According to the specifics of the structure-dynamics control problems, they belong to the class of the AMCS structure-functional synthesis problems and the problems of program construction for AMCS development [3-7].

The main disadvantage of the problems belonging to the above class is that, optimal control programs for AMCS main elements and subsystems can be implemented only when the functions and algorithms for control and information processing in these subsystems and elements are known [5-6].

We propose a dynamic interpretation of monitoring situation in local waters, which is based on results of classical control theory [8-15].

2 Conceptual model of monitoring situation in local waters

The preliminary investigations confirm that the most convenient concept for formalization monitoring situation in local waters is the concept of an active moving object (AMO). In general case, it is an artificial object (a complex of devices) moving in space (in our case in local waters) and interacting (by means of information, energy, or material flows) with other AMO and objects-in-service (OS) [1, 2].

* Corresponding author: sokolov_boris@inbox.ru

Figure 1 shows a general structure of AMO as an object being controlled. It is seen that AMO consists of four subsystems relating to four processes (functioning forms): moving, interaction with OS and other AMO, functioning of the main (goal-oriented) and auxiliary facilities, resources consumption (replenishment).

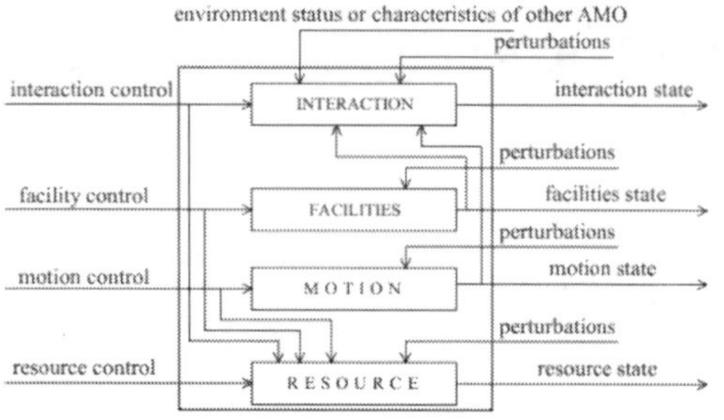

Fig. 1. General block-diagram of AMO

The four functions of AMO are quite different, though the joint execution of these functions, the interaction being the main one, provide for AMO new characteristics. Thus, it becomes a specific object of investigation, and AMO control problems are strictly different from classical problems of mechanical-motion control. The proposed structure of AMO can be widely interpreted. It gives a common basis for description ships on inland waters. In this case we use AMO for describing both ships and their automated monitoring and control system (AMCS) including ships (in our terms – AMO), ground-based radar observation stations (GRS), data processing center (DPC).

To construct the models of AMO AMCS we should firstly formulate the goal of its functioning. This goal is to be related to the interaction between other AMO and OS. Secondly, the sequence of operations that lead to the specified goal should be determined. Each operations are characterized by their parameters. Some of parameters describe the results of the operation performance (amount of work, quality, duration, resource consumption, etc), the others present the characteristics of material or information flows necessary to perform the operation.

Thus, at the conceptual level, the process of AMO AMCS functioning can be described as a process of operation execution, while each operation can be regarded as a transition from one state to another one. Meanwhile, it is convenient to characterize the AMO AMCS state by the parameters of operations.

The particular control models are based on the dynamic interpretation of operations and the previously developed particular dynamic models of AMO AMCS functioning.

3. Multiple-model description of AMO monitoring and control processes

The main aim of our paper is to prove the need of integrated modelling and simulation for parallel structural-functional synthesis of AMO AMCS under dynamic conditions. Moreover the main idea of our approach is to use fundamental results of structural-dynamic control theory [1,2-6] for multiple-model description of AMO AMCS planning and scheduling tasks. The dynamic interpretation of structure-dynamics control processes allows for application of the results, previously received in the theory of dynamic systems stability, ability, failure tolerance, effectiveness and sensitivity, for AMO AMCS analysis and synthesis problems [7-14].

Our investigation have shown that joint use of diverse models allows one to improve the flexibility and adaptability of AMO AMCS, as well as to compensate the drawbacks of one class of models by the advantages of the other.

We propose four interconnected dynamics models for description both ships (AMO) and their automated monitoring and control system (AMCS) including ground-based radar observation stations (GRS), data processing center (DPC).

3.1 Dynamic model of AMO motion control (Mg model)

$$M_g = \Big\{\mathbf{u}^{(g)}(t) \mid \dot{\mathbf{x}}_i^{(g)} = f_i^{(g)}\big(\mathbf{x}_i^{(g)}, \mathbf{u}_i^{(g)}, t\big), \\ \mathbf{q}^{(g)}(\mathbf{x}^{(g)}, \mathbf{u}^{(g)}, t) \le \mathbf{O}, \mathbf{h}_0^{(g)}(\mathbf{x}^{(g)}(T_0)) \le \mathbf{O}, \\ \mathbf{h}_1^{(g)}(\mathbf{x}^{(g)}(T_f)) \le \mathbf{O},\ i \in \widetilde{M}\Big\} \quad (1)$$

where $\mathbf{x}^{(g)} = \left\| \mathbf{x}_1^{(g)\mathrm{T}}, \mathbf{x}_2^{(g)\mathrm{T}}, ..., \mathbf{x}_{m+\overline{m}}^{(g)\mathrm{T}} \right\|^{\mathrm{T}}$ is a vector characterizing the state of motion for elements and subsystems of AMO AMCS, including ships (in our terms – AMO) and ground-based radar observation stations (GRS); $\mathbf{u}^{(g)} = \left\| \mathbf{u}_1^{(g)\mathrm{T}}, ..., \mathbf{u}_{m+\overline{m}}^{(g)\mathrm{T}} \right\|^{\mathrm{T}}$ is a vector of control inputs including, in a general case, program and regulative inputs; $\mathbf{q}^{(g)}(\mathbf{x}^{(g)}, \mathbf{u}^{(g)}, t)$, $\mathbf{h}_0^{(g)}(\mathbf{x}^{(g)}(t_0))$, $\mathbf{h}_0^{(g)}(\mathbf{x}^{(g)}(t_f))$ are known vector functions used for motion constraints and end conditions. An interaction operation can be performed when each communicating object is in the interaction zone (IZ) of the other one. These zones are described by the matrix function $E(t) = \|\varepsilon_{i,j}(t)\|$; $i,j \in \widetilde{M} = \{M \cup \overline{M}\}$.

This function was called contact potential [1, 2]. Under the above conditions, the contact potential for a pair of objects <AMO$_i$, GRS$_j$> can be obtained as follows [1]:

$$\varepsilon_{i,j}(t) = \gamma_+ \Big\{R_j^{(g)} - \left|\mathbf{r}_i^{(g)}(t) - \mathbf{r}_j^{(g)}(t)\right|\Big\}, \quad (2)$$

where $i,j \in \widetilde{M}$, $\gamma_+(\widetilde{\alpha}) = 1$ if $\widetilde{\alpha} \ge 0$, $\gamma_+(\widetilde{\alpha}) = 0$ if $\widetilde{\alpha} < 0$; $R_j^{(g)}(t)$ is IZ radius for the object B_j, $\mathbf{r}_i^{(g)}(t)$, $\mathbf{r}_j^{(g)}(t)$ are, in a general case, three-dimensional radius-vectors for B_i, B_j and are sub-vectors (together with $\dot{\mathbf{r}}_i^{(g)}(t)$, $\dot{\mathbf{r}}_j^{(g)}(t)$) of the vectors $\mathbf{x}_i^{(g)} = \| \mathbf{r}_i^{(g)}(t), \dot{\mathbf{r}}_i^{(g)}(t) \|^{\mathrm{T}}$,

$\mathbf{x}_j^{(g)} = \| \mathbf{r}_j^{(g)}(t), \dot{\mathbf{r}}_j^{(g)}(t) \|^{\mathrm{T}}$.

Obviously, for the stationary (fixed) objects we have $\mathbf{r}_i(t) = \mathbf{r}_i(t_0) = \mathbf{r}_{i0}$, where r_{i0} is a position of an object.

3.2 Dynamic model of AMO AMCS interaction-operations (IO) control (Mo model)

$$M_o = \Big\{ \mathbf{u}^{(o)}(t) \mid \dot{x}^{(o)}_{iæ} = \sum_{j=1}^{m} \varepsilon_{ij}(t) \cdot u^{(o)}_{iæj};\ x^{(o)}_{iæ}(t_0) = 0,$$
$$x^{(o)}_{iæ}(t_f) = a^{(o)}_{iæ}, \sum_{i=1}^{m}\sum_{æ=1}^{s_i} u^{(o)}_{iæj} \le c^{(o,1)}_j,$$
$$\sum_{j=1}^{m}\sum_{æ=1}^{s_j} u^{(o)}_{iæj} \le c^{(o,2)}_i,\ u^{(o)}_{iæj}(t) \in \{0,1\}, \qquad (3)$$
$$u^{(o)}_{iæj}\left[\sum_{\tilde{\alpha}\in\Gamma_{iæ1}} (a^{(o)}_{i\tilde{\alpha}} - x^{(o)}_{i\tilde{\alpha}}) + \prod_{\tilde{\beta}\in\Gamma_{iæ2}} (a^{(o)}_{i\tilde{\beta}} - x^{(o)}_{i\tilde{\beta}}) \right] = 0,$$
$$i, j = 1,\dots,m; i \ne j; æ = 1,\dots,s \Big\},$$

where $x^{(o)}_{iæ}$ is a variable characterizing the state of interaction operation $D^{(i)}_{æ}$; $a^{(o)}_{æ}, a^{(o)}_{i\tilde{\alpha}}, a^{(o)}_{i\tilde{\beta}}$ are given amounts of work for operations $D^{(i)}_{æ}$, $D^{(i)}_{\tilde{\alpha}}$, $D^{(i)}_{\tilde{\beta}}$; $u^{(o)}_{iæj}(t)$ is a control input, $u^{(o)}_{iæj}(t)=1$ if the operation $D^{(i)}_{æ}$ is being performed, $u^{(o)}_{iæj}(t)=0$ if not; $\Gamma_{iæ1}, \Gamma_{iæ2}$ are the sets of operations numbers for the operations that involve the object B_i and precede the operation $D^{(i)}_{æ}$; here $\Gamma_{iæ1}$ is used for the operations connected with $D^{(i)}_{æ}$ by the logical "$\wedge$" - ("and"-conjunction), $\Gamma_{iæ2}$ is used for the operations connected with $D^{(i)}_{æ}$ by the logical "$\vee$" ("or" - disjunction); $c^{(o,1)}_j, c^{(o,2)}_i$ are constants used for technical conditions of AMO AMCS functioning.

3.3 Dynamic model of AMO AMCS channels control (Mk model)

$$M_k = \Big\{ \mathbf{u}^{(k)}(t) \mid \dot{x}^{(k)}_{ij} =$$
$$= \sum_{l=1}^{m} \Theta_{lj}(t) \frac{h^{(j)}_{il} - x^{(k)}_{ij}}{x^{(k)}_{lj}} \cdot u^{(k)}_{lj};$$
$$x^{(k)}_{ij}(t_0), x^{(k)}_{ij}(t_f) \in \mathrm{R}^+, \qquad (4)$$
$$x^{(k)}_{ij} \cdot u^{(o)}_{iæj} = 0; \sum_{l=1}^{m} u^{(k)}_{lj} \le 1,$$
$$\sum_{j=1}^{m} u^{(k)}_{lj} \le 1, u^{(k)}_{lj}(t) \in \{0,1\}; i,j,l \in \tilde{M} \Big\},$$

where $x^{(k)}_{ij}$ is the state of readjustment process for the channel $C^{(i)}_{\lambda}$; $h^{(j)}_{il}$ is a given value of the readjustment duration; it is supposed that the channel was ready to service the object GRS_l at the beginning of the readjustment and must be ready to service the object AMO_i at the end of readjustment process; R^+ is a set of positive real numbers; $u^{(k)}_{lj}(t)$ is a control input; $u^{(k)}_{lj}(t)=1$ if C_j is in the readjustment state, $u^{(k)}_{lj}(t)=0$ if not; $\Theta_{lj}(t) \in \{0,1\}$ is a given value characterizing potential ability of the channel C_j to interact with the object B_i (in our case AMO_i).It can be seen that the operation $D^{(i)}_{æ}$ can be started when appropriate channel is ready [$x^{(k)}_{ij}(t)=0$], while each channel C_j at each time t can be prepared for only one object GRS_l ($l \ne j$, $l = 1,\dots m$).

3.4 Dynamic model of AMO AMCS operation parameters control (Me model)

$$M_e = \Big\{ \mathbf{u}^{(e)}(t) \mid \dot{\mathbf{x}}^{(g)}_i = F_i(t)\mathbf{x}^{(g)}_i;$$
$$y^{(i)}_j(t) = \mathbf{d}^{\mathrm{T}}_j(t)\mathbf{x}^{(g)}_i + \xi^{(e)}_j; \qquad (5)$$
$$\dot{Z}_i = -Z_iF_i - F_i^{\mathrm{T}}Z_i - \sum_{j=1}^{n}\sum_{\tilde{æ}\in\Gamma_i} u^{(e)}_{i\tilde{æ}j} \frac{\mathbf{d}_j\mathbf{d}^{\mathrm{T}}_j}{\sigma^2_j};$$
$$i \ne j; i,j \in \tilde{M}; 0 \le u^{(e)}_{i\tilde{æ}j} \le c^{(e)}_{j\tilde{æ}} u^{(o)}_{iæj} \Big\}$$

where $\mathbf{x}^{(g)}_i = \left\| \mathbf{r}^{(g)\mathrm{T}}_i \dot{\mathbf{r}}^{(g)\mathrm{T}}_i \right\|^{\mathrm{T}}$ is a motion state vector of the AMO B_i the parameters $y^{(i)}_j(t)$, $j \in \tilde{M}$ of motion are cinematically associated with $x^{(g)}_j(t)$; $F_i(t)$ is a given matrix; $\xi^{(e)}_j$ denotes uncorrelated measurement errors over the channel C_j, it is asserted that $\xi^{(e)}_j$ follows normal distribution with zero expectation and the variance equal to σ^2_j; $D^{(i)}_{\tilde{æ}} \in D^{(i)}$; $u^{(e)}_{i\tilde{æ}j}(t)$ is a control input defining the intensity of the parameter $y^{(i)}_j(t)$ measurements; $c^{(e)}_{i\tilde{æ}}$ is a given value of the channel C_j technical capacity for the operation $D^{(i)}_{\tilde{æ}}$; Z_i is the inverse matrix for the matrix $K_i(t)$ that is a correlation matrix for the estimation errors of the motion state variables; Γ_i is a set of subscripts for measurement operations of the AMO B_i; $\boldsymbol{d}_j(t)$ is a given vector defining technical features of the measuring device (of the channel C_j); $\boldsymbol{b}_{\gamma}=\|0\ 0\dots1\dots0\ 0\|^{\mathrm{T}}$ is an auxiliary vector used to mark a particular element number «γ» of the matrix K_i; K_{i0} is the initial value of the matrix K_i at time $t = t_0$; $\sigma^2_{\gamma i}$ is a specified measurement accuracy for the γ-th element of the motion state vector.

The operations considered in the proposed models are sometimes executed under an interrupt prohibition. Such conditions can be stated in the additional model namely the model of auxiliary-operation control (**M**$_b$ model). The detailed description of this model can be found in [1, 2].

The quality of AMO AMCS control processes can be evaluated through various goal functions [1, 2, 5, 6]. Let us consider some examples of these functions:

$$J_1^{(o)} = \frac{1}{2}\sum_{i=1}^{m}\sum_{æ=1}^{S_i}\left\{ \left[a_{iæ}^{(o)} - x_{iæ}^{(o)}(t_f)\right]^2 + \sum_{j=1}^{m}\int_{t_0}^{t_f}\xi_{iæ}(\tau)u_{iæj}^{(o)}(\tau)d\tau\right\}, i \neq j; \quad (6)$$

$$J_5^{(e)} = \vec{b}_\gamma^{\mathrm{T}} K_i(t_f)\vec{b}_\gamma, J_{6j}^{(e)} = \sum_{i=1}^{m}\sum_{\tilde{æ}\in D^{(i)}}\int_{t_0}^{t_f} u_{i\tilde{æ}j}^{(e)}(\tau)d\tau, \quad (7)$$

where $\xi_{iæ}(\tau)$, $i \in \{1,\ldots,m\}$, $æ \in \{1,\ldots,S_i\}$ are monotone functions that state the penalties for the overdraft of operations' scheduled times. The functional (6) is used, firstly, when the degree of end-conditions violation should be evaluated and, secondly, when the total losses caused by the excess of operations durations are to be measured. The value of the functional $J_5^{(e)}$ characterizes the accuracy of measurement achieved for the γ–th element of the vector $\mathbf{x}_i^{(g)}$.

All the presented models can be interrelated into a generalized model.

3.5 Generalized dynamic model of AMO AMCS control processes (M model)

$$M = \{\mathbf{u}(t) \mid \dot{\mathbf{x}} = \mathbf{f}(\mathbf{x},\mathbf{u},t); \mathbf{h}_0(\mathbf{x}(t_0)) \le \mathbf{O}, \mathbf{h}_1(\mathbf{x}(t_f)) \le \mathbf{O}, \mathbf{q}^{(1)}(\mathbf{x},\mathbf{u}) = \mathbf{O}, \mathbf{q}^{(2)}(\mathbf{x},\mathbf{u}) < \mathbf{O}\} \quad (8)$$

$$J_\theta = J_\theta(\mathbf{x}(t),\mathbf{u}(t),t) = \varphi_\theta(\mathbf{x}(t_f)) + \int_{t_0}^{t_f} f_\theta(\mathbf{x}(\tau),\mathbf{u}(\tau),\tau)d\tau, \theta = 1,\ldots,\Theta, \quad (9)$$

where $\mathbf{x}$ is a vector of AMO AMCS generalized state, $\mathbf{u}$ is a vector of generalized control; $\mathbf{h}_0, \mathbf{h}_1$ are known vector-functions that are used for the state $\vec{x}$ end conditions at the time points $t = t_0$ and $t = t_f$; the vector-functions $\mathbf{q}^{(1)}, \mathbf{q}^{(2)}$ define the main spatio-temporal, technical and technological conditions for the SF functioning process.

On the whole, the constructed model **M** (8) is a deterministic nonlinear non-stationary finite-dimensional differential system with a reconfigurable structure.

4. Algorithm

Computational scheme of the scheduling algorithm is as follows [1, 2, 16-18].

Step 4.1. Setting the rough decision (any valid plan) $\vec{u}_g(t), t \in (t_0, t_f]$. In the special case empty plan ($\vec{u}_g(t) \equiv \vec{0}$) can be selected.

Step 4.2. Integrating basic system of equations (1)–(4) with initial conditions and the $\vec{u} = \vec{u}_g(t)$. Vector $\vec{x}_{(0)}(t)$ obtained as a result of integration. In addition, determine (9) at $t = t_f$, which is taken as a record. Calculating the transversality condition which could be found in [8,13,14].

Step 4.3. Integrating the conjugate system of equations

$$\dot{\psi}_l = -\frac{\partial H}{\partial x_l} + \sum_{\alpha=1}^{I_1}\eta_\alpha(t)\frac{\partial q_\alpha^{(1)}(\vec{x}(t),\vec{u}(t))}{\partial x_l} + \sum_{\rho=1}^{I_2}\rho_\beta(t)\frac{\partial q_\beta^{(2)}(\vec{x}(t),\vec{u}(t))}{\partial x_l}, l = 1,\ldots,\tilde{n} \quad (10)$$

Where, H – Hamiltonian [13,14], η_α and ρ_β could be found from

$$\rho_\beta(t)q_\beta^{(2)}(\vec{x}(t),\vec{u}(t)) \equiv 0, \beta = 1,\ldots,I_2,$$

$$grad_{\vec{u}} H(\vec{x}(t),\vec{u}(t),\vec{\psi}(t)) = \sum_{\alpha=1}^{I_1}\eta_\alpha(t) grad_{\vec{u}} q_\alpha^{(1)}(\vec{x}(t),\vec{u}(t)) + \sum_{\beta=1}^{I_2}\rho_\beta(t) grad_{\vec{u}} q_\beta^{(2)}(\vec{x}(t),\vec{u}(t)),$$

where $\vec{\psi}(t)$ is a general vector of the conjugate system of equations, $q_\alpha^{(1)}(\vec{x}(t),\vec{u}(t))$ and $q_\beta^{(2)}(\vec{x}(t),\vec{u}(t))$ – components of given system of constraints, from $t = t_f$ to $t = t_0$ with $\vec{u} = \vec{u}_g(t)$. At $t = t_0$ we get a first approximation of $\vec{\psi}_i(t_0)$. This completes the iteration $r = 0$.

Step 4.4. From the moment t_0 control $\vec{u}^{(r+1)}(t)$ is sought ($r = 0,1,2,\ldots$ – number of iterations) based on the maximization of the Hamiltonian $H(\vec{x}^*(t),\vec{u}^*(t),\vec{\psi}^*(t))$ [2]. Simultaneously the main system of equations is integrated. Thus at any given time there is a dynamic decomposition of the main tasks for several mathematical programming problems: linear programming, assignment problems [4].

The iterative optimization process ends when the following conditions are satisfied: $\left|J^{(r+1)} - J^{(r)}\right| \le \varepsilon$. Thus $\vec{\psi}(t_0)$ defines vector of associative values of dynamic priority $\vec{\zeta}$ for each operation $D_\kappa^{(i)}$ [2,3].

5 Conclusion

The constructed general model has a form of linear (bilinear as regards the model **Mk**) nonstationary finite-dimensional differential dynamic system with reconfigurable structure. The solutions obtained in the presented multiple-model complex are coordinated by the control-inputs vector $\mathbf{u}^{(o)}$ of the model **Mo**. This vector determines the sequence of interaction operations and fixes AMO AMCS resources allocation. The applied procedure of solution adjustment was called in [5,10-12] resource coordination.

The model complex M evolves and generalizes the dynamic models of scheduling theory [2-4,6,15]. The main distinctive feature (as apposed to [3]) of the complex is that nonlinear technological constraints are actualized in the convex domain of allowable control inputs rather than in differential equations. Therefore, Lagrangian coefficients keeping the information about technical and technological constraints can be defined explicitly using the local-sections method [2]. In [5-6] more complicated iterative procedure was suggested to obtain Lagrangian coefficients. Furthermore instead of relay constraints $u_{i\alpha j}^{(o)}(t) \in \{0,1\}$, less strict ones $u_{i\alpha j}^{(o)}(t) \in [0,1]$ can be considered. This substitution allows for the use of fundamental scientific results of the modern control theory [8,13,14] in various AMO AMCS control problems (including scheduling theory problems).

Computational investigation [2] showed that the use of the AMO AMCS dynamic models entails considerable dimensionality decrease for control problems to be solved in a real-time operation mode. Recurrence description of models allows parallel computations accelerating problem solving.

The proposed original description of AMO AMCS control processes establishes dependence relations between control technology and the goals of AMO AMCS. For example, the methods of optimal-control theory applied to the models **Mo, Me** help to estimate the degree of interdependency between quality of spacecraft operating according to a specified purpose and such technological aspects of AMO management as AMCS resource allocation, trajectory measurement schemes, and information-flow routing methods. Consequently, the optimal programs for resource allocation, for flow routing, and trajectory measurements can be obtained.

Various combinations and interactions of particular control models forming the general model M is the basis for detailed multi-criteria analysis of the factors influencing upon the objective results of AMO AMCS operating.

The research is supported by the Russian Science Foundation (project № 17-11-01254).

References

1. V.N. Kalinin, B.V. Sokolov, International Journal of Difference Equations, **21**(5) (1985)
2. V.N. Kalinin, B.V. Sokolov, Automation and Remote Control, **48**(1-2) (1987)
3. I.N. Zimin, Yu.P. Ivanilov, Zh. Vychisl. Mat. Mat. Fiz., **11(3)** (1971)
4. M.Y. Okhtilev, B.V. Sokolov, R.M. Yusupov, *Intellectual Technology of Complex-Technical-Object Structural-Dynamic Monitoring and Control* (Moscow, "Nauka", 2014)
5. D. Ivanov, B. Sokolov, International Journal of Production Research, **51(9)** (2013)
6. D. Ivanov, B. Sokolov, Journal of Scheduling, **15(2)** (2012)
7. M. Athans, P.L. Falb. *Optimal control: An introduction to the theory and its applications* (New York San Francisco, CA, Sidney, NSW: McGraw-Hill 1966).
8. F. Werner, Y. Sotskov (Eds.), *Sequencing and Scheduling with Inaccurate Data.* (Nova Publishers, 2014)
9. E. Khmelnitsky, K. Kogan, O. Maimom, International Journal of Production Research, **10(35)** (1997)
10. J.G. Kimemia, T.I. Gershwin, IIE Transactions, **15** (1983)
11. K. Kogan, E. Khmelnitsky, International Journal of Production Research, **3** (1996)
12. K. Kogan, E. Khmelnitsky. *Scheduling: control-based theory and polynomial-time algorithms.* (Dordrecht, Kluwer 2000)
13. E.B. Lee, L. Markus, *Foundations of optimal control theory.* (Wiley & Sons, New York 1967)
14. D. Tabak, B.C. Kuo, *Optimal control by mathematical programming.* (Prentice Hall, NY 1971)
15. H. Ye, R. Liu, Transportation Research Part B: Methodological, **93** (Part A) (2016)
16. F.L. Chernousko, A.A. Lyubushin, Optimal Control Applications and Methods, **3(2)** (1982)
17. A. Alabyan, V. Zelentsov, I. Krylenko, S. Potryasaev, B. Sokolov, R. Yusupov, SPIIRAS Proceedings, **41**(4) (2015)
18. I.A. Krylov, F.L. Chernousko, Zh. Vychisl. Mat. Mat. Fiz., **12(1)** (1972)
19. F. Werner, Y. Sotskov (Eds.), *Sequencing and Scheduling with Inaccurate Data.* (Nova Publishers, 2014)
20. E. Khmelnitsky, K. Kogan, O. Maimom, International Journal of Production Research, **10(35)** (1997)
21. J.G. Kimemia, T.I. Gershwin, IIE Transactions, **15** (1983)
22. K. Kogan, E. Khmelnitsky, International Journal of Production Research, **3** (1996)
23. K. Kogan, E. Khmelnitsky. *Scheduling: control-based theory and polynomial-time algorithms.* (Dordrecht, Kluwer 2000)
24. E.B. Lee, L. Markus, *Foundations of optimal control theory.* (Wiley & Sons, New York 1967)
25. D. Tabak, B.C. Kuo, *Optimal control by mathematical programming.* (Prentice Hall, NY 1971)
26. H. Ye, R. Liu, Transportation Research Part B: Methodological, **93** (Part A) (2016)
27. F.L. Chernousko, A.A. Lyubushin, Optimal Control Applications and Methods, **3(2)** (1982)
28. A. Alabyan, V. Zelentsov, I. Krylenko, S. Potryasaev, B. Sokolov, R. Yusupov, SPIIRAS Proceedings, **41**(4) (2015)
29. I.A. Krylov, F.L. Chernousko, Zh. Vychisl. Mat. Mat. Fiz., **12(1)** (1972)

Ontology-based indirect interaction of mobile robots for joint task solving: a scenario for obstacle overcoming

Mikhail Petrov[1,2] and *Alexey Kashevnik* [1,2]

[1]ITMO University, Saint Petersburg, Russian Federation
[2]SPIIRAS, Saint Petersburg, Russian Federation

Abstract. This paper describes an ontology-based approach to interaction of users and mobile robots for joint task solving. The use of ontologies allows supporting semantic interoperability between robots. The ontologies store knowledge about the tasks to be performed, knowledge about the functionality of robots and the current situation factors like a robot location or busyness. Ontologies are published in a smart space which allows indirect interaction between participants. On the basis of the knowledge, a robot can define a task that is to be performed and get the current status of other robots. The paper presents a reference model of the approach to indirect interaction between mobile robots for joint task solving, an ontology model for the knowledge organization, and application of the presented approach for the scenario for obstacle overcoming.

1 Introduction

Nowadays, mobile robots become more and more popular in many areas [1-3]. They are actively used for different tasks such as scouting [1, 2], technological accidents and catastrophes [3], counterterrorism operations [3] and patrolling [1, 2].

The utilization of robots is useful in case of the impossibility of employment of human resources, in the presence of threats to health and human life, as well as to perform consuming tasks in adverse conditions.

Performing a task a robot tends to face obstacles on its path because of both heterogeneity and complexity of a landscape. Depending on their design and the task to perform as well as the nature of these obstacles robots can avoid or, if it is impossible, overcome them. The implementation of both of these options requires an appropriate approach to problem solving and interaction with users, which takes into account various factors such as an obstacle size and shape.

This paper proposes ontology-oriented approach to indirect interaction of mobile triaxial robots between themselves and users. These robots have to be able to work and move on the ground having uneven terrain, and perform various tasks remotely given by one or more users or other robots. Each robot consists of several mobile units (blocks) with one or two pairs of wheels and is equipped with ultrasonic sensors. Robots are able to detect obstacles, and overcome not only low ones (the height is less than the radius of the wheel), which can be run over without making additional operations, but also obstacle height 1-4 times bigger than the radius of the wheel.

Robots are supposed to distribute tasks among themselves effectively, to take into account promptly events occurring during the execution of tasks. For example, a robot should try to overcome encountered obstacles automatically.

The interaction and information exchange between robots and users is conducted using smart space technology. This technology allows organizing the information and knowledge sharing for various actors due to an ontology-based publication subscription mechanism. Smart space, in fact, is an information model that describes the interaction of mobile robots and the tasks they perform. This technology allows the use of robots and user devices of different vendors within a single scenario.

The rest of the paper is organized as follows. Section II describes related work in the area of mobile robots interaction. Section III presents the reference model of the approach to indirect interaction between users and mobile robots. Section IV describes the implementation. Finally, Conclusion summarizes the paper.

2 Related works

Paper [4] deals with an intelligent system of robot and human interaction. This system uses adaptive learning mechanisms to account for the behavior and preferences of a human. It includes algorithms for speech and face recognition, natural language understanding, and other ones so that the robot could recognize human actions and commands. A robot used in this paper (Pioneer 3-AT) is equipped with multiple sensors, scanners and a camera. The paper focuses on the process of teaching the robot and its mechanisms of interaction with a human.

Paper [5] describes a coordinated process for the agents (in particular, mobile robots) jointly performing a task, using the knowledge about the environment, their abilities and possibilities for communication with other agents. The effective use of knowledge about the agents capabilities is implemented by using suitability rates which allow agent to choose the most appropriate action. This approach was tested during a football match between mobile robots. The interaction in this article is considered in terms of the choice of an agent to perform an action.

Paper [6] proposes a methodology for coordination of autonomous mobile robots in jams in congested systems. The methodology is based on reducing a robot's speed in jams and congested areas, as well as in front of them. Thus, the robots slow down detecting other robots in front of themselves. The effectiveness of the methodology was tested in the simulation of congested system with a jam.

Paper [7] discusses combining wireless sensor networks and mobile robots. Authors suggest layered swarm framework for the interaction of decentralized self-organizing complex adaptive systems with mobile robots as their members. This framework consists of two layers (robots and wireless LAN) and communication channels within and between layers. During the experimental verification of framework mobile robots independently reached the destination.

Paper [8] deals with a system of robots that allows garbage collection in the territories without the involvement of human. The system consists of a group of the robots responsible for the recognition of objects and bypassing the area. While collecting garbage robots can interact with each other via communication channels.

Paper [9] is devoted to the problem of robots group control in non-deterministic, dynamically changing situations and, above all, in conditions of organized resistance. The paper contains a formal statement of the robots group management task for these conditions, as well as a model of collective interaction of robots in groups. Simulation of the robots group control task in a military counteraction showed the effectiveness of the approach.

Paper [10] shows a system of robots interaction based on BML – Battle Management Language. This language is intended for expressing concise, unambiguous orders read by both humans and robots. The orders are transmitted from the system of multiple robots and processed for further distribution specific commands between them.

The paper [11] describes designing human-robot interaction, improving human situational awareness on the basis of an agent-oriented approach. This approach is intended to apply at movement of mobile agents (autonomous robots and cosmonauts in a protective gear) on the Moon's surface in the controlled territorial area.

Paper [12] discusses the use of multi-agent systems for the organization of industrial robots interaction. On the basis of them a system architecture of robotic systems interaction was built. During the interaction of agents within designed architecture the programs are creating which govern behavior of the agents. These programs are stored in knowledge base.

Paper [13] considers the problem of program agents' access to the smart space based on Smart-M3 platform. This paper suggests a model of concurrent sessions to work with multiple smart spaces. The model allows an agent to use multiple connections and control objects in a single session. Moreover, it provides local storage, divided between session connections. The storage is synchronized using subscriptions.

The approaches used in papers [4] and [6] require a well-defined functionality of robots, so the set of undertaken tasks is restricted. In paper [5] the knowledge of the capabilities of robots is generated dynamically during the execution of tasks what allows managing the robots effectively. Thus, no limitation to robots functionality and dynamic determination of the options available can provide the widest range of tasks to execute. Paper [7] shows the effectiveness of the indirect interaction of robots, that is, the interaction through an intermediary that is Smart-M3 platform in the paper. Paper [8] demonstrates an additional module necessity for the efficient interaction of robots and users. Paper [9] provides a formal model that can be used for interaction. Paper [10] shows the effectiveness of using specialized language for robots control. Papers [11] and [12] demonstrate the applicability of the multi-agent systems approach for robots and humans interaction in various fields. Finally, paper [13] proposes a model of interaction between agents and a smart space.

3 Interaction of mobile robots

The reference model of the approach to indirect interaction mobile robots is presented in Figure 1. It is conducted on the basis of ontology-based publish / subscribe mechanism implemented in Smart-M3 platform [14, 15]. Smart-M3 platform includes a semantic information Broker (SIB) and knowledge processors (KP).

The SIB broker contains the information provided in the form of a knowledge base; KP processors provide an access to the information and its changes and ensure notifications of new information. The interaction between agents and broker operates according to Smart Space Access Protocol (SSAP) [8]. All communications between the smart space and users, robots, additional services are conducted via a LAN or the Internet.

Functions of the robot can be classified into two types: basic and additional (Fig. 2). The basic functions are necessary to move the robot. These ones include moving in two directions and overcoming obstacles by lifting, lowering and pulling units.

Additional features are necessary to expand the number of tasks performed by robots. An obstacle recognition allows storing a sequence of actions to overcome obstacles. The movement to a given point involves automatic overcoming encountered obstacles or manual control request if automatic overcoming is impossible. The "Explore a point" combines moving to a given point, and filming and sending a video.

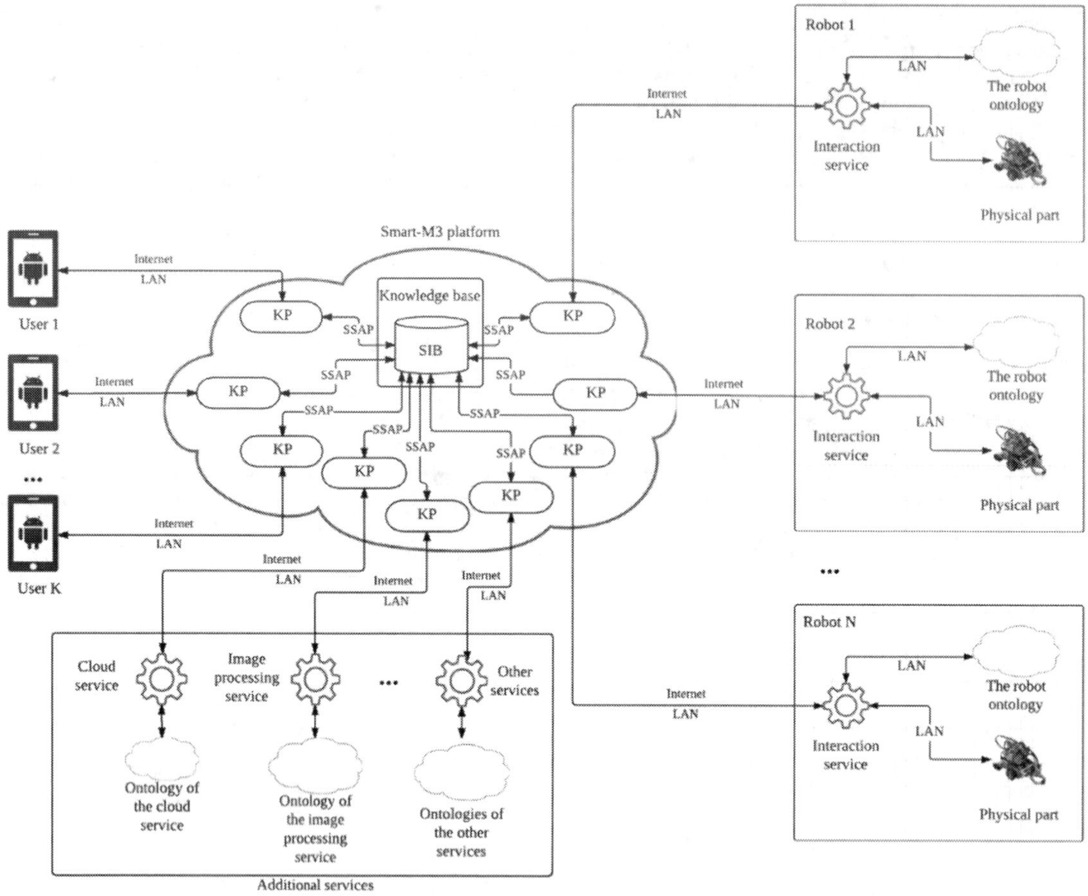

Fig. 1. Reference model of the approach to indirect interaction of mobile robots.

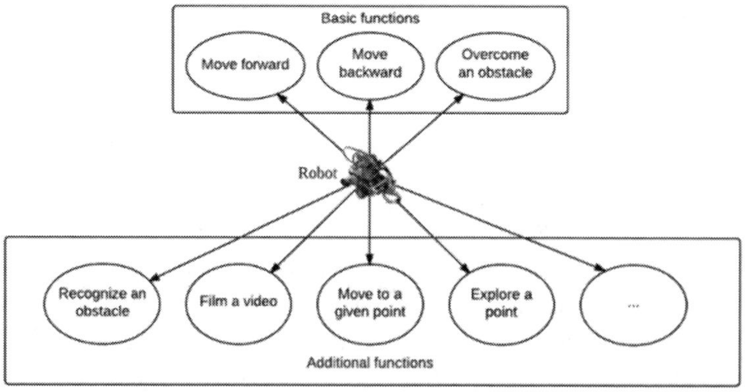

Fig. 2. Robots functions.

Smart-M3 platform involves storing information in the form of RDF-ontologies, in which information is presented as a set of triples of "subject-predicate-object". A triples subject describes an entity such as an object, a device or a person. This entity has a property or can perform an action described in the triple. A triples object describes an entity, which the subject is associated with, that is its function or property. The object of one triple may be the subject of another one.

A triples predicate describes the type of relationship between subject and object. It can determine the type of functions performed by the subject, or the correlation between subject and object.

A simple example of a triple stored in the ontology is <"robot", "task", "goToLocation">. The subject "robot" describes a robot, which information in the triple relates to, the predicate "task" defines what this information is, and the object "goToLocation" is a particular "task" for the subject "robot". At the same time "goToLocation" is also a subject in the triple <"goToLocation", "coordinates", "40; 200">.

Devices connected to the smart space based on Smart-M3 platform are able to "subscribe" to a certain pattern triples to respond timely to new information published in the smart space. In the case of a new triple satisfying pattern all devices, which are subscribed to triples of this pattern, are notified.

A typical example of the ontology provided in the smart space at the time of publication of a task for the robot is shown in Figure 3. Triples contained in the ontology are presented in the figure in the form of a set of ovals and arrows connecting them. An oval, which is a source for the arrow, denotes the subject; the arrow denotes the predicate, and an oval, which is a target for the arrow, denotes the object.

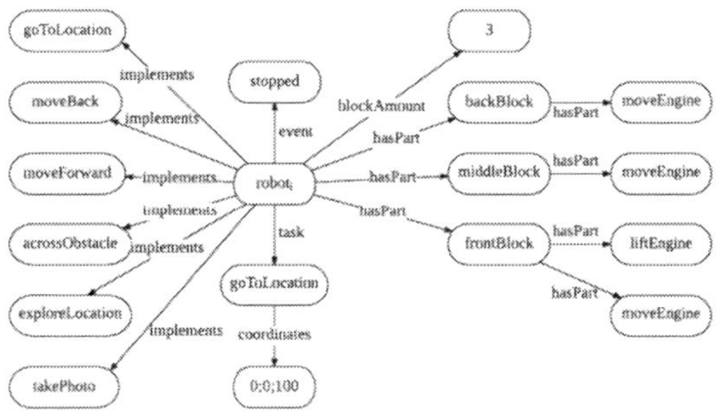

Fig. 3. Ontology at the time of the task publication for the robot in the smart space.

This example shows information about a robot consisting of three blocks (<"roboti", "blockAmount", "3">). Each block has its main engine ("moveEngine"). The front block is also equipped a lift engine (<"frontBlock", "hasPart", "liftEngine">) that is used to raise this block. The robot can move backward (<"roboti", "implements", "moveBack">) or forward (<"roboti", "implements", "moveForward">), overcome an obstacle (<"roboti", "implements", "acrossObstacle">), and snapshot (<"roboti", "implements", "takePhoto">). Besides, the robot can go to a given point (<"roboti", "implements", "goToLocation">) and explore a location (<"roboti", "implements", "exploreLocation">). Fig. 3 shows the moment the robot is given the task to move back (a triple with "task" predicate). In addition, triples with the "event" predicate can be published in the ontology describing the events that occurred during the execution of robotic tasks such as stopping due to detecting obstacles (<"roboti", "event", "stopped">).

When starting the robot "subscribed" to the triples with the "task" predicate. Therefore, when triple <"roboti", "task", "goToLocation"> appeared in the ontology, the robot analyzed the subject of the triple. It determined whether to perform the task (on the basis of the coordinates of the point and the location of its own and other robots), and then went to the point.

4 Implementation

To assess the approach efficiency the scenario of a point exploring and encountered obstacles overcoming is chosen, as it uses most of the robot functions. During this scenario, if a robot finds an obstacle using ultrasonic sensors, it makes a request to the cloud service to get a sequence of actions to overcome such an obstacle. If the information about obstacles of this type is not found in the cloud service, the control over the robot is transmitted to a user. A prototype of a mobile robot has been developed for this scenario. The robot prototype is able to move forward, to detect obstacles and overcome them. It is based on a set of robotic Lego Mindstorm EV3. This set allows one to easily design a robot with required functionality for training purposes. At the same time there is the possibility to use electronic units, motors and sensors and program them in Java. Fig. 4 shows the design of the robot constructed for the scenario of forward moving and encountered obstacles overcoming.

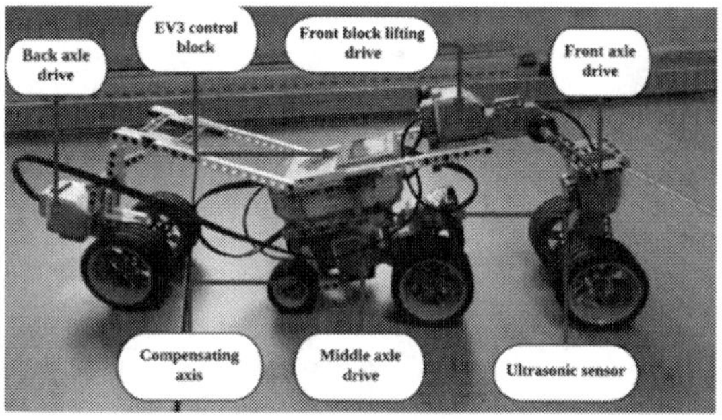

Fig. 4. Design of a mobile robot prototype.

In order to overcome obstacles in accordance with the present scenario, the robot was constructed from several blocks. Due to the design the robot could climb an obstacle gradually. First, it lifted and fixed the front block, then basing on the front and back blocks it raised the middle one. All blocks were equipped with a pair of drive wheels; in addition, the middle block had a pair of wheels without a drive for balance. An ultrasonic sensor for measuring the distance to objects was also mounted on the central block.

The sensor detected obstacles in the path of the robot, automatic stop occurred. At this stage, only basic functions of the robot were necessary, so it was not equipped with a camera or other devices to receive any information about the nature of the terrain.

Fig. 5 shows the scenario of forward moving and encountered obstacles overcoming.

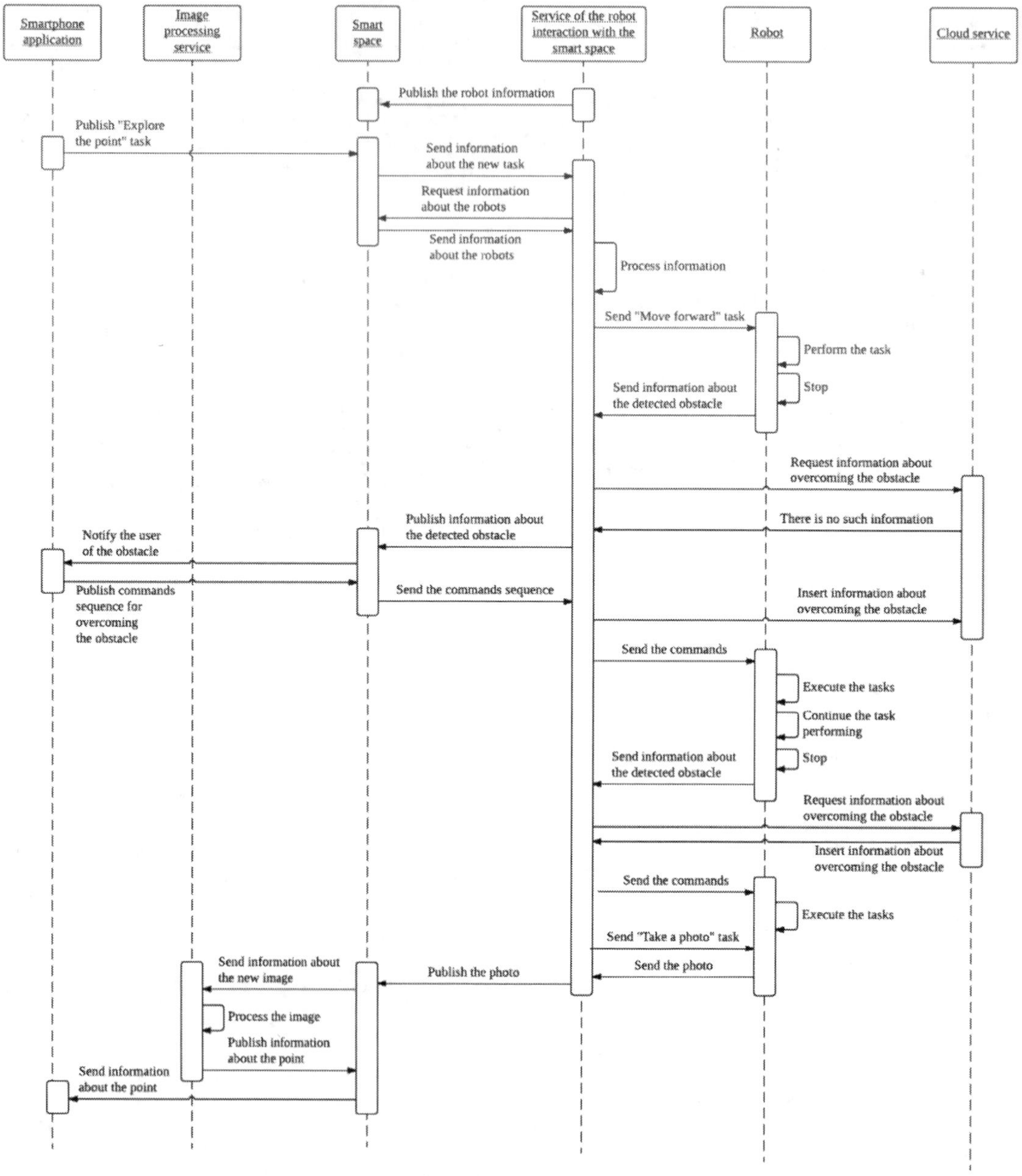

Fig. 5. The scenario of forward moving and encountered obstacles overcoming.

The diagram shows six main components of the system.

- Smartphone application publishes tasks to perform and shows the status of performance.
- Image processing service is designed to retrieve information about an object or area from a photo.
- Smart space is a repository of information on the basis of Smart-M3 platform.
- Service of the robot's interaction with the smart space is a program which receives, processes and transmits information to the robot from the smart space. At the same time the service receives information about the events and results of task performance from the robot and publishes this information in the smart space .
- The robot executes commands received from the interaction service.
- Cloud service is a knowledge base that stores information about various obstacles in the form of a sequence of actions to overcome these obstacles.

When starting each robot publishes information about its functions and limitations from its own ontology in the smart space. Moreover, while executing tasks a robot publishes information about its location and the current task.

For example, a user publishes task "Explore a point" in the smart space. Since the robot is free, that is, it does not perform any other tasks, its interaction service determines whether the robot is to perform published task. For this purpose, the service follows the information about the task and about other robots. If a robot is the closest one to the point, it proceeds to the task.

When a robot detects an obstacle in the way, it automatically stops, informing the interaction service about it. For further performance of the initial problem the robot needs to overcome this obstacle. Therefore, it sends a request to the cloud service on how to do it. If there is no information about such an obstacle, the interaction service publishes information about the impossibility of overcoming the obstacle automatically in the smart space.

After receiving a notification a user accepts the control over the robot. As a result, the commands that are to be done in order to overcome this obstacle are consistently being published in the smart space. The interaction service inserts the sequence of instructions and information about obstacles in the template database for further use. After overcoming the obstacle the robot continues to perform the task and finds the second obstacle that is similar to the first one. Since the information about it is already in the obstacles templates database, the robot overcomes it without the user's help.

When the robot reaches the destination, and receives command "Take a photo", it takes a snapshot of terrain and sends it to the interaction service, which in its turn publishes an image in the smart space. Then, the image processing service receives information about the new image and processes it. After receiving information about the point, the service publishes it in the smart space, and then information is sent to the user.

5 Conclusion

The approach to the indirect interaction between mobile robots for the joint problem solving, described in the paper, provides an effective solution to a wide range of tasks due to the indirect interaction of not only the robots and users, but also the robots with each other. The use of ontologies minimizes the dependence of the approach on the manufacturers as well as software of robots and user devices: only possibility of working with ontologies support is necessary. To evaluate the working capacity and effectiveness of the approach a robot prototype has been designed.

Further development involves adding computing services, and equipping the robot with new types of sensors to expand the set of tasks to perform. Moreover, the implementation of additional robot functions is planned. Such modifications include, for example, equipping the robot with cameras and developing the image processing service that can publish the nature of the terrain, the type of encountered obstacles, the presence or absence of any given object in sight. This service is likely to make the robot control easier and more convenient (to make the robot itself be able to identify and overcome obstacles), and make the use of a robot more efficient (due to new tasks performed). In addition, it is necessary to evaluate the effectiveness of the joint problem solving. For this task it is planned to design the second robot prototype.

The presented results are part of the research carried out within the project funded by grants ## 16-29-04349 and 16-07-00462 of the Russian Foundation for Basic Research.

References

1. A.V. Vasiliev, Izvestija JuFU. Tehnicheskie nauki, **3**(104), 119-124 (2010)
2. A.V. Vasiliev, *Robototehnika i tehnicheskaja kibernetika*, 67-71 (2014)
3. S. Phani Teja, S. Harsha, A. Siravuru, S.V. Shan, K. Madhava Krishna, *AIR '15 Proceedings* (2015)
4. A. Sekmen, P. Challa, Knowledge-Based Systems, **42**, 49-59 (2013)
5. S. Ibarra-Martínez, J.A. Castán-Rocha, J. Laria-Menchaca, J. Guzmán-Obando, E. Castán-Rocha, Ingeniería, Investigación y Tecnología, **14**, 37-42 (2013)
6. S. Hoshino, H. Seki, Robotics and Autonomous Systems, **61**, 808-820 (2013)
7. W. Li, W. Shen, Journal of Network and Computer Applications, **34**, 1398-1407 (2011)
8. P.L. Sajapin, V.G. Pak, Informatika i Kibernetika (ComCon-2016), 113-115 (2016)
9. S.G. Kapustjan, L.Z. Usachev, Izvestija JuFU. Tehnicheskie nauki, **12**, 26-36 (2008)
10. T. Remmersmann, A. Tiderko, U. Schade, F.E. Schneider, Robototehnika i tehnicheskaja kibernetika, **3**, 48-52 (2014)
11. A. Karpov, B. Kryuchkov, A. Ronzhin, V. Usov, Jekstremal'naja robototehnika, **1**, 71-81 (2016)
12. I.O. Protalinskiy, I.A. Shcherbatov, Vestnik AGTU. Serija: upravlenie, vychislitel'naja tehnika i informatika, **2**, 76-82 (2009)
13. A.A. Lomov, Uchenye zapiski Petrozavodskogo Gosudarstvennogo Universiteta, **8**, 118-121 (2013)
14. J. Honkola, H. Laine, R. Brown, O. Tyrkkö, International Symposium on Computers and Communications, 23-25 (2010)
15. S. Mikhailov, A. Kashevnik, SPIIRAS Proc., **52**(3), 180-203 (2017)

Underwater floating robot-fish: a comparative analysis of the results of mathematical modelling and full-scale tests of the prototype

Sergey Jatsun*, *Boris* Lushnikov, *Evgeny* Politov and *Sergey* Knyazev

SouthWest State University, Department of Mechanics, Mechatronics and robotics, 305040 94, 50 let Oktyabrya, Kursk, Russia

Abstract. The article presents a comparative analysis of the results of computer mathematical modelling of the motion of the underwater robot-fish implemented by using the MATLAB / Simulink package and full-scale tests of an experimental model developed in the laboratory of mechatronics and robotics of the SouthWest State University.

1 Introduction

At present, unmanned autonomous underwater vehicles (AUV) that possess high autonomy, modern sensory and navigational equipment and intelligent control have become the main robotic mobile device for solving various tasks of human activity in the hydrosphere of the ocean [1-11]. The main leaders in the field of autonomous underwater vehicles are the United States, China and the United Kingdom. They produce about 50% of the total number of underwater robots, and also have about the same relative volume of scientific publications on this topic. An even more significant predominance of these countries is the financing of research and development work related to various aspects of underwater robotics.

The main tasks of such devices are: monitoring the hydrosphere and the technical condition of underwater man-made objects, exploration of underwater resources, conducting survey, search and rescue operations, as well as military reconnaissance operations that make high demands on the noiselessness and secrecy of the underwater robot.

Papers [12-20] considers the new principles of motion of mobile robots interacting with the water environment, as well as the development of underwater vehicles of various types and the principle of motion, including bionic.

In this paper, we present a comparative analysis of the results of computer mathematical modelling of the movement of an underwater floating robot-fish in the MATLAB / Simulink and full-scale prototype tests.

2 Design and principle of action of the robot-fish prototype

A three-dimensional model of the developed design of the bionic robot is shown in Figure 1. The main structural elements of it are the body, peduncle 6, caudal fin 7 and lateral fins (in Figure 1 - not shown). In the body of the robot we have: a system of static immersion and ascent of the robot 1 with drive 2; video camera 3; rechargeable battery 4; drive of the fin control system 5; drive of the precursor control system 8 and the lighting, navigation and autonomous control subsystems (not shown in Figure 1). In the body of the antechamber there is a drive 8 of the caudal fin 7.

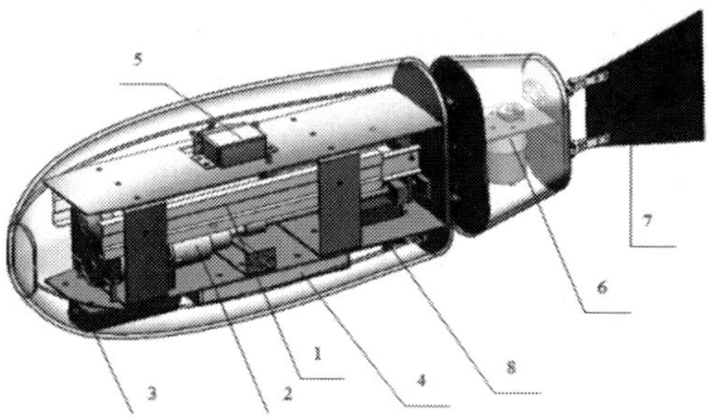

Fig.1. The three-dimensional model of the developed design of the bionic robot: 1 - the robot immersion system; 2 - the drive of the immersion system; 3 - video camera; 4 - the storage battery; 5 - drive of the fin control system; 6 - tail control system drive; 7 - tail; 8 - drive of the control system.

The general view of the prototype of the robot fish is shown in Figure 2. The forward rectilinear movement of the robot occurs due to the oscillating motion of the caudal fin, while turning in the horizontal plane - due to rotation of the forecourt relative to the robot body. Vertical diving or ascent is implemented by regulating Archimedes force with the help of intake or displacement of seawater. The regime of dynamic immersion or ascent is possible by changing the angle of inclination of the lateral fins with simultaneous operation of the caudal fin. If the premounting is included in the

* Corresponding author: teormeh@inbox.ru

work, then the robot can move along a complex spatial trajectory.

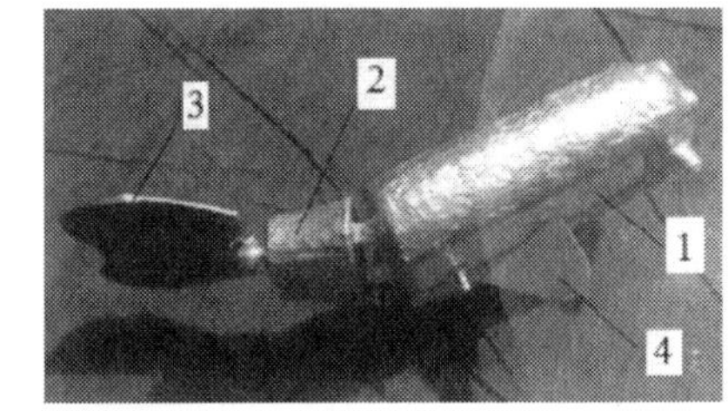

Fig. 2. General view of the prototype of the robot-fish during the tests in the shallow basin: 1 - body, 2 - peduncle, 3 - caudal fin, 4 - lateral fins.

The main technical characteristics of the developed prototype of the robot with a bionic principle of motion are presented in Table 1.

Table 1. The main technical characteristics of the robot fish.

№	Characteristic	Value
1.	Dimensions (length×height×width), m	0.70×0.23 ×0.28
2.	Displacement, kg	5.9
3.	Payload, kg	1.8
4.	Battery capacity, mA/hour	3500
5.	Maximum speed underwater, m/s	0.42
6.	Maximum vertical diving (ascent) speed, m/s	0.08
7.	Maneuverability (minimum turning radius), m	0.45
8.	Swing angle of the caudal fin, degrees	-30…30
9.	Turning angle of peduncle, degrees	-25…25
10.	Angle of rotation of the lateral fins, degrees	-30…30
11.	Frequency of oscillations of the caudal fin, Hz	0…8
12.	Power consumption at maximum speeds, W	55

The operating principle of the underwater robot control system is described in [16].

3 Mathematical model of robot-fish

During the design stage, a mathematical model of the robot-fish was developed to determine and study the basic laws of motion, the choice of optimal design parameters according to the quality criteria, which determines the minimum power consumption, maximum speed, maneuverability and other characteristics.

A three-link kinematic scheme shown in Figure 3 was adopted as an analytical model of the mathematical model of motion of a bionic floating robot in the horizontal plane.

The system of algebraic differential equations describing the dynamics of the system under consideration will have the following form:

$$\begin{cases} m_{tX}\ddot{X} = F_{TX}(X, Y, \theta, \Psi_1, \Psi_2, t) + F_{sopr}(\dot{X}); \\ m_{tY}\ddot{Y} = F_{TY}(X, Y, \theta, \Psi_1, \Psi_2, t) + F_{sopr}(\dot{Y}); \\ I_\theta\ddot{\theta} = M_C(F_t) + M_{sopr}(\dot{\theta}) + M_t(\dot{X}, \dot{Y}, \theta, \Psi_1, \Psi_2); \\ \Psi_1(t) = \pi + \theta(t) + \theta_1(t); \\ \Psi_2(t) = \Psi_1(t) + \theta_2(t). \end{cases} \quad (1)$$

Using the provisions of the aerodynamic theory of the wing, the lifting (traction) force acting on the moving flat plate can be determined according to the following expression [7]:

$$\overline{F_T} = \pi\rho S(\overline{v_q} \times \overline{l_B}) \times \overline{v_q}, \quad (2)$$

where ρ is the density of the medium (water); S is the surface area of the plate; $\vec{v}_q$- velocity vector of the point of the plate, located at a distance of a quarter of its length from the axis of rotation; $\vec{l}_e$- a unit vector located in the plane of the plate and determined by the coordinates:

$$l_e = [-\cos(\psi_2), \ -\sin(\psi_2), \ 0] \quad (3)$$

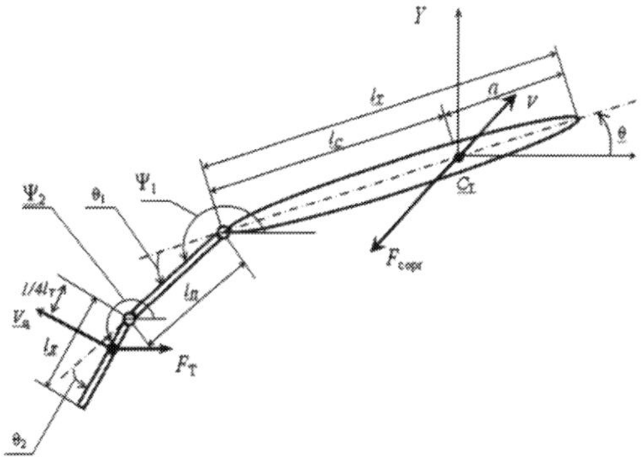

Fig. 3. Analytical diagram of the bionic floating robot in the horizontal plane: X-Y - rectangular inertial coordinate system; *l* - length of the "body" of the robot; l_Π is the length of the caudal peduncle; l_T is the length of the caudal fin; C_T is the point corresponding to the center of mass of the "body" of the robot; V is the vector of the absolute velocity of the center of mass; F_{copr} is the vector of the drag force of the robot body. V_q is the linear velocity vector of the point of the caudal fin at a distance equal to *1/4* of its length from the axis of rotation; θ - angular coordinate of the central longitudinal axis line of the robot body; θ_1 and θ_2 are the relative rotation angles of the leading and tail fins respectively around their rotation axes; ψ_1 and ψ_2 are the absolute angular coordinates of the position of the caudal peduncle and caudal fin, respectively; F_T - the vector of the traction force, which arises when rotational motion of the caudal fin.

The coordinates of the speed vector $\overrightarrow{v_q}$ are determined by the following expressions:

$$V_{qX}(t) = V_X + L_{copt} \cdot \sin(\theta(t)) \cdot d\theta(t) -$$

$$-L_P \cdot \sin(\Psi_1(t)) \cdot d\Psi_1(t) - \frac{L_t}{4}\sin(\Psi_2(t)) \cdot d\Psi_2(t); \quad (4)$$

$$V_{qY}(t) = V_Y - L_{copt} \cdot \cos(\theta(t)) \cdot d\theta(t) +$$

$$+L_P \cdot \cos(\Psi_1(t)) \cdot d\Psi_1(t) + \frac{L_t}{4}\sin(\Psi_2(t)) \cdot d\Psi_2(t); \quad (5)$$

$$V_{qz}(t) = 0\,. \quad (6)$$

Then the projections of the vector F_T on the X and Y axis have the form:

$$F_{TX}(t) = \pi\rho S(V_{qX}(t) \cdot \sin(\Psi_2(t)) - V_{qY}(t) \cdot \cos(\Psi_2(t))) \cdot V_{qY}(t); \quad (7)$$

$$F_{TY}(t) = \pi\rho S(-V_{qX}(t) \cdot \sin(\Psi_2(t)) + V_{qY}(t) \cdot \cos(\Psi_2(t))) \cdot V_{qX}(t). \quad (8)$$

Besides thrust a torque, given by the expression below, also acts on the tail plate:

$$M_t = -\pi\rho \frac{L_t^2}{4}\left(\dot{x}_m \dot{y}_m \cos(2\Psi_2(t)) + \frac{1}{2}\left(\dot{y}_m^2 - \dot{x}_m^2\right)\right)\sin(2\Psi_2(t)), \quad (9)$$

where $\dot{x}_m$, $\dot{y}_m$ – projection of the velocity vector of the midpoint of the tail plate on the X and Y axes.

The body of the robot-fish will also be acted upon by the torque vector of the thrust force, the drag force of the fluid and the viscous drag moment. To determine the last two factors, it is necessary to integrate the forces distributed over the surface of the robot-fish body. For preliminary calculations, we will consider these force factors to be proportional to the corresponding velocities:

$$F_{soprX} = \mu_X V_X\,; \quad (10)$$

$$F_{soprY} = \mu_Y V_Y\,; \quad (11)$$

$$M_{sopr} = \mu_\theta \dot{\theta}\,. \quad (12)$$

The moment of the traction force acting on the body of the robot relative to its center of mass is determined in accordance with the vector product:

$$M_C(F_T) = [X_{Cq}, Y_{Cq}] \times [F_{TX}, F_{TY}], \quad (13)$$

where X_{Cq}, Y_{Cq}– coordinates of the vector connecting the center of mass of the body of the robot-fish C_T and the point of application of the traction force on the caudal fin.

As a result we obtain

$$M_C(F_T) = Y_{Cq}F_{TX} - X_{Cq}F_{TY}. \quad (14)$$

For carrying out numerical studies, the mathematical model described above was implemented in the MATLAB / Simulink environment, and a comprehensive verification of its adequacy was performed [13].

In the course of computer simulation the effect of the parameters of the mathematical model of the robot-fish design on its functional and dynamic characteristics was studied.

One of the most important kinematic characteristics of a mobile robot in terms of assessing its dynamic efficiency is the swimming speed. The greatest influence on the swimming speed is exerted by the vibration frequency of the caudal fin, the mass of the robot and the area of the lateral surface of the caudal fin: the greater the frequency of oscillations and the tail area and lesser the mass of the robot, the greater the speed of its translational movement in the horizontal plane.

At the same time, it is desirable to achieve the best conditions for monitoring the environment, in particular, video and photography, measurements of physical properties of the medium, etc when the robot moving. In this regard it is therefore important to minimize the lateral displacement of the center of mass of the robot and the angular "wobble" of the body of the robot relative to the vertical axis passing through its center of gravity. In the course of numerical experiments it was found that the reduction of these two parameters is possible by increasing the frequency of tail angular vibrations and decreasing the area of the caudal fin with simultaneous elongation and an increase in mass.

As an example, Figure 4 shows graphs of the dependence of the swimming speed of the robot V on the frequency f of the tail fin for the three amplitude values θ2 obtained as a result of mathematical modelling.

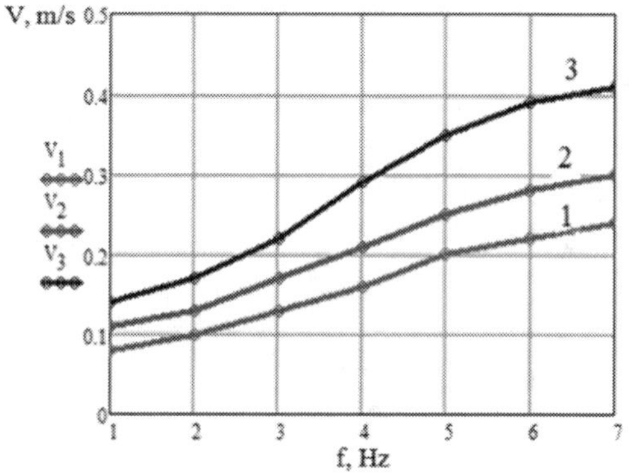

Fig. 4. The graph of the swimming speed V against the oscillation frequency f of the tail in for three amplitude values θ2: 1 - θ2=10°, 2 - θ2=20° and θ2=30°.

From the analysis of the presented results of mathematical modelling it follows that in order to increase the efficiency of the robot (increasing speed and improving stability) it is desirable to create oscillations of the caudal fin with a sufficiently high frequency (above 3-5 Hz). However, an increase in the frequency of tail angular oscillations leads to a significant increase in the consumption of power, which obviously is a limited resource.

The size of the tail and mass of the robot adversely affect speed and lateral "wobble", and with an increase in the tail area and a decrease in body weight, the velocity decreases, but the amplitude of lateral movements increases. Therefore, in order to increase the efficiency of the robot's movement, the task of multicriteria optimization of constructive and control parameters by known methods was set and realized [12, 15].

4 Comparative analysis of the results of mathematical modelling and full scale tests of the experimental prototype

Experimental studies of the prototype fish robot were carried out in three types of reservoirs: in two types of basins (shallow water as shown in Figure 2 and deep water) and in full-scale conditions at the Kursk reservoir.

Quantitative and qualitative assessments are as follows:

• quality of functioning of lighting, video surveillance, and power supply subsystems.

• steady position of the robot during swimming and diving;

• hermeticity of the housing;

• range of controllability depending on the depth of immersion;

• range of speeds of horizontal movement in the underwater position;

• speed of dynamic immersion and ascent;

• turning radius in the horizontal plane;

• speed of static (vertical) immersion and ascent;

• depth of immersion, with which a robot can safely surface;

• performance of lighting, video surveillance, and power supply subsystems.

Figure 5 shows a graph of the average velocity V (m/s) of the horizontal movement of the robot in a completely submerged position against the frequency of the tail fin f (Hz) at a swing angle θ2 = 30°, recorded from the results of the full-scale experiment.

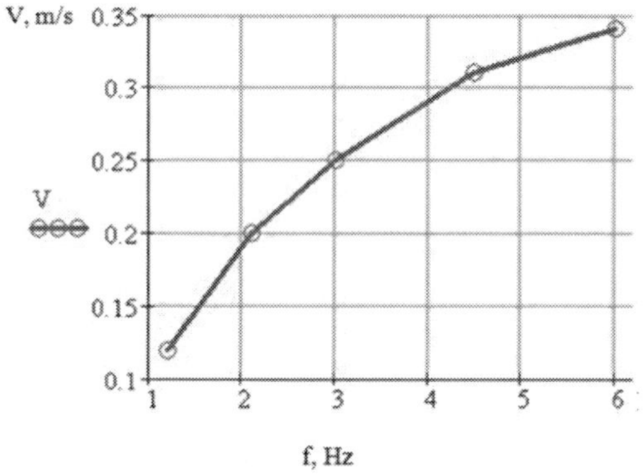

Fig. 5. Graph of the average velocity V (m/s) of the horizontal movement of the robot in a completely submerged position against the frequency of the tail fin f (Hz).

Based on the analysis of the presented graph obtained in the course of full-scale tests, we can see that the nature of the influence of the frequency of the oscillations of the caudal fin on the speed of movement of the robot is similar to that obtained in mathematical computer simulation (see Fig. 3). At the same time, an increase in the frequency f leads to a nonlinear increase in the average speed of the robot's motion and the shape of the graph takes the form of a "saturation" curve. This is caused by an increase in resistance to the robot's movement of the liquid with increasing speed. Therefore, for an economical mode of the robot's motion it is necessary to use the vibration frequencies of the caudal fin in the interval from 1 to 5 Hz, which corresponds to a speed range of 0.1 to 0.3 m/s.

The turning radius of the robot in the horizontal plane (turning radius) that characterises maneuverability is 0.45 m, which also satisfies the requirement specification.

The rate of dynamic immersion and ascent depends on the speed of the horizontal movement of the robot and is determined by the deflection angle of the lateral fins. The maximum values of the speed of ascent (ascent) during the tests reached 0.08 m/s with a maximum angle of attack of the lateral fins of 45 degrees.

The range of the radio control system reached 80-100 m with an immersion depth of up to 1 m and for large immersion depths it is necessary to use a floating antenna or switch to a stand-alone program control mode. The conducted tests confirmed the results of mathematical modelling, showing that the robot's movement pattern corresponds to the calculated one since the divergence of the main characteristics does not exceed 15%. The tests on the produced prototype showed that it can move in the water environment for a distance of up to 150 m. This device makes it possible to investigate the influence of design parameters (tail area and length, frequency and amplitude of the oscillations of the tail) on the dynamic characteristics of the mobile robot, such as the speed of navigation and maneuverability.

5 Conclusions

The article describes the design and basic elements of an underwater floating robot device based on bionic principles of motion. The robot is able to perform both movement at a given constant depth and vertical ascent and ascent, as well as movement along a complex spatial trajectory. A mathematical model of the developed robot-fish implemented in MATLAB/Simulink for numerical computer simulation was presented.

The conducted full-scale tests of the prototype of the device of a bionic floating robot for monitoring natural and technogenic objects in the hydrosphere showed its effective operation and normal functioning of all its subsystems. The technical characteristics of the developed robot-fish fixed during the tests, basically correspond to the laid requirements and mathematical modelling. Comparative analysis of the results of computer mathematical modelling of the underwater robot-fish movement and full-scale tests of the experimental prototype showed that the discrepancy between their main characteristics does not exceed 10-15%.

References

1. A.Yu. Bocharov, Underwater research and robotics, **2**, 36 (2006)
2. R.J. Mason, W. Burdick, *Proc. 2000 ICRA*, 428 (2000)

3. E.J. Kim, Y. Youm, *Proc. 2004 IEEE Int. Conf. ICRA*, 488 (2004)
4. D. Lachat, A. Crespi, A.J. Ijspeert, *Proc. 1st IEEE RAS & EMBS Int. Conf. on Biomedical Robotics and Biomechatronics*, 643 (2006)
5. L. Wang, S. Wang, Z. Cao, M. Tan, C. Zhou, H. Sang, Z. Shen, *Proc. IEEE International Conference on Industrial Technology*, 1263 (2005)
6. J. Yuh, Autonomous Robots, **8(1)**, 7 (2000)
7. K.A. Morgansen, V. Duindam, R.J. Mason, J.W. Burdick, *Proc. of IEEE Int. Conf. on Robotics and Automation*, 427 (2001)
8. E.V. Vetchanin, Yu.L. Karavaev, A.A. Kalinkin, A.V. Klekovkin, E.N. Pivovarova, Bulletin of the Udmurt University. Mathematics. Mechanics. Computer science, **25. 4**, 544 (2015)
9. L.V. Kiselev, A.V. Medvedev, Underwater research and robotics, **1 (13)**, 24 (2012)
10. M.D. Ageev, *Autonomous underwater robots. Systems and technologies* (Nauka, Moscow, 2005)
11. V.F. Filaretov, A.V. Lebedev, D.A. Yukhimets, *Devices and control systems for underwater robots* (Nauka, Moscow, 2005)
12. B.V. Lushnikov, S.F. Jatsun, E.N. Politov, E.S. Tarasova, Izv. Samar. Sci. Center of the Russian Academy of Sciences, **13, 4 (4)**, 1193 (2011)
13. B.V. Lushnikov, S.F. Jatsun, E.N. Politov, E.S. Tarasova, Izv. Samar. Sci. Center of the Russian Academy of Sciences, **12, 4 (3)**, 562 (2010)
14. B.V. Lushnikov, S.F. Jatsun, E.S. Tarasova, E.N. Politov, *Controlled Vibration Technologies and Machines: Sat. Sci.*, **2**, 111 (2012)
15. B.V. Lushnikov, S.I. Savin, K.G. Kazaryan, A.S. Yatsun, A.V. Malchikov, *Controlled Vibration Technologies and Machines: Sat. Sci.*, **2**,107 (2012).
16. B.V. Lushnikov, E.N. Politov, E.S. Tarasova, K.G. Kazaryan, Electronic Journal of the Cloud of Science. **1(1)**, 61 (2014)
17. S.F. Jatsun, B.V. Lushnikov, E.N. Politov, *Proc. Perspective systems and control tasks*, 208 (2016)
18. S.F. Jatsun, E.N. Politov, P.A. Bezmen, K.G. Kazaryan, *Proc. Vibration technologies, mechatronics and controlled machines*, **2**, 277 (2016)
19. P.A. Bezmen, E.N. Politov, K.G. Kazaryan, Natural and technical sciences, **6**,111 (2016)
20. B.V. Lushnikov, S.F. Jatsun, S.I. Savin, K.G. Kazaryan, A.V. Malchikov, E.S. Tarasova, E.N. Politov, A.S. Yatsun, Underwater floating robot with bionic principle of movement Russian Federation Patent 124656, Feb. 10, (2013)

Towards heterogeneous robot team path planning: acquisition of multiple routes with a modified spline-based algorithm

Roman Lavrenov* and *Evgeni* Magid

Intelligent Robotics Department, Higher Institute for Information Technology and Information Systems (ITIS), Kazan Federal University, 35 Kremlyovskaya street, Kazan, 420008, Russian Federation

Abstract. Our research focuses on operation of a heterogeneous robotic group that carries out point-to point navigation in GPS-denied dynamic environment, applying a combined local and global planning approach. In this paper, we introduce a homotopy-based high-level planner, which uses a modified spline-based path-planning algorithm. The algorithm utilizes Voronoi graph for global planning and a set of optimization criteria for local improvements of selected paths. The simulation was implemented in Matlab environment.

1 Introduction

Currently, one of the actual problems of robotics deals with interaction between the unmanned ground robots (UGV) and unmanned aerial vehicles (UAVs). A heterogeneous group of robots, which work as a team, can significantly improve information acquisition about environment and solve various tasks of transportation logistics, reconnaissance, and search-and-rescue due to different sensory system features of the robots and different operational space [1-3].

Our research is targeted to investigate collaboration aspects between a heterogonous group comprising a UGV and a number of small-size UAVs, focusing on operation in uncertain and dangerous environments [4]. The latter could be completely unknown, include dynamic scene or objects, or represented by an outdated and imprecise map. To facilitate a reliable autonomous operation, robots will collaboratively perceive the surrounding environment and infer their own states, and plan high-quality actions and paths while taking different sources of uncertainty into account. In this research, we investigate a multi-robot collaborative framework, with the aim to advance the state of the art in multi-robot autonomy. Our objective is to develop collaboration strategies between robots equipped with various sensors such that these can autonomously operate given partial and possibly uncertain information regarding the environment. To that end, we will build upon recent progress in simultaneous localization and mapping (SLAM, [5]) and belief space planning [6], and explore multi-robot belief space planning approaches that take into account the uncertainty in the environment and are expected to facilitate efficient collaboration between robots. In particular, such a framework will allow UAVs and the UGV to devise proper motion to improve localization and mapping even in lack of GPS signal.

2 Path planning approach

Path planning is the act of finding a path from some starting configuration S to goal configuration G. The objective of a path planner is to provide a continuous path from S to G, which will avoid all obstacles and minimizes some positive cost metric [7]. Classical path planning methods usually involve a global or a local path planner. A global path planner exploits a precise map of the environment, which is available to the planner a priori, and the path planning process generates a hazard-free path in the off-line mode; its main concern is to provide a computationally effective scheme for the particular cost metric (e.g., [8]). Global planning allows optimizing the solution with regard to various user-defined criteria in order to obtain a path, which is close to the optimal or at least sub-optimal path. Thus, the algorithm performance requirements are mainly dictated by such qualities as ability to generate a collision free path (soundness), time and space performance (complexity), and a guaranteed ability to discover a path if it does exist (completeness). However, in realistic settings, a robot cannot rely on complete a priori knowledge of the environment, but has to continuously collect sensory data from its potentially dynamic environment and plan accordingly. The two main sensor based approaches use either global or local path planning. On the opposite, a local path planner is a purely reactive on-line planner that collects sensory information and acts based on its analysis (e.g., [9]).

Each of these two classical approaches has its own strong and weak points and, while in the past robotics researchers were mainly developing them separately,

* Corresponding author: magid@it.kfu.ru

a more effective approach would combine them together in order to strengthen their advantages and compensate for the pitfalls (e.g., [10]). Moreover, the recent appearance of affordable swarm-type UAV and UGV robots on the market made it possible to develop a more reliable global (e.g., [11]) and local path planning (e.g., [12]) approaches by efficient use of multiple agents working toward a common goal.

In uncertain environments of urban search and rescue (USAR) or surveillance scenarios pure global path planning is of a little use because previously detailed maps become outdated and imprecise due to dynamic changes, while precise positioning becomes impossible due to the lack of GPS signal. However, often it is the case that some global data about the environment is still available: plans of the buildings, a bird-eye view map of the scene (e.g., obtained by a fleet of UGVs over the site) or abstract sketches by survivors. In such circumstances, the team of UAVs and UGVs could re-build the outdated map dynamically while moving toward the specified target and initially use this outdated map.

We are developing Homotopy based high level planner [13] that utilizes helpful global data of an outdated map and select pareto optimal trajectory classes based on user-tuned minimization criteria. Further, the selected class trajectory will be applied by the pilot system and actively combined with sensor based local path planning. The robotic team will create a detailed map and further update it on the fly as the team moves toward its target. The global re-planning will take place in real time taking into account dynamic changes of the environment and objective function, which contains various optimization criteria such as path length, safety, stealth, energy efficiency, path curvature, etc. combined together and allowing dynamic selection of coefficients for each function criterion. Some of such criteria are discussed in more details in the next section. To construct a global road map we apply Hierarchical Generalized Voronoi Graph method [14]. Next, the map is updated locally in real-time in order to introduce newly discovered obstacles [15] utilizing UAVs and UGV updates.

One of popular approaches for dynamical solution of a collision avoidance problem is a potential field approach [16]. It is based on an artificial potential field concept with an attractive pole at a target and obstacles that are represented with repulsive surfaces, while a robot follows the potential function gradient toward its minimum. The major advantages of potential field methods are simplicity and a capability of applying them reactively for obstacles avoidance. Its typical drawbacks include oscillations for certain obstacle setup and presence of local minima that attracts the robot and then keeps it captured inside unless the robot succeeds to detect the local nature of the minima and to execute some special escape procedure.

In [8] we had proposed an iterative path planning algorithm for a car-like mobile robot that simultaneously optimizes (collision free) path length and smoothness. Moreover, while typical path planning algorithms apply path-planning smoothing in last stages only [17], our algorithm, which we briefly overview in Section IV, considered smoothness criterion from its first iterations. In order to improve the original algorithm performance, to add flexibility for path optimization and a possibility for simultaneous exploration with a team of UAVs that would afford a fast re-planning for UGV path in a case the initially off-line selected path becomes unavailable, we integrated Voronoi Diagram approach into our algorithm.

3 Cost function criteria selection

This section presents a number of valuable criteria for path optimization. Some particular criteria that are selected for our algorithm are introduced in more details through the prism of the spline-based navigation algorithm [8]. This algorithm navigates an omni-directional point robot in a planar known 2D configuration space environment populated with static obstacles. Each obstacle is presented with a finite set of intersecting circles of different sizes, assuming that any arbitrary obstacle could be well approximated with such set. Other criteria, which are not integrated into our path-planning algorithm, are only briefly overviewed in this section due to the lack of space.

Collision avoidance, which we further refer as path topology function is a basic requirement for any path to generate a collision free path. With potential function based approach, to provide a collision free path, a repulsive potential function should have a high value inside an obstacle (and on its boundary) and a small value in free space. Its high value in an obstacle center pushes a path outside in order to minimize path cost during local optimization procedure. The potential field begins to decrease drastically on obstacle's boundary, keeps decreasing with distance from the boundary and becomes zero in free space rather fast.

Assuming the robot's position at $q(t)=(x(t), y(t))$ at time t, a contribution of a single circle (which is a part of an obstacle) repulsive potential to a global potential function is defined with the equation:

$$U_{rep}(q) = 1 + \tanh(\alpha(\rho - R_{\inf})) \ ,$$

$$R_{\inf} = \sqrt{(x(t) - x_c)^2 + (y(t) - y_c)^2} \ , \qquad (1)$$

where ρ is the radius of the obstacle with the center at (x_c, y_c) and α is an empirically defined parameter that is responsible for pushing a path outside of an obstacle. Figure 1 demonstrates the examples of two different selections of α parameter for the environment with two obstacles, shown in Figure 4. Figure 2 demonstrates the examples of $\alpha = 0.5$ and $\alpha = 0.2$ for the environment in Figure 5 with two complicated concave obstacles that are formed by finite sets of intersecting circles.

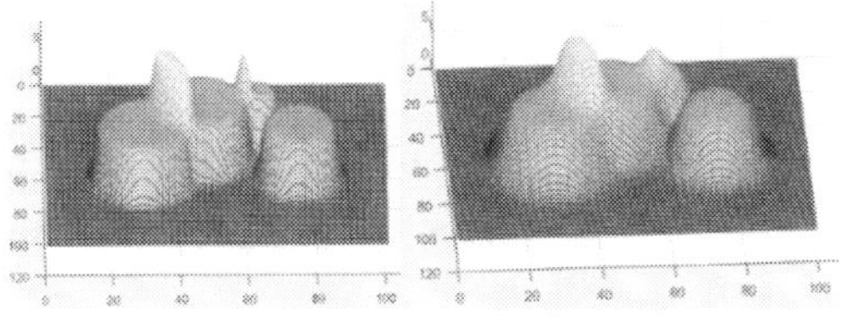

Fig. 1. Repulsive potential function of eq.1 for α = 0.5 (left) and α = 0.2 (right) that corresponds to the obstacle in Fig.4.

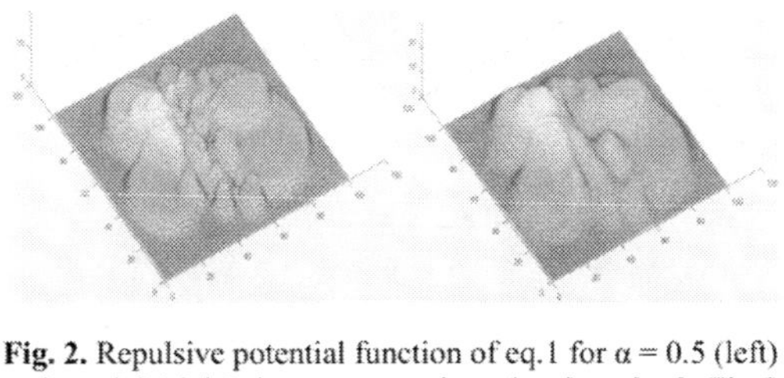

Fig. 2. Repulsive potential function of eq.1 for α = 0.5 (left) and α = 0.2 (right) that corresponds to the obstacles in Fig.5.

Topology function $T(q)$ takes into an account all N obstacles of the environment and their influence on the robot along the whole path, which is defined as a parametric function within [0,1]:

$$T(q) = \sum_{j=0}^{N-1} \int_{t=0}^{1} U^{j}_{rep}(q) \cdot \delta l(t) \cdot dt, \qquad (2)$$

where $\delta l(t)$ is simply a length of a short segment:

$$\delta l(t) = \sqrt{(x'(t))^2 + (y'(t))^2}. \qquad (3)$$

Path smoothness, which we further refer as path roughness function is responsible for absence of abrupt changes in the trajectory. It is particularly important for wheeled robots, and allows to maximize robot speed along the path and to avoid extreme accelerations and decelerations while following the trajectory. Roughness function $R(q)$ is also integrated along the path:

$$R(q) = \sqrt{\int_{t=0}^{1} (x''(t))^2 + (y''(t))^2 dt}. \qquad (4)$$

Path length is a rather typical criterion for all path-planning algorithms. It simply sums up the lengths of all path segments:

$$L(q) = \int_{t=0}^{1} \delta l(t) \cdot dt. \qquad (5)$$

Safety or minimal distance from obstacles criterion is responsible for preserving some minimal distance from obstacle boundaries [18]. While this feature is partially integrated into topology function, in hostile environments, there may be an explicit requirement to stay at least at some minimal distance from obstacle boundaries - these may happen in semi-structured debris of USAR scene, which risk to collapse further, or a danger of short-range emissions from the obstacles, etc. In such cases, the use of some paths that may be optimal with regard to other criteria should be completely forbidden due to violation of minimal distance requirement, and a different path should be selected. A minimal distance from all obstacles of environment is calculated in each robot configuration $q(t)$ along the parametrically defined path:

$$m(C) = \min_{\forall t \in [0,1]} d_c(q(t)), \qquad (5)$$

where $d_c q(t)$ is a minimal distance from obstacles in configuration $q(t)$:

$$d_c(q(t)) = \min_{\forall c \in C} \sqrt{(x(t) - x_c)^2 + (y(t) - y_c)^2} - r_c, \quad (6)$$

where C is a set of all circular obstacles c with the center at (x_c, y_c) and radius r_c; further, these elementary circular obstacles may intersect to form compound obstacles. The user is required to specify a particular minimally acceptable distance to the boundaries of any obstacle d_m, and them optimization criterion D(q) is defined as follows:

$$D(d) = \omega^{d_m - m(C)}, \qquad (7)$$

where ω is a sufficiently large empirically defined value that prevents a path from approaching obstacles beyond the permitted limit. Value of this function is one when minimal distance to obstacles $m(C)$ is equal to safe value d_m, while it gains large positive value when the limit is violated, and diminishes to a small positive value when the robot keeps safe distance from an obstacle.

Maximal distance from obstacles criterion is an opposite for safety criterion and it is responsible for preserving some maximally allowed clearance of a path with regard to obstacles. The situation of keeping obstacles at particular distance may become necessary when visual localization requires the obstacles that, for example, serve as environment features for SLAM to be within a limited sensor range.

Start and target point visibility time refers to the path length where the robot keeps the start point (or target point respectively) within its direct line of sight without any obstacle occlusions [19]. These two criteria are important if a robot needs to maximize time of a direct visual or radio contact with a monitoring device or a router, which supports path planning, localization or any other functionality of the robot. The criterion considers a ratio of visible and invisible from start S (or target T) points segments of a path:

$$I_S = 1 - \lim_{\delta t, \delta l(t) \to 0} \frac{\sum_{t=0}^{u} dist(A(t), A(t+\delta t))}{\int_{t=0}^{1} \delta l(t) dt} \qquad (8)$$

$$\forall t \in [0, u + \delta t] : [A(t), S] \cap \left(\bigcup_{j=1}^{N} Obs_j\right) = 0.$$

Equation (10) defines start point S visibility time and the equation for target point T is similar.

Path passing time determines importance of passing a route in a fastest way. The value correlates with trajectory length, smoothness, and a number of other criteria.

If there are critical or danger points within a map [20] or, on the contrary, attraction points that always should be kept at a certain distance (e.g. for communication routers), then we might set several additional parameters:

Maximum distance to critical points, which is necessary to be preserved in order to avoid losing contact between a robot and a monitoring device, e.g., while inspecting a map or following a path via control points that should stay within sensors range.

Safe path length rate, which is defined as a percentage of a sheltered from hazards path segments (i.e., their length). For example, if there is a radiation source or another source of danger in a critical point on a map, then a mobile robot should follow the safest way (e.g., behind obstacles for protection).

Minimum permitted distance to danger points is the minimal distance to a particular dangerous object or location, which a robot should always maintain.

4 Spline-based planning with potential field approach

Given a complete information about the environment, start S and target T positions of a robot, the robot searches for a collision-free path, which is guided by a pre-determined cost function. The original algorithm of spline-based path planning works iteratively [8]. An initial path is suggested as a straight line between S and T points, which defines a first spline with three points: S, T and an equidistant point that lies on the straight line between S and T. Equation 9 sets a path cost, and the path is further optimized with Nelder-Mead Simplex Method to minimize the cost:

$$F(q) = \gamma_1 T(q) + \gamma_2 R(q) + \gamma_3 L(q) \qquad (9)$$

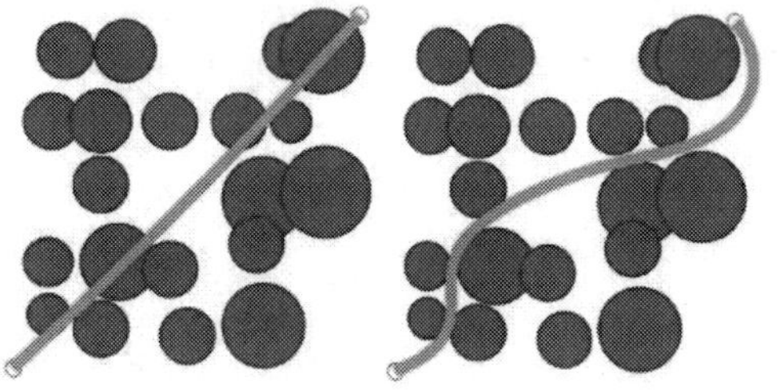

Fig. 3. Simple convex obstacles: the initial state (left), the final path after 6 iterations (right).

A resulting better path serves as an initial guess for the next iteration. The optimization procedure operates only with those points of the path, which define its spline, while path evaluation accounts for all points of the path. The path-defining spline is rebuilt at each iteration using information from a previous stage, increasing the number of spline's points by one and adjusting parameters of the target cost function. The algorithm terminates if iteration count exceeds a user-defined limit or a new iteration fails to improve a previous one. Figure 3 demonstrates an example of a successful execution of the original algorithm with cost function from eq. 9. The initial path was obtained in six iterations that took 9 minutes and succeeded to reduce total cost of the path, while further increase of via points number did not provide any improvement. Further details of the algorithm could be found in [8].

The original spline-based method succeeds to obtain a collision free smooth path for any environment under assumption that each obstacle is approximated with a single circle, there are no circle intersections and some minimal distance between the obstacles is preserved. In this case, since the potential function diminishes rapidly as we move away from the obstacle boundary, we could neglect the probability of getting stuck in a local minima. When a number of circles intersect to form a compound concave obstacle, intersections introduce potential field local maxima. Moreover, if in such settings an initial spline passes through intersection of several obstacles, the cost function $F(q)$ priority is to push the spline out of the intersection area. Upon successful pushing out, a further solution often sticks in a local minima and a next iteration spline cannot overcome it due to a local nature of the optimization.

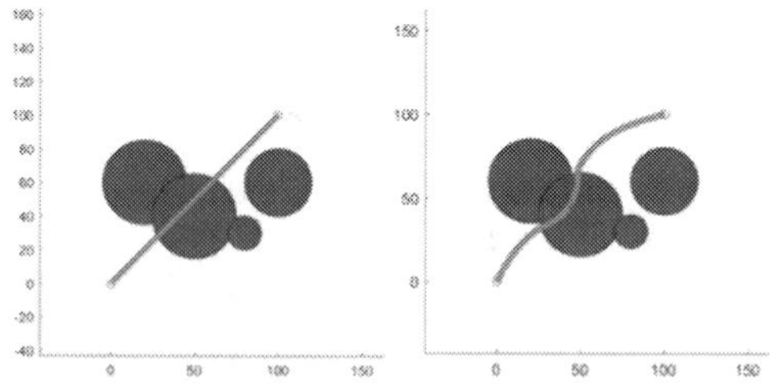

Fig. 4. Simple concave obstacle: the first iteration path (left) and the final path after 18 iterations (right).

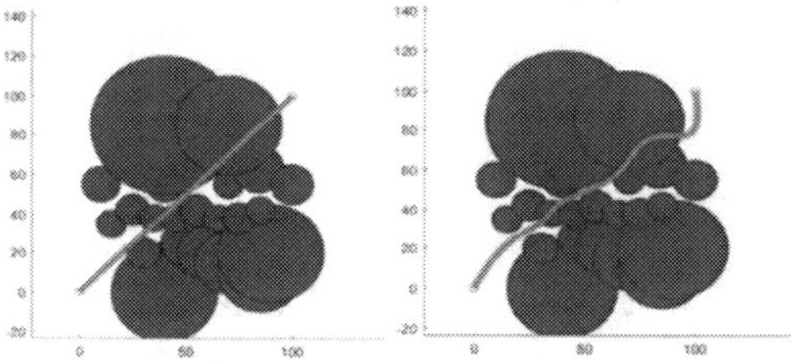

Fig. 5. Two complicated concave obstacles: the first iteration path (left) and the final path after 37 iterations (right).

While in some cases the original algorithm succeeds to negotiate with concave obstacles (that are formed with a set of intersecting circles) as well, it is not a generic case. Figure 4 presents an example of two obstacles (the corresponding potential field is in Figure 1). The algorithm stops after 18 iterations as no further improvements are possible - while a spline succeeds to optimize the path locally so that it avoids a local maximum, it is still stuck at a local minimum and the resulting path could not be used for navigation due to obstacle collision. Figure 5 demonstrates an example with two complicated concave obstacles and a narrow pass in between. Due to multiple intersecting circles, which generate their own local repulsive potentials, the corresponding global field conceals the pass (Figure 2) and permits only its local optimization. The local optimization successfully avoids local maxima, but even after 37 iterations that continued for 44 minutes, the final path is still occluded with obstacles and thus useless for navigation.

5 Voronoi diagram based solution for multi-robot exploration

We recognize that the local nature of the optimization procedure creates a strong dependence of algorithm success or failure on initial spline. In order to provide a good initial spline that could be further improved locally with regard to user selection of the cost weights, we apply Voronoi Diagram approach. Voronoi graph VG is constructed (Fig. 6, top left) based on a classical brushfire approach [14] and then VG nodes and edges, which are a part of some selected path from S to T, help to define an initial spline. Voronoi graph helps to avoid the spline seizing at local minima.

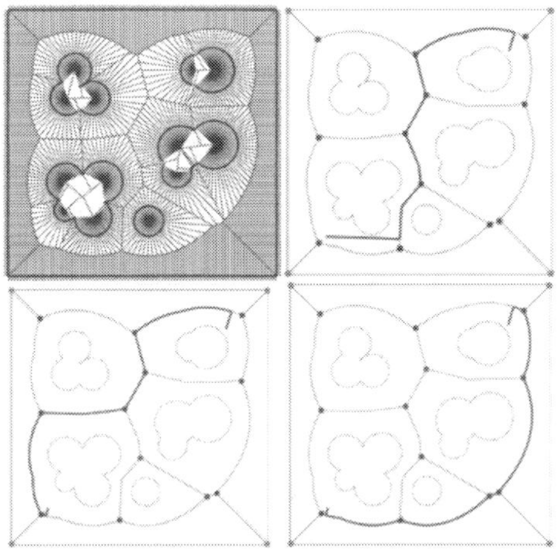

Fig. 6. Voronoi graph VG construction (top left): blue circles denote obstacles, blue lines are emitted at construction stage by obstacles rays, red lines form VG. Three other sub-figures demonstrate examples of path calculation within different homotopies: grey figures are obstacles, green lines are edges of VD, red dots are nodes of VD, red lines are the selected paths in different homotopies.

Voronoi graph VG contains multiple homotopy class paths and their variety depends on map complexity: the more distinct obstacles appear within the map, the larger is the amount of homotopy classes. For example, Figure 6 demonstrates three different paths that belong to different homotopy classes. Having multiple options for initial spline selection within various homotopies is of special interest for USAR: for example, USAR mission control centre has a city plan before an earthquake and thus could create a Voronoi graph VG offline for further use by a rescue robot. It is worth mentioning that off-line stage does not limit us with valuable on-board resources, as it could be performed far in advance using high performance computational clusters. VG contains a large number of paths, one for each homotopy class. This feature provides a possibility for a simultaneous exploration of a number of prospective paths with a team of UAVs and affords discovering blocked passages that were previously available for a main rescue UGV. In turn, such approach permits a fast dynamic re-planning for UGV path within a different homotopy class in a case that the initially off-line selected path becomes unavailable.

We emphasize that it is not feasible to use all possible homotopies and most of them would obviously offer non-optimal paths. Thus, in our simulated experiments, we limit the algorithm to 5-7 homotopies. Next, these selected homotopies served as initial spline for a new smart spline-based method. We have tested this strategy with a number of environments.

6 Simulations and results

In order to verify our approach, the new algorithm was implemented in Matlab environment and an exhaustive set of simulations was performed. Particular attention was paid to the cases where the original algorithm fails. As a cost function we used the following equation:

$$F(q)=\gamma_1 T(q)+\gamma_2 R(q)+\gamma_3 L(q)+\gamma_4 D(d), \quad (10)$$

where the empirical parameter selection was: $\gamma_1=1$, $\gamma_2=1$, $\gamma_3=0.5$, $\gamma_4=1$, $d_m=3$ and $\omega=20$ (in eq. 7). We included additional safety criteria into original algorithm's cost function (eq. 9). The algorithm succeeded to provide collision-free paths in all cases, which was a natural consequence of applying initial Voronoi-based path as an input for iterative optimization algorithm.

We compute the cost function value (eq. 10) for each homotopy. As a result, optimization takes place in several homotopies simultaneously. For example, in Figure 7 the resulting path in the left image is more safe than the path in the right image. Since path optimization is performed only locally, the effect of the additional parameter(s) is also local. The path is constrained to remain within the same homotopy group even in the case when it is possible to bypass any obstacle from the other side.

Figure 7 environment was selected to demonstrate the advantage of the new algorithm with regard to the original algorithm. In such environment the original spline-based method failed to find an existing path within the map - the algorithm had spent 44 minutes and completed 18 iterations before reporting a failure. On the contrary, our new smart spline-based method succeeded to provide high quality paths for both initial spline selections from two different homotopies.

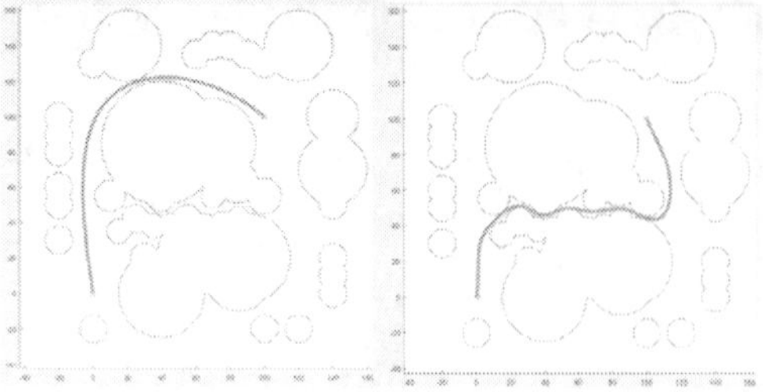

Fig. 7. Path planning results with a new spline-based algorithm: two corresponding optimal for a particular homotopy paths.

In the first homotopy example (Fig. 7, left), Voronoi graph VG calculation took only 5 seconds, and another 8 minutes were spent for two iterations of local optimization of the path with regard to user-defined criteria. In the second case, the same VG was utilized and a further optimization spent 21.5 minutes in order to complete the process within five iterations (Fig. 7, right).

7 Conclusions

Our research focuses on effective operation of a heterogeneous robotic group that could carry out point-to point navigation in dynamic environments, applying a combination of local and global planning approaches. This paper presents a homotopy-based high-level planner that uses a modified spline-based iterative algorithm, which constructs Voronoi graph off-line and runs local path optimization with a potential function approach. Voronoi graph supports a selection of an initial spline, which is further optimized with regard to path length, curvature and safety criteria. Furthermore, Voronoi graph provides an opportunity to perform simultaneous exploration of several prospective paths with a team of UAVs varying paths from different homotopy classes. This way UAVs discover blocked passages that were previously available for a main UGV, and a fast dynamic re-planning of the UGV path within a different homotopy class could be performed.

The algorithm was implemented in Matlab environment and its results were explicitly compared to our original algorithm. The new approach requires less optimization iterations than the original algorithm due to a smart selection of an initial spline. While the original algorithm fails to find an existing path in complicated environments with concave obstacles, its smart version was successful in all simulated tests.

This work was partially supported by the Russian Foundation for Basic Research (RFBR) and Ministry of Science Technology & Space State of Israel (joint project ID 15-57-06010). Part of the work was performed according to the Russian Government Program of Competitive Growth of Kazan Federal University.

References

1. A. Rosenfeld, N. Agmon, O. Maksimov, A. Azaria, S. Kraus, *24th Int. Conf. on Artificial Intelligence*, 1902-1908 (2015)
2. A. Ronzhin, I. Vatamaniuk, N. Pavluk, *Int. Conf. and Exposition on Electrical and Power Engineering*, 675–680 (2016)
3. A.I. Panov, K. Yakovlev, *Robot Intelligence Technology and Applications, 4,* 3–20 (2017)
4. E. Magid, T. Tsubouchi, E. Koyanagi, and T. Yoshida, Journal of Robotics and Mechatronics, **23(4)**, 567 (2011)
5. A. Buyval, I. Afanasyev, and E. Magid. *9th Int. Conf. on Machine Vision* (2016).
6. V. Indelman, L. Carlone, F. Dellaert, The International Journal of Robotics Research (2015)
7. A. Stentz, Optimal and efficient path planning for unknown and dynamic environments (1993)
8. E. Magid, D. Keren, E. Rivlin, I. Yavneh, *IEEE/RSJ Int. Conf. on Intelligent Robots and Systems*, 2296-2301 (2006)
9. E. Magid, E. Rivlin, *IEEE/RSJ Int. Conf. on Intelligent Robots and Systems* (2004)
10. H. Zhang, J. Butzke, M. Likhachev, *IEEE Int. Conf. on Robotics and Automation* (2012)
11. M. Whitty, J. Guivant, *IEEE Int. Conf. on Robotics and Automation* (2011)
12. L. Rodriguez, J. A. Cobano, A. Ollero, *XV Workshop of physical agents: book of proceedings* (2014)
13. M. Kuderer, Ch. Sprunk, H. Kretzschmar, W. Burgard, *IEEE Int. Conf. on Robotics and Automation* (2014)
14. H. M. Choset, Principles of robot motion: theory, algorithms, and implementation. (MIT press, 2005)
15. B. Lau, and C. Sprunk, W. Burgard, *IEEE Int. Conf. on Intelligent Robots and Systems* (2010)
16. L. Tang, S. Dian, G. Gu, K. Zhou, S. Wang, X. Feng. *3rd IEEE Int. Conf. on Computer Science and Information Technology*, 9, 633-637 (2010)
17. S. Fleury, P. Soueres, J.-P. Laumond, R. Chatila, IEEE Transactions on robotics and automation, **11(3)**, 441-448 (1995)
18. R. Lavrenov, E. Magid, *2nd Int. Con. on Interactive Collaborative Robotics*, LNCS, Springer (2017, to be published)
19. E. Magid, R. Lavrenov, A. Khasianov, *Int. Conf. on Informatics in Control, Automation and Robotics*, MATEC Web of Conferences (2017, to be published)
20. E. Magid, R. Lavrenov, I. Afanasyev, *Int. Conf. on Mechanical, System and Control Engineering*, MATEC Web of Conferences (2017, to be published)

Trajectory generation for a walking in-pipe robot moving through spatially curved pipes

Sergei Savin*, *Sergey* Jatsun and *Lyudmila* Vorochaeva

Southwest State University, Department of Mechanics, Mechatronics and Robotics, 305040 Kursk, Russia

Abstract. In this paper, a walking in-pipe robot is studied. The robot has six legs, each consisting of 3 links connected via rotary joints. The legs are attached to the robot's body. The work is focused on the problem of generating desired position and orientation for the robot's body, using a given footstep sequence. An iterative geometric algorithm for generating orientation sequence is proposed. The problem of finding the desired position of the center of mass of the robot's body is formulated as a problem of minimizing stretching of the robot's legs during steps. Also, an analytical solution for inverse kinematics problem has been given. All proposed algorithms do not require extensive calculation and use basic algebraic operations.

1 Introduction

The main use of in-pipe robots is the transport of specialized equipment, which, in its turn, is used for diagnostics and maintenance of the pipeline [1, 2]. Having the equipment inside the pipeline is sometimes the only option other than excavation of the pipe, which is associated with significant costs. Additionally, being able to access the inner surface of the pipe can provide for a better detection of defects that occur there. It also provides a direct way to monitor the deposits on the inner surface of the pipe. Sensor systems employed in such inspection tasks are discussed in paper [3]. The use of in-pipe robots for cleaning the pipelines was also proposed in [4].

There are several types of in-pipe robots. Many of the proposed classifications distinguish seven or more different types [5, 6]. Paper [7] is devoted to the study of motion of worm-like robot; the work [8] describes the parallel-link robot with flexible structure. Papers [9,10] present the results of numerical simulations in-pipe robot based on the vibration principle of movement. All of the above robots are used for inspection of pipelines. In this paper we focus on walking in-pipe robots. This type of robots is particularly interesting because of its ability to "step over" obstacles, potentially allowing such robots to navigate pipelines with branches, matter deposits and other types of obstacles.

Some of the walking robot designs are shown in papers [11-17]. In works [11-13] an eight legged in-pipe robot and its sensory system are studied. In paper [14] a six legged in-pipe robot designed for motion in spatially curved pipes is discussed. Articles [15-17] focus on planar in-pipe robots with four legs, considering designs with a solid and segmented main link.

We should note that there are a number of problems associated with controlling walking in-pipe robots. That includes step sequence generation, trajectory generation, inverse kinematics, controller design that takes into account information about contact with the inner surface of the pipe, observer design and others. This is similar to the set of problems associated with other types of walking robots. Many of the solutions for these problems found for different types of walking robots are viable for in-pipe walkers as well; this is discussed in paper [14] in more detail.

Some of the listed above problems have been studied specifically for the case of in-pipe walking robots. In papers [14,15] step sequence generation algorithms are proposed for planar and spatial in-pipe walking robots. In work [15] an analytic solution for inverse kinematics problem is given for a 9 link planar robot. In [16] a quadratic programming-based controller design for an in-pipe walking robot is proposed, able to take into account actuator torque limits and mechanical constraints, preventing loss of contact and slipping of the leg. Additionally, inverse kinematics algorithms, control designs and state-estimator designs were studied in [18] for a bipedal walking robot, using approaches that can be translated to in-pipe robots.

In this paper we focus on trajectory generation problem for the robot's body (the main link that all robot's legs are connected to). Knowing position and orientation of robot's body and robot's footstep sequence allows us to solve inverse kinematics problem for each leg individually, obtaining all joint angles. If there are three joints in each leg and the end effector of the leg (its foot) has three translational degrees of freedom when the robot's body is fixed, then in this case the inverse kinematics problem has a finite number of solutions that may be found analytically. For such a case, it should be possible to write an efficient inverse kinematics algorithm that can be executed using on-board computer without producing excessive computational load. Also,

* Corresponding author: savinswsu@mail.ru

knowing robot's body position and orientation allows us to estimate the feasibility of the robot's trajectory with respect to the geometry of the pipeline.

This paper can be considered a follow-up on the work [15], where the problem of body position and orientation generation was solved for a planar robot. The goal of the paper is to derive a systematic approach for generating center of mass trajectory for the robot's main link, as well as time functions defining its orientation.

2 Robot description

In this paper, we consider a six legged in-pipe robot with 19 individual links. Each leg includes three links connected via rotary joints and is attached to the robot's body. The construction of the robot is shown in Figure 1.

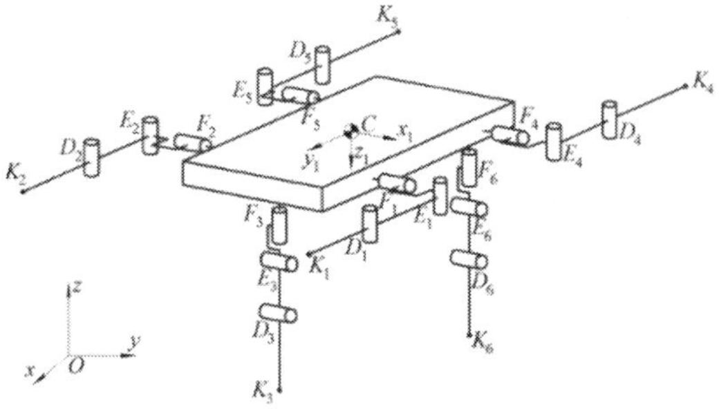

Fig. 1. Construction of the six-legged walking in-pipe robot: 1 – robot's body, 2 – robot's legs.

In Figure 1 point *C* denotes the center of mass of the body link, F_i, E_i and D_i denote robot's joints and points K_i denote contact pads on the end of the robot's legs, $i=\overline{1,6}$. We denote the position of these points as $\mathbf{r}_C$, $\mathbf{r}_{Fi}$, $\mathbf{r}_{Ei}$, $\mathbf{r}_{Di}$ and $\mathbf{r}_{Ki}$ accordingly. The orientation of the body is defined by a directional cosines matrix $\mathbf{T}$.

If $\mathbf{r}_C$ and $\mathbf{T}$ are given, then we can determine $\mathbf{r}_{Fi}$ using the following expression:

$$\mathbf{r}_{Fi}^{(0)} = \mathbf{T}\mathbf{r}_{Fi}^{(1)} + \mathbf{r}_C^{(0)}, \quad (1)$$

where the superscript denotes the reference frame, $\mathbf{r}^{(0)}$ means the vector is expressed in world frame $Oxyz$ and $\mathbf{r}^{(1)}$ means it is exressed in local frame $Cx_1y_1z_1$.

Furthermore, we assume that links F_iE_i have such shape, that rotations in joints F_i don't change the positions of points E_i.

3 Determining desired position and orientation of the robot's body

The algorithm described in this section discretizes robot's trajectory and aims to find $\mathbf{T}$ and $\mathbf{r}_C$ for every discrete point in time t_j. We assume that we are given vectors $\mathbf{r}_{Ki}(t_j)$ for the current time point t_j, and we are given matrix $\mathbf{T}(t_{j-1})$ and vector $\mathbf{r}_C(t_{j-1})$ for the previous time point t_{j-1}. The algorithm for generating footstep sequence $\mathbf{r}_{Ki}(t_j)$ is described in detail in [14]. The algorithms includes two steps. First, we find the desired orientation of the robot's body, and then the position of its center of mass.

3.1. Determining desired orientation

First, we define directional vector σ_K for the footsteps $\mathbf{r}_{Ki}$ as follows:

$$\sigma_{K,i} = \frac{\mathbf{r}_{K,i+3} - \mathbf{r}_{K,i}}{\left\| \mathbf{r}_{K,i+3} - \mathbf{r}_{K,i} \right\|}, \quad (2)$$

$$\sigma_K = \frac{\sigma_{K,1} + \sigma_{K,2} + \sigma_{K,3}}{\left\| \sigma_{K,1} + \sigma_{K,2} + \sigma_{K,3} \right\|}, \quad (3)$$

Then, using $\mathbf{T}(t_{j-1})$ and formula (1) we can find $\mathbf{r}_{Fi}(t_{j-1})$, the positions of the body joints during the previous step. We define directional vector σ_F for robot's body the same way as we defined σ_K, only instead of $\mathbf{r}_{Ki}(t_j)$ we use $\mathbf{r}_{Fi}(t_{j-1})$.

Finally we can find rotation matrix that takes σ_F to σ_K. To do this we find axis of rotation $\mathbf{e}_{FK}$ and angle of rotation α_{FK}:

$$\mathbf{e}_{FK} = \frac{\sigma_F \times \sigma_K}{\left\| \sigma_F \times \sigma_K \right\|}, \quad (4)$$

$$\alpha_{FK} = \cos^{-1}(\sigma_F \cdot \sigma_K). \quad (5)$$

The directional cosines matrix $\mathbf{T}(t_j)$ is then given by the following expression:

$$\mathbf{T}(t_j) = \mathbf{T}(\mathbf{e}_{FK}, \alpha_{FK})\mathbf{T}(t_{j-1}), \quad (6)$$

where $\mathbf{T}(\mathbf{e}_{FK}, \alpha_{FK})$ is the matrix representing rotation around axis $\mathbf{e}_{FK}$ by angle α_{FK} [19].

We should note that this method can accumulate error associated with roll angle (in roll-pitch-yaw representation of rotations). This can be avoided by performing an additional rotation, using σ_K as an axis.

3.2. Determining desired position

To find desired position of the center of mass $\mathbf{r}_C(t_j)$ we find the vector $\mathbf{p}$ that shifts the previous its position $\mathbf{r}_C(t_{j-1})$ to the new one, such that the resulting new positions of points F_i minimize the distance between them and points K_i. This condition aims at minimizing stretching of the robot's legs, allowing them to stay further away from kinematic singularities.

Then the vector $\mathbf{p}$ given by the following expression:

$$\mathbf{p} = \operatorname{argmin} \sum_{i=1}^{6} \left\| \mathbf{T}(t_j)\mathbf{r}_{Fi}^{(1)} + \mathbf{r}_C^{(0)}(t_{j-1}) + \mathbf{p} - \mathbf{r}_{Ki}^{(0)} \right\|. \quad (7)$$

This can be formulated as a quadratic program or can be solved directly using least squares method. To do the latter, we define a difference vector $\mathbf{d}$ and matrix $\mathbf{C}$:

$$\mathbf{d} = \begin{bmatrix} \mathbf{T}(t_j)\mathbf{r}_{F1}^{(1)} + \mathbf{r}_C^{(0)}(t_{j-1}) - \mathbf{r}_{K1}^{(0)} \\ \vdots \\ \mathbf{T}(t_j)\mathbf{r}_{F6}^{(1)} + \mathbf{r}_C^{(0)}(t_{j-1}) - \mathbf{r}_{K6}^{(0)} \end{bmatrix}, \quad \mathbf{C} = \begin{bmatrix} \mathbf{I} \\ \vdots \\ \mathbf{I} \end{bmatrix}, \quad (8)$$

where $\mathbf{I}$ is a 3 by 3 identity matrix, and both $\mathbf{C}$ and $\mathbf{d}$ have 18 rows. Then the solution for least squares can be found as:

$$\mathbf{p} = \mathbf{C}^{+}\mathbf{d} = (\mathbf{C}^{\mathrm{T}}\mathbf{C})^{-1}\mathbf{C}^{\mathrm{T}}\mathbf{d}. \quad (9)$$

Thus, the new position of the center of mass is given as:

$$\mathbf{r}_C(t_j) = \mathbf{r}_C(t_{j-1}) + \mathbf{p}. \quad (10)$$

3.3. Generating continuous trajectories

In order to generate continuous trajectories from the obtained sequence of positions and orientations, we can use interpolation. Sequence $\mathbf{r}_C(t_j)$ can be directly interpolated using splines. In order to interpolate orientations, they need to be converted to a different representation, because interpolating sequence of directional cosines matrix can produce matrices that would not be orthonormal and thus would require orthogonalization. Instead, we can replace matrices $\mathbf{T}(t_j)$ with quaternion $\mathbf{q}(t_j)$ describing equivalent rotations [19]. This sequence of quaternions can then be interpolated using special algorithms, such as Slerp [20]. This interpolation can be used to recover rotation matrices $\mathbf{T}$ for any moment in time, or directly, by modifying formula (1) and other kinematics expressions accordingly.

Figures 2 and 3 shows a sequence of positions of robot's body inside a pipeline (legs are not drawn, pipeline is opaque in Figure 2 and transparent in Figure 3). Figure 2 shows that the proposed method allows generating a sequence of positions of the center of mass of the robot. The robot always stays roughly at the center of the pipe, and keeps its orientation "parallel to the pipeline's direction".

4 Inverse kinematics algorithm

Having $\mathbf{T}(t)$ and $\mathbf{r}_C(t)$ we can find $\mathbf{r}_{Fi}(t)$ using (1). Furthermore, we can find $\mathbf{r}_{Ei}(t)$ with the next formula:

$$\mathbf{r}_{Ei}^{(0)} = \mathbf{T}\mathbf{r}_{FiEi}^{(1)} + \mathbf{r}_{Fi}^{(0)}, \quad (11)$$

where $\mathbf{r}_{FiEi}$ is vector drawn from point F_i to E_i. Figure 4 shows the leg of the robot. In the Figure 4, l_1 and l_2 and lengths of the links E_iD_i and D_iK_i and ψ_i, $\varphi_{i,1}$ and $\varphi_{i,2}$ are joint angles.

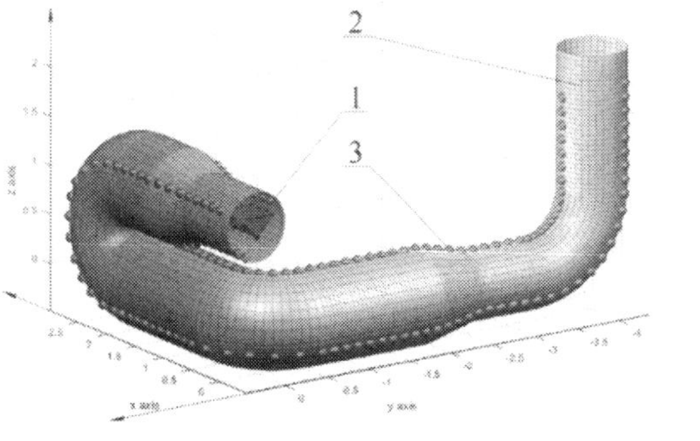

Fig. 2. Spatially curved pipeline with robot in it; 1 – initial robot's position, 2 – pipeline, 3 – points where the robot steps to.

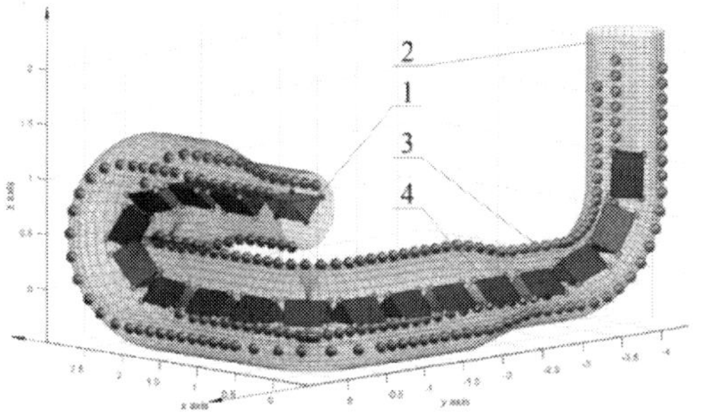

Fig. 3. Sequence of positions for robot's body inside a pipeline; 1 – initial robot's position, 2 – pipeline, 3 – points where the robot steps to, 4 – intermediate positions.

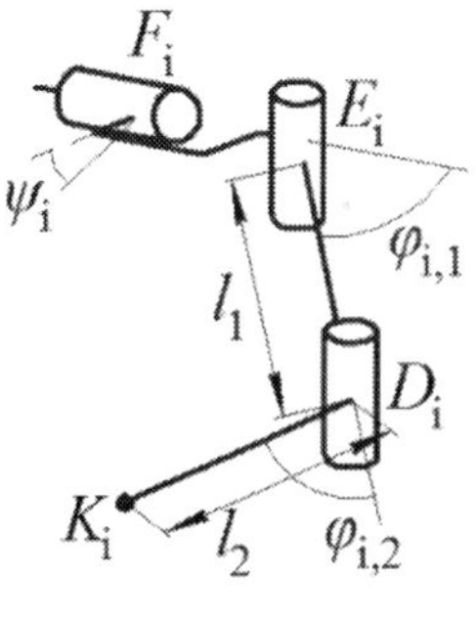

Fig. 4. Construction of the robot's leg.

Let us consider vector $\mathbf{r}_{EiKi}$ drawn between points E_i and K_i. Using formula (11) and the given sequence $\mathbf{r}_{Ki}(t_j)$ (or its continuous version) we can find the value of this vector. Then we can calculate the length of this vector L_i and the azimuth angle ψ_i in the following way:

$$L_i = \left\|\mathbf{r}_{EiKi}\right\|, \quad \psi_i = \tan^{-1}\frac{y_1 \cdot \mathbf{r}_{EiKi}}{z_1 \cdot \mathbf{r}_{EiKi}}, \tag{12}$$

Then we rotate the vector $\mathbf{r}_{EiKi}$ so it would lie parallel to Ozx plane :

$$\overset{*}{\mathbf{r}}_{EiKi} = \mathbf{T}_x(\psi)\mathbf{T}^{-1}\mathbf{r}_{EiKi}, \tag{13}$$

With this we can consider a simple planar inverse kinematics problem, illustrated in Figure 5.

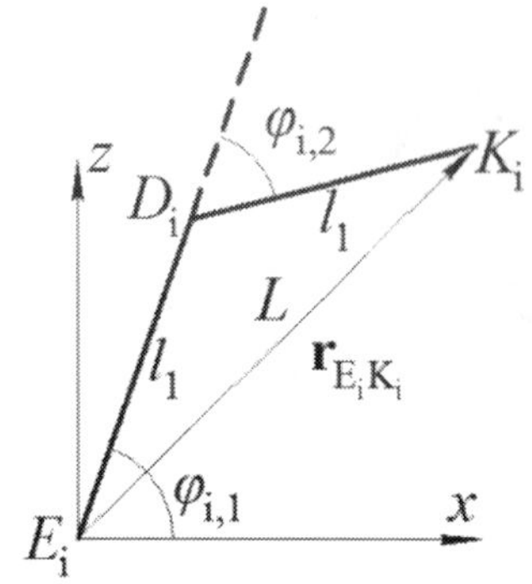

Fig. 5. Inverse kinematics diagram.

The solution for the planar inverse kinematics problem is given by the next expressions:

$$\varphi_{i,1} = \tan^{-1}\left(\frac{\overset{*}{z}_{EiKi}}{\overset{*}{x}_{EiKi}}\right) + \cos^{-1}\left(\frac{l_1^2 + L^2 - l_2^2}{2l_1 L}\right), \tag{14}$$

$$\varphi_{i,2} = \pi - \cos^{-1}\left(\frac{l_1^2 + l_2^2 - L^2}{2l_1 l_2}\right), \tag{15}$$

Using these expressions we can solve for ψ_i, $\varphi_{i,1}$ and $\varphi_{i,2}$ triplets for each leg, thus solving inverse kinematics problem for the robot. Figure 6 shows the result of applying these algorithms, demonstrating obtained orientations of robot's legs while it moves through a curved pipe.

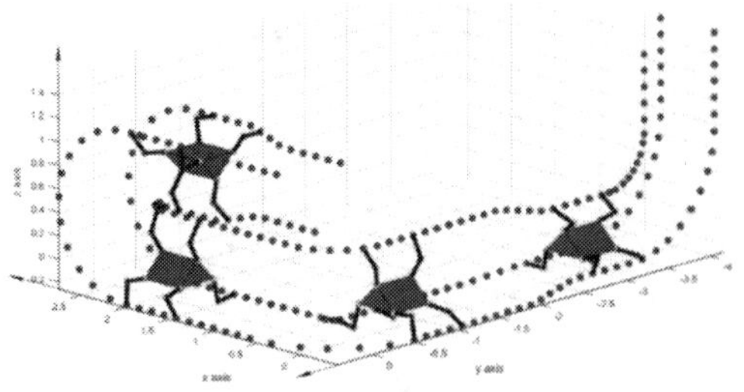

Fig. 6. Robot with legs drawn, their orientation found by the inverse kinematics algorithm (the pipe is not shown).

We can see that the robot bends its "knees" in the same direction, which is the result of the chosen approach. This can be controlled by changing signs in the expressions (14) and (15). Similar problem was discussed in [15, 21].

We should note that this algorithm does not handle singularities, and it would fail in the case when a point can't be reached. There are easy ways to check if the algorithm will fail, for example by using the following condition:

$$L_i \le l_1 + l_2. \tag{16}$$

If the leg segments are not of equal length, we can add another condition, requiring that the length of the longer segment would be less or equal to the length of the other segment plus L_i. If the conditions are not satisfied, then the inverse kinematics problem for the leg has no solution.

6 Conclusions

In this paper a six legged in-pipe walking robot was studied. The aim of the study was to provide an efficient algorithm for generating positions and orientations of the robot's body, and to show how it can be used to derive an analytic solution for inverse kinematics problem.

The proposed method for generating center of mass trajectory is based on interpolation of the sequence of positions of the center of mass. That sequence is in its turn found by considering the problem of minimizing the distance between the footsteps locations (points of contact where the robot needs to step to) and the bases of robot's legs. This problem can be solved algebraically for the case when robot's orientation is fixed.

The orientation is found by a heuristic-based approach, which uses so-called directional vectors, which indicate where the robot should be orientated. The procedure of generating a sequence of orientations is iterative and requires initialization, although the initial orientation supplied to the algorithm does not need to be very accurate. Then the sequence of orientations can be represented as a sequence of quaternions and interpolated.

The proposed inverse kinematics solution is based on simple geometric manipulations. The advantage of this approach, same as with the other methods proposed in this paper, is its low computational requirements.

This work is supported by the Presidential grant MK-2577.2017.8.

References

1. H.T. Roman, B.A. Pellegrino, W.R. Sigrist, *IEEE Transactions on Energy Conversion*, **8**(3), 576-583 (1993)
2. J.M. Mirats Tur and W. Garthwaite, Journal of Field Robotics, **27**(4), 491-508 (2010)
3. O. Duran, K. Althoefer, L.D. Seneviratne, IEEE Sensors journal, **2**(2), 73-81 (2002)
4. C.D. Jung, W.J. Chung, J.S. Ahn, M.S. Kim, G.S. Shin, S.J. Kwon, *International Conference on*

Mechatronics and Automation (ICMA), 1327-1332 (2011)

5. N.S. Roslin, A. Anuar, M.F.A. Jalal and K.S.M . Sahari, Procedia Engineering, **41**, 1456-1462 (2012)
6. S.G. Roh, H.R. Choi, IEEE transactions on robotics, **21**(1), 1-17 (2005)
7. Jatsun S., Malchikov A., *Handbook of Research on Advancements in Robotics and Mechatronics,* 168-218 (2014)
8. S. Jatsun, A. Yatsun, S. Savin. *15th International Conference on Climbing and Walking Robots and the Support Technologies for Mobile Machines, CLAWAR 2012*, 713-719 (2012)
9. G. Carbone, M. Ceccarelli, S. Jatsun, A. Malchikov, *10th IFToMM International Symposium on Science of Mechanisms and Machines,* 103-114 (2010)
10. A.N. Grankin, S.F. Yatsun. Journal of Computer and Systems Sciences International, **48**(1), 155-163 (2009)
11. J.A. Gálvez, P.G. De Santos, F. Pfeiffer, IEEE/ASME Transactions on mechatronics, **6**(1), 26-35 (2001)
12. A. Zagler, F. Pfeiffer, *IEEE International Conference on Robotics and Automation ICRA'03,* 2954-2959 (2003)
13. F. Pfeiffer, Philosophical Transactions of the Royal Society of London A: Mathematical, Physical and Engineering Sciences, **365**, 109-131 (2007)
14. S. Savin, L.Yu. Vorochaeva, *13th Siberian conference SIBCON-2017* (to be published)
15. S. Savin, L. Vorochaeva, *International Conference on Industrial Engineering 2017,* (to be published)
16. S. Savin and L.Yu. Vorochaeva, *International Conference On Industrial Engineering 2017*, (to be published)
17. S. Jatsun, A. Malchikov, O. Loktionova, *XX CISM-IFToMM Symposium on Theory and Practice of Robots and Manipulators ROMANSY 2014,* 341-348 (2014)
18. S. Kuindersma, R. Deits, M. Fallon, A. Valenzuela, H. Dai, F. Permenter, T. Koolen, P. Marion, R. Tedrake, Autonomous Robots, **40**(3), 429-455 (2016)
19. J. Diebel. Matrix, **58**(15-16), 1-35 (2006)
20. K. Shoemake. ACM SIGGRAPH computer graphics **19**(3), 245-254 (1985)
21. L.Yu. Vorochaeva, A.S. Yatsun, S.F. Jatsun. SPIIRAS Proceedings. **3**(52), 70-94 (2017)

AUTHOR INDEX